Order and Disorder

Science Essentials for the Non-Scientist

Order and Disorder

Science Essentials for the Non-Scientist

Myron Kaufman
Emory University, USA

Imperial College Press

Published by

Imperial College Press
57 Shelton Street
Covent Garden
London WC2H 9HE

Distributed by

World Scientific Publishing Co. Pte. Ltd.
5 Toh Tuck Link, Singapore 596224
USA office: 27 Warren Street, Suite 401-402, Hackensack, NJ 07601
UK office: 57 Shelton Street, Covent Garden, London WC2H 9HE

British Library Cataloguing-in-Publication Data
A catalogue record for this book is available from the British Library.

ORDER AND DISORDER
Science Essentials for the Non-Scientist

Copyright © 2011 by Imperial College Press

All rights reserved. This book, or parts thereof, may not be reproduced in any form or by any means, electronic or mechanical, including photocopying, recording or any information storage and retrieval system now known or to be invented, without written permission from the Publisher.

For photocopying of material in this volume, please pay a copying fee through the Copyright Clearance Center, Inc., 222 Rosewood Drive, Danvers, MA 01923, USA. In this case permission to photocopy is not required from the publisher.

ISBN-13 978-1-84816-574-8
ISBN-10 1-84816-574-9

Typeset by Stallion Press
Email: enquiries@stallionpress.com

Printed in Singapore by Mainland Press Pte Ltd.

Contents

Preface — vii

1. The Nature of Science — 1
2. Order — 15
3. Particles — 31
4. Forces and Motion — 45
5. Waves — 61
6. Energy — 75
7. Particle-Waves — 93
8. The New Physics — 113
9. Atoms — 127
10. The Elements — 139
11. Compounds — 157
12. Organic Compounds — 175
13. Entropy — 193
14. Equilibrium and Steady State — 207
15. Introduction to Life — 221

16	The Components of Life	241
17	Molecules of Life	257
18	Information Storage and Transmission in Life	275
19	Beyond the Cell	293
20	The Nervous System	317
21	The Earth	333
22	Ecology and Evolution	355
23	Astronomy	369
Appendix		395
Glossary		397
Credits		433
Index		437

Preface

As C. P. Snow, the British physicist and novelist, noted a half-century ago, intellectual society is divided into two cultures, literary and scientific, and these groups are each incapable of appreciating the other, much to the detriment of society and individuals. With the advancement of science, Snow's observation is probably even more valid today. While scientists, if they are so inclined, can probably remedy some of their literary shortcomings in their post-university life, it is challenging for literary intellectuals to make much progress learning science on their own.

Science has become increasingly important in the lives of all citizens. Every day we have to make medical, investment and political decisions for which a modicum of scientific knowledge is required. Science is too important to be left to the scientists, doctors and stockbrokers. In addition, our lives are filled with technological devices, whose use is facilitated by just a little scientific understanding. And wouldn't it be great if we could understand the exciting scientific advances reported in the media? Scientific ideas have also made their way into our vocabulary. When terms such as symmetry, entropy, gender, genes, mind or sustainability, are used in conversation, it would be helpful to know just what we and others are talking about.

From my experience as a faculty member of Emory University and from perusing the catalogs of other institutions of higher learning, I doubt that American universities are preparing their non-science students to function in our scientific-oriented society after they graduate. This shortcoming affects even the most elite of U. S. educational institutions, as indicated by 22 of 25 graduating Harvard students being unable to correctly answer the question "Why is it warm in summer and cold in winter?" Meanwhile the American public remains exceedingly open to mystical ideas. Polls have reported that a large fraction of Americans believe in ghosts and astrology. Can our society be at once the most technically advanced and the least scientifically knowledgeable among the developed nations, and continue to prosper? This doesn't seem possible.

I believe a major cause for the scientific illiteracy of non-science graduates of major universities is the lack of good courses designed for them in this area. Offerings are often "watered-down" versions of regular science courses or special topic courses taught at a superficial level. One reason for this is that science is built on a pyramidal structure. To understand biology, you need chemistry; to understand chemistry, you need physics; underlying the entire structure is mathematics.

Over a decade ago, I started to teach a course at Emory University that I hoped would circumvent this difficulty. By teaching physics in an intuitive, non-mathematical way, I hoped to cover sufficient background, so students could understand the basic ideas of chemistry and then go on and apply their physics and chemistry to biology, geology and astronomy. Since I couldn't find an appropriate textbook for the course, I began by providing the students with handouts on each of the topics. Those handouts have provided the foundation for this book. Since the ultimate goal of science is to understand the universe, I decided to tie the topics together with the theme of order and disorder — explaining the tendencies of Nature to move in each of these directions. Realizing that the vocabulary of science is quite extensive, I have included a glossary at the end of the book.

I must admit that I also had another goal in developing my course. As a practicing scientist for many years, I began to be aware of how limited my own scientific training had been. Frustrated at not being able to follow seminars whose titles interested me, it was exciting to go back and learn the basics of the various sciences. I suspect many other scientists have similar feelings, but are not at a stage of their careers where they can invest the time to do this. Perhaps this book would also be of some interest to them.

I am indebted to faculty members at Emory University who have helped me out by leading classes in my course, while I was learning their subjects. Edmund Day and Richard Williamon of Physics, Stephan Lutz of Chemistry (with a biochemistry specialty), Chris Beck and Walter

Escobar of Biology, Darryl Neill of Psychology, Paul Lennard of Neuroscience and William Size and Anthony Martin of Environmental Studies contributed over the years. In addition Amy Benson Brown of Emory's Manuscript Development Program read an early version of the book and provided very useful feedback. I always solicited comments from my class on the handouts, and one year even offered extra credit for these. Many of the student comments were very helpful.

Most of all, I would like to thank my wife, June, who read all the chapters, some more than once. In addition to her fine editorial eye, she continually encouraged me to make the book simpler and more relevant. I tried to take her advice, but for some of the topics I just didn't feel it was possible. Where the book is difficult, obscure or boring, the fault is all mine.

1 The Nature of Science

The great tragedy of Science — The slaying of a beautiful hypothesis by an ugly fact.

Thomas Huxley

Science is useful, science is mysterious, but one thing for sure: science is just too important to be left to the scientists. We all have to make decisions involving lifestyles, medical treatments and politics that are intimately involved with science. Listening to the experts is not very satisfying, since sometimes their advice is contradictory or changes with time. The need to understand the world around us is a basic human drive, and at the very least we should know what sort of questions can yield to science and what sort are better answered by our religious or intuitive beliefs.

This book is organized around the theme of *order and disorder*, a concept that pervades the universe at every scale, from the smallest elementary particle to the largest grouping of galaxies. Our objective will be to see how the very simple principles on which the universe is based combine to produce the limitless complexity that we observe in the world around us, i.e., the approach is **reductionist**. Before approaching this task, in this chapter we will briefly describe science and how it differs from other ways in which humans approach their environment.

The Ways of Science

What is this concept called "science" that has so enhanced and threatened our lives, so empowered our imagination? Why is it that, while stock market theories oscillate back and forth and philosophers and theologians continue to ponder age-old questions, science marches on, making ever-widening gains in the understanding and use of Nature? Is it something in the scientist's approach to comprehension that leads to these advances, something that perhaps could be applied to other enterprises and would be a worthy topic for study by nonscientists? If, as said by C. P. Snow, "The steam engine helped to shape the modern world at least as much as Napoleon and Adam Smith," non-scientists can little afford to remain ignorant of technological developments and their scientific underpinnings. In this chapter we will find out a little about the scientist's approach to "knowing" about the universe.

We learn about the cosmos by observation. Even at this first step of understanding, there are differences between the approach of a scientist and a layperson. (By laypeople I mean everyone besides scientists, including many of the world's greatest thinkers.) A scientist prefers observations to be repeatable and, if possible, **quantitative** (represented by a number). Thus, while a layperson might be satisfied with the observation: *Vegetarians are thinner,*[1] a scientist considers this a **hypothesis**. A hypothesis is an assumption that has to be tested, in this case by weighing a group of vegetarians and a group of meat eaters. In weighing, an immediate question is: how good are the measurements? To answer this, a scientist wants to weigh at least some of the participants a number of times. Statistical analysis of these repeated measurements gives an estimate of their **precision**; repeated measurements that are very close are precise. A competent scientist, however, knows there is more to observations than how precise they are. It is also critical that the measurements are **accurate**, i.e., record the actual value of the property of interest. As an example of good precision but poor accuracy, the participants in our study could be weighed with an inaccurate scale. One way around this is to calibrate the measuring instrument: for a scale — using certified weights spanning the range of weights of the participants.

Just what do we mean by an **observation**? The only real things are the forces that influence our sense organs. Between these forces and our

---

[1] For a scientific study of this hypothesis see *American Journal of Clinical Nutrition,* June 2005.

interpretation of them are a number of filters, both conscious and unconscious, that influence our understanding of the external world. These filters are located in our sense organs as well as our brain. For example, it is likely that when observing a forest, the retinas of our eyes deemphasize the vertical pattern of trees, allowing us to focus on more significant information, such as a lurking tiger. In science, we additionally interpose instruments between our sense organs and the external world. We use electronic measurement to very accurately determine the time it takes a sprinter to cross the finish line. Some observations may seem rather indirect, when in fact they are as direct as they can be. For example, we learn about the Earth's core, 1500 miles below its surface, by observing the bending of seismic (sound) waves traveling through the Earth. Like our observation of the tiger in the forest, this is an observation at a distance; we no more want to reach down and touch the Earth's molten core than we want to reach out and pet the tiger. However, interpretation of the seismological measurements, while very well founded, is still less direct than recognition of the tiger. Observations are made through time, as well as through space. At the simplest level, my dog knows that there have been other dogs, or cats, in the neighborhood by the scents they have left on the grass or cement. In the same way, the paleontologist knows about the existence of ancient creatures by the fossils they have left in sediments.

Even with accurate measurements, it can be difficult to test a hypothesis, since only a subset of a population is observed. (By testing a hypothesis we mean determining the probability that it can be rejected. A hypothesis cannot be proven.) For example, we can only weigh a small fraction of all vegetarians (subjects) and meat eaters (controls). It's necessary to make sure that these two groups are equally representative of different sexes and ages. With human subjects, however, there are very many other factors that can influence weight—we call these *variables*, and it is impossible to match the two groups for all of these. Once again scientists rely on statistical analysis to test whether these ignored variables can affect the conclusions. It is the difficulties in matching subject and control groups that often result in changing recommendations in health and nutritional fields. No wonder those diets don't always work!

In some ways, experiments are simpler in the physical sciences, which deal with inanimate matter. In this case, there are fewer inherent variables that must be controlled. Two samples of pure copper differ only in the mass of the sample, and the mass indicates the number of atoms in the sample. (We say that the mass is **directly proportional** to the number of atoms.) In some experiments the shape or isotopic constitution of the sample (Don't worry if you don't know what this means right now, we'll get to it later.) might also be important. Copper atoms don't have hidden variables that have to be controlled. On the other hand, measuring the properties of interest in the physical sciences often involves very sophisticated equipment. The experimenter must be careful to insure that results are not influenced by factors such as vibrations, contamination or atmospheric disturbances.

Observations, aided by statistics, are the "stuff" that *all* science is made of. Unfortunately, there is just too much stuff to observe in the world around us, and if scientists are to make any sense of it, they must find ways of organizing and making generalizations about the world.

Scientific laws are generalizations of observations.

They are usually expressed mathematically. Examples of scientific laws are Ohm's law: electric current is ***directly proportional*** to voltage, as one doubles so does the other; and Boyles law: the volume of a gas is ***inversely proportional*** to its pressure, as one doubles the other halves. The name "Scientific Law" is unfortunate, since it conjures up the idea that the generalizations it implies cannot be violated. In fact, a scientific law often holds only with limited accuracy and can be applied to only a restricted range of phenomena. For example, Ohm's law "breaks down" at very high current densities (current flowing through unit area) and Boyle's law at very high pressures.

There is another way that we can try to understand scientific observations, and that is to try to explain them at a more fundamental level. When such an explanation is first proposed, it is a hypothesis, i.e., an assumption that must be tested. In this case, the testing is usually by the test of time. What happens is that when the explanation is publicized, scientists think of the various consequences that should occur or should not occur, if the explanation were correct. They do experiments to try to observe these consequences. If the results of their experiments are in accord with the hypothesis, they gain confidence in it.

A scientific theory is a well-tested hypothesis.

If, on the other hand, the results of their experiments do not agree with the predictions of the hypothesis, they tend to reject it. Note that a hypothesis is never proved, because there may be observations incompatible with it that lie just beyond the horizon. Hypotheses and theories are answers to a four-year-old's favorite question, "Why?". Scientists are among the adults who never stop asking this question.

Supplemental Reading: The Kinetic Theory of Gases

Boyle's law will be used as an example. The most famous scientist of Boyle's era, Isaac Newton (1642–1727), proposed a theory to account for Boyle's measurement of the spring of the air.[2] Newton envisioned a gas as comprised of stationary particles that repelled each other. The nature of this repulsive force was such, that as the molecules were pushed closer together, the repulsion increased (hence behaving like a spring). As the volume of such a gas is decreased, its pressure would increase, in accord with Boyle's observations. Such was Newton's reputation, that this theory was almost universally accepted, and an alternative theory, proposed by Daniel Bernoulli (1700–1782), received very little attention. In Bernoulli's theory, the pressure produced by a gas confined in a container is explained as due to the incessant bombardment of the walls of the container by the very small particles constituting the gas — much as the way a soccer ball bouncing off a players chest drives him backwards. This is called the **kinetic theory** of gases. In the simplest form of Bernoulli's theory, all interactions between the constituent particles were neglected, giving results in exact agreement with Boyle's law. The kinetic theory, however, had some definite advantages, most importantly in relating the temperature of the gas to the average kinetic energy of the gas particles. In its more advanced form, which included interactions between the gas particles, the kinetic theory also predicted observed deviations from Boyle's law and allowed the calculation of other properties of gases. Notwithstanding Newton's reputation, over time, because of its clear advantages, the scientific community overwhelmingly adopted Bernoulli's theory.

---

[2] *New Experiments Physico-Mecanicall, Touching the Spring of the Air and its effects*, was Boyle's title for his 1660 manuscript.

The above example of an early struggle between two scientific theories illustrates the conflicting tendencies in science. On the one hand, scientists, like other professionals, demonstrate a degree of conservatism. They don't like to change the way they think about something, nor do they want to offend the powers-that-be. Since in the peer-review process, other scientists review scientific papers and grant applications, new ideas often struggle to get an adequate hearing. On the other hand, in science there are considerable rewards for introducing a valuable new paradigm. In addition, experimental scientists are always looking for better ways to explain their data. Rarely can inadequate entrenched ideas hold out for too long against such forces. While things may not be as simple as stated by Huxley at the beginning of this chapter, in science, better and more useful theories eventually win out. Thus we see that an established scientific theory is not something to be taken lightly. It has stood up to the test of time, and in the realm of its applicability, has *not* made predictions that disagree with observations. In essence,

> **a scientific theory must be testable,**

or according to the philosopher Karl Popper, it must be *falsifiable* — capable of being proven wrong. In fact, since it may change the way that scientists think about a phenomenon, an experiment giving a negative result may ultimately be more important than one giving a positive result. It is the accumulation of negative results, of observations that cannot be explained by the current theory, that opens scientific minds to accept a new way of looking at things — a new paradigm.[3]

For some reason, among the general public, a scientific theory is often regarded more like a hypothesis. We hear expressions such as, "It's only a theory." Such expressions devalue scientific theories, which, after all, have provided a powerful framework for organizing and developing a body of knowledge. Although we must always admit that at some later date, something may be observed that will disprove a theory, in the realm of science it serves as the best explanation that we currently have. On the other hand, explanations of observations that are incapable of being disproved (such as, "because God made it that way") might be comforting to many people, but they do not fall within the realm of science. This does

---

[3] Thomas Kuhn, *The Structure of Scientific Revolutions*, 1962.

not mean that science is the only way to look at Nature. It is, for example, useless in describing the beauty of a sunset or the awe we feel when gazing down into the Grand Canyon,[4] nor can it answer questions such as "Is there existence after death?" or "What is reality?" Questions involving morality, legality, good and evil also fall outside of the realm of science.

Although the ultimate test for a scientific theory is how well it agrees with observation, scientists do have a predilection for theories that satisfy their aesthetic, and even perhaps their religious, sense. As stated by Einstein, "My religion consists of a humble admiration of the illimitable superior spirit who reveals himself in the slight details we are able to perceive with our frail and feeble minds." Scientists often ask themselves questions like: Is the theory beautiful? Does it connect a number of different phenomena in the simplest possible manner? Is this the way God would do it? Of course, some of these are matters of intuition. Intuition can lead us up the wrong tree, as exemplified by Einstein's comment, "God does not throw dice," which summed up his opinion of the probabilistic interpretation of quantum mechanics—the physics of very small particles. However, this probabilistic interpretation *has* survived the test of time and is now accepted by the great majority of physicists. It appears that God does throw dice!

This is a book about Nature, which is the subject matter of science. Since we are all observers of Nature in our everyday lives, we might think that we would have accumulated considerable intuition concerning its operation and organization. However, we must realize that our experience encompasses only a limited range of physical parameters. Does our experience with mass ranging from that of a flea to a planet have much relevance for dealing with super-massive black holes at the centers of galaxies or particles as light as an electron? Is a lifetime traveling at speeds up to that of a jet plane relevant for dealing with speeds approaching the velocity of light? Do observations at durations as short as a "blink of the eye" or over our lifetimes apply to processes that occur at durations as short as a billionth of a billionth of a second or as long as a billion years? Do all creatures that exist or have existed resemble those with which we are familiar? We should not be surprised if some of our common-sensical ideas, developed from our limited vantage point in the universe, must be modified to deal with phenomena outside our realm of experience.

A limited perspective can affect scientific theories as well as intuition. At any one time the scientific literature encompasses a range of

---

[4] Neuroscientists have made some progress in understanding our appreciation of beauty.

experiences, a range of familiarity with scientific variables. The scientific theories of the time might be perfectly adequate to deal with this range, but as experience expands, they may be found lacking. In this book, we will review a number of examples of scientific theories, which were once thought to be universally applicable, but later had to be modified or dropped entirely in order to deal with newly observed phenomena.

Vignette — Spontaneous Generation of Life

By the seventeenth century, observations, such as the appearance of maggots on meat left exposed or mice found in stored grain, led to the general acceptance of *abiogenesis*, or the spontaneous generation of life. In 1668, Francesco Redi decided to test this hypothesis by an experiment. By using experiment, rather than just observations, Redi was able to employ *control* groups, in addition to the test group of open flasks containing meat. In one set of controls, Redi sealed the flasks, while in another set he covered the flasks with gauze. After several days, flies were observed in the test group and in the flasks covered with gauze. No maggots were observed in the sealed flasks. Noting that there were flies buzzing around in the open flasks and above the gauze-covered flasks, Redi concluded that the maggots are not spontaneously generated, but rather formed from eggs laid by flies, either directly on the meat or dropping onto the meat through the gauze.

The results of Redi's experiment seriously damaged the abiogenesis hypothesis, at least as far as it applied to maggots and meat. However, the idea persisted for another two centuries, until Pasteur showed in 1859, that it was micro-organisms in the air that were responsible for things growing in sterilized cultures. Pasteur's results disproved one of the last existing "examples" of abiogenesis. It should be noted, however, that the experiments of Pasteur, Redi and others applied to specific instances of abiogenesis. They did not disprove abiogenesis under conditions far from those in which it had previously been suggested to occur. Specifically, in all cases discussed at the time abiogenesis was thought to have occurred over days, weeks or years. Such experiments have little to say about whether life can be spontaneously generated from inanimate matter over millions or billions of years.

The Languages of Science

For science to advance rapidly, scientists must be able to communicate with each other efficiently. One problem is that scientists from different countries have different native languages. Although there have been attempts to develop languages that are syntheses of many languages,[5] they have never become important for scientific communication. Over time, English has gradually become the dominant language in science. At international scientific conferences almost all papers are currently presented in English. In addition, the most important journals are either published in English, or have English translations readily available.

Many observables involve a unit in addition to a number. It is meaningless to say that something or someone weighs 140, without indicating whether we are talking about pounds, ounces or kilograms. The units in which quantities are expressed are an important aspect of scientific language. Measuring quantities in a single set of units facilitates communication. In 1960 an international conference adopted the set used in France, the SI system — we call it the metric system, as the units to be used for scientific communication. Its primary advantage over the system used in English-speaking countries is that it is a decimal system, in which derived units are obtained by multiplying or dividing by powers of 10. Only a minimum set of SI units must be defined; the others can be obtained from these from basic laws of science. The basic SI units are given in Table 1.1.

Table 1.1. Basic SI units

Quantity	Unit	Abbreviation
Mass	kilogram	kg
Length	meter	m
Time	second	s
Electric charge	Coulomb[a]	C
Temperature	Kelvin	K
Amount of substance	mole	mol

[a]We will use the unit for electric charge, rather than that for electric current, the Ampere.

---

[5] Esperanto, for example, is a synthetic language, synthesized from many Western languages.

Table 1.2. Some useful numeric prefixes

Prefix	Abbreviation	Multiple
giga	G	10^9 — a billion
mega	M	10^6 — a million
kilo	k	10^3 — a thousand
centi	c	10^{-2} — a hundredth
milli	m	10^{-3} — a thousandth
micro	μ	10^{-6} — a millionth
nano	n	10^{-9} — a billionth

A problem with scientific units is that the range of quantities of interest in different areas of science is very large. How can one system of units accommodate discussions of viruses, elementary particles and galaxies? There are two methods for doing this in science. The first is by the use of a group of prefixes affixed to the scientific units. The most commonly used of these, and the only ones used in this book, are given in Table 1.2.

Even more powerful is the use of *scientific notation*. This involves multiplying or dividing a quantity by an appropriate power of 10 (10 raised to an appropriate *exponent* — negative exponent indicates division). By these methods, 0.000000003 meters could be simply written as 3 nanometers or as 3×10^{-9} meter. The latter method is particularly useful when an appropriate prefix is not available. For example, a much-used quantity in science that relates the macroscopic and microscopic world is Avogadro's number, 6.022×10^{23} molecules per mole.

Each time a quantity is multiplied by a factor of 10 we say it is increased by an *order of magnitude*. Quantities known to a factor of ten are said to be known to an order of magnitude. An order of magnitude that comes up in number of contexts is 10^{11}, a hundred billion. It is a good approximation to the number of stars in our galaxy (the Milky Way), the number of galaxies in the universe and the number of neurons in a human brain.

Due to the wide range of distances that they deal with, astronomers use some specialized distance units: the astronomical unit (the average distance between the sun and Earth), ca. 1.5×10^{11} m; the light year (the distance light travels in a year), ca. 9.5×10^{15} m; and the parsec, ca. 3.1×10^{16} m.

Probability in Science

Many ideas in science are based on probability (a field called statistics). Probabilities of events are often independent (the occurrence of one event does not depend on the occurrence of the other). In these cases the probability of combined events is the product (they are multiplied) of the probabilities of the occurrence of the individual events. This is equivalent to adding the exponents of the probabilities expressed in scientific notation.

Example 1. What is the probability of drawing the queen of spades from a deck of cards?

Discussion: Since there is only one queen of spades and 52 cards, the probability is obviously 1/52. However, this question can also be answered by realizing that there are four suites and 13 types in each suite. The combined probability of the independent events of drawing a spade and of drawing a queen is $1/4 \times 1/13 = 1/52$.

Example 2. If one person in 10,000 gets a disease and one of every 100 who get the disease dies, what is the probability of dying of the disease?

Discussion: As written, these are independent events, so the combined probability (dying of the disease) is $1/10{,}000 \times 1/100 = 1/1{,}000{,}000$. In scientific notation, $10^{-4} \times 10^{-2} = 10^{-6}$. Note that when you multiply powers of a quantity, you add its exponents.

Science employs a number of languages besides native languages and English to communicate ideas. Among these are mathematics, chemical equations, molecular models and genetic maps. More than the others, mathematics permits manipulation as well as presentation of quantities. What mathematics manipulates, however, are symbols, rather than ideas. Mathematics is not a science.[6] Its theorems are not subject to refutation based on comparison with experiment. A mathematical equation is not a scientific theory. However, when we give physical significance to the

---

[6] Although in universities, it is often placed in the same division as science departments.

symbols in the equation and assign values to its parameters, we say that we have made a ***mathematical model*** of some phenomenon based upon a theory. Mathematics can then be used to explore and present the logical consequences of the theory. When one of these consequences does not agree with experiment, the scientist looks at the adequacy of the theory, the correctness of assignments in the model or the accuracy of the measurements as the source of the problem, instead of questioning the mathematics employed.

Most of the difficulties in understanding science results, not from comprehending its laws and theories, but rather from having facility in using one of its languages. It is the premise of this book, that for understanding science, in contrast to practicing it, this difficulty can be circumvented, and scientific ideas can be presented in a manner understandable to non-scientists.

Organization of the Book

The general order in which the material is presented is:

physics → chemistry → biology → Earth sciences → astronomy.

This order allows discussion of topics before they are used. In some instances a concept may have become so familiar to me, I didn't realize that it would give the reader difficulty. My wife has tried to catch me on these, but I take complete responsibility for them. Realizing that it might be difficult for the reader to remember all the terms used, I have included a glossary at the end of the book.

Summary

In order to assess precision and accuracy, scientists prefer observations that are repeatable and quantitative and which compare experimental and control groups. Observations are summarized in laws, which are explained at a deeper level by theories. A scientific theory must be testable and is only accepted over time, after it has been not disproved by many tests.

Scientific communication is usually in English, using the SI set of units, with prefixes indicating large or small numbers. Alternatively, scientific notation may be used. Expressing laws mathematically allows novel insights to be gained by mathematical manipulation, but is not essential for understanding basic concepts.

Questions

1. Why in the vegetarian study, would the accuracy of the scale be less important, if both the vegetarians (subjects) and the meat eaters (controls) were weighed with the same scale? Even if this were done, would the accuracy of the scale be completely unimportant? E.g., what if the scale was off by 1000 pounds?
2. If a given amount of gas at room temperature and 1.0 atmosphere pressure occupies a volume of 8.0 liters, what volume would the gas occupy, if the pressure were increased to 2.0 atmospheres?
3. You go to the doctor with a sore throat and fever. The doctor prescribes an antibiotic. You take it and get better.

    (a) What was the doctor's hypothesis?
    (b) What experiment did this hypothesis lead to?
    (c) Did the results of the experiment prove the hypothesis?

4. Give an example of direct proportionality and inverse proportionality.
5. Explain how the kinetic theory of gases could justify the observation that as the volume of a given amount of gas is decreased (keeping the temperature constant), the pressure of the gas is increased.
6. Is it part of the scientific method that scientists plan their experiments without any preconceived ideas?
7. Most universities have a committee for prior review of experiments involving human subjects. Do you think that the membership of such a committee should be limited to scientists? Defend your answer.
8. Are there any vegetarians in your class? Are they thin? What does this prove?
9. An identical twin claims that at the instant that his twin, who resided in a distant city, had a heart attack, he felt a sharp pain in his chest. As a scientist, would you dismiss this assertion or take it as

evidence of a new type of telepathy between twins? What experiments would you propose to test the latter?

10. Is each of the following a law, theory or hypothesis?

    (a) Crop circles are produced by alien beings.
    (b) When the supply of a commodity rapidly increases, its price drops.
    (c) Investor behavior is determined by fear or greed.

11. If one quantity is a billion times as large as another, by how many orders of magnitude do they differ?

12. Devise an experiment to test the efficacy of magnetic therapy — namely the ability of magnets to relieve pain. Be sure to consider the placebo effect, experimenter bias and matching experimental and control groups.

13. Gordon Moore, cofounder of Intel, observed that computer power doubled every year. Is this Moore's law, Moore's hypothesis or Moore's theory? (Recently the doubling time is closer to two years.)

2 Order

Symmetry,… is one idea by which man through the ages has tried to comprehend and create order, beauty, and perfection.

Hermann Weyl

Order at Different Scales

The ideas of order and disorder are dominant themes in the universe. We can imagine exploring order in the universe at different scales[7] using a super telescope–microscope combination instrument. First, with the telescope, we find that we are bathed in microwave radiation (more about this in Chapter 5). This radiation was created not long after the cataclysmic formation of the universe in what is called the *"**Big Bang**."*[8] The microwaves impinge upon us in nearly equal intensity from all directions, as would be expected from the disorder at the incredibly high temperature of the early universe. On closer inspection, however, the intensity distribution of radiation shows structure, showing that even in the very young universe some order was present.

---

[7] A number of web sites called Powers of Ten allow one to visually explore the universe at different scales.

[8] We see the radiation from the time that the universe first became transparent, approximately 400,000 years after its formation.

With visible light, we see that galaxies (each containing billions of stars) are not arranged randomly, but are organized in everything from clusters to giant walls. The galaxies themselves have different shapes, among them spiral (like our Milky Way) and elliptical. Individual stars range from hardly visible brown dwarfs to massive red giants. Some stars have collapsed to neutron stars or even black holes, while others are more like our own sun. Planets surround many stars.

In our solar system, we see remarkable differences among the planets. The inner planets consist primarily of rocky material, while the outer ones are much larger, surrounded by rings of matter and are composed mostly of highly compressed gases. Many smaller bodies also orbit our sun. Our own planet, Earth, is highly structured both internally and on its surface, with high mountain ranges and deep ocean trenches persisting, in spite of continual erosion by water and wind. Below the surface, temperature and pressure rise, until at its core the Earth is mainly solid iron at nearly 6000 degrees centigrade — about the same temperature as the surface of the sun.

Nothing is more organized than life. Our planet is rich in liquid water and its atmosphere in oxygen gas — two substances indispensable for most life on earth. We are most familiar with two types of life: plants that draw their sustenance from solar energy and have roots that suck up nutrients from the soil, and animals that survive by consuming the plants or each other. But fungi and microscopic forms of life also exist, and their variety exceeds that of plants and animals. It almost seems that there is no outward form, color or size that is or has not been represented in Earth's biosphere. However, at the molecular level, all life is constructed, functions and stores information in a similar manner.

With our super microscope, we can observe molecules, which demonstrate remarkable beauty in their symmetry and linking of form and function. For example, most protein molecules only function when they are folded in a particular manner. The versatility of one element, carbon, is more than can be comprehended in a lifetime. In combination with only hydrogen, carbon forms millions of different hydrocarbon molecules. These compounds, modified with oxygen, nitrogen, sulfur, phosphorous and trace amounts of other elements, comprise almost all of life that is not water.

When we look at or within atoms, the microscope images get fuzzy. Adjusting the instrument doesn't help. It seems that the fuzziness is something fundamental. Our inability to directly observe sub-atomic structures does not indicate a lack of understanding of atoms. In fact, we can make predictions regarding them with outstanding accuracy. For example, we know that electrically neutral atoms are composed of charged components — negatively charged electrons and a positively charged nucleus. The nucleus itself is found to have structure, being made up of protons, which supply its positive charge and electrically neutral particles called neutrons.

We are not through yet, as protons and neutrons are not the ultimate bits of matter, since they are composed of more fundamental particles, called "quarks," in various combinations. Quarks are very difficult to isolate, but knowing they exist allows physicists to make sense of the otherwise bewildering array of particles that are produced in high-energy accelerators. Some physicists now tell us that quarks, and everything else, may have another level of structure made up of vibrating loops of string. This "string theory," or the theory of everything (TOE) is not universally accepted. Don't ask what the vibrating strings are made of — there are fundamental reasons why this is an unanswerable question.

With such variety of form in the universe, are there any general principles that we can use to organize our ideas of order (and its converse, disorder)? First we will consider just how we might define order and disorder.

What is Order?

In defining order, it's helpful to think back to the first time that we heard this word. I am sure that few of us are capable of doing this, but for me, a likely possibility was when my kindergarten teacher said, "Now class, come to *order*." Presumably she meant — well, stop running around and take your seat. We were in school now, not in our permissive home environment. In other words, she was placing a restriction on the class. For the purpose of this book:

> **Order is restriction; disorder is freedom from restriction.**

We will also talk about order in terms of *information*. When the class comes to order, we know that the students are sitting in their seats — when they are in *disorder*, they can be anywhere in the room — we have much less information about their position.

We can classify the three different types of order in the following manner:

- limits,
- symmetry and
- functional order.

Limits

Limits are the permissible range of variables, such as position, velocity and time that members of a system can adopt. In kindergarten, our class was the system and the students the members. Even before class started, the students were restricted to the classroom (position). They had more order in this state than they would have if they were dispersed. Dispersal is disorder, the opposite of order. Students were also required not to run around (velocity) and to get to class promptly (time). These are the limits on their variables. As an example from the physical world, we can consider water, which when in a gaseous state (steam) might have its molecules confined to a container. In liquid water, there is more restriction, more order; molecules must stay in close proximity to each other. Moreover, there are limits on the velocities of liquid water molecules. If these velocities are too large, molecules will escape from the liquid into the gas — a process called vaporization.

Symmetry

Symmetry is more difficult to define than limits. Our appreciation of a beautiful piece of pottery, a flower, or an attractive face shows that our aesthetic senses are very much attuned to symmetry. An appreciation of symmetry is found in other species, as well. For example, there is evidence that bees are attracted to more symmetric flowers — such symmetry indicating health and abundant nectar.

How can we give an all-inclusive definition of symmetry? Thinking of the kindergarten class, we will make things simple by assuming it has only four students. Suppose that the teacher calls for the disorganized class of four students to form a circle to play a game. The system (the class) now has more symmetry than it did when in free play. How does a scientist describe this symmetry? First, to be able to grasp the essence of the new arrangement of the class, the teacher idealizes it by assuming the students are identical and form a perfect circle. Since there are 360 degrees in a circle, in the idealized situation the children are spaced at 360/4 = 90 degree intervals around the circle. "Hogwash!" you might say, "Students aren't identical and their circle isn't perfect." True, but unless you idealize the system, you miss the basic difference between the class in free play and in the circle. Scientists often perform such *idealization* to grasp the essence of a situation.

Now that we have idealized the system, we characterize its symmetry in the following manner:

> **Symmetry is specified by symmetry operations.**

A *symmetry operation* is any procedure that does not produce a distinguishable change in the system. The system appears identical before and after the procedure.

Rotation symmetry

One symmetry operation for the students in a circle is a rotation of 90 degrees (or some integer multiple of 90 degrees) around the center of the circle (more exactly, rotation around a *symmetry axis* perpendicular to the plane of the circle at its center). If we perform this symmetry operation, the idealized system is indistinguishable from it before the operation. The result of this symmetry operation is shown in Figure 2.1. Since the numbering of the students is something we have done for convenience (the students are identical) — the system is unchanged by the rotation.

In biology, body symmetry is one way of classifying life forms and is often related to function. For example, *rotational symmetry*, called *axial symmetry*, is useful for species such as jellyfish that take in food from all directions.

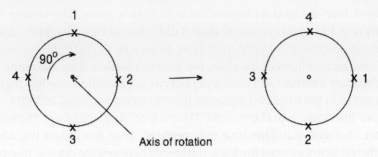

Figure 2.1. Rotation around an axis.

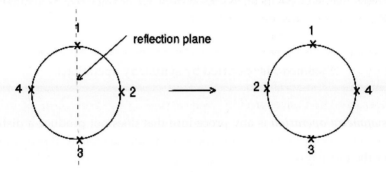

Figure 2.2. Reflection in planes.

Reflection symmetry

Other operations leaving the system indistinguishable that can be performed on the four students are *reflections through planes*. Reflection leaves all points at the same distance from the plane, but on the opposite side. For example, we can imagine a plane perpendicular to the plane of the circle, including its center and two students on opposite sides of the circle. As shown in Figure 2.2, upon reflection, the two students in the plane don't move and the other two students change places. There are two such planes, depending on the two students that we select to be in the plane. The two planes that bisect the angles between two students are also symmetry planes. Since no students are in these planes, they all change places upon this reflection.

In biology, reflection symmetry, such as found more or less in the outward appearance of humans, is called *bilateral symmetry*. Bilateral symmetry is useful for species that spot and chase their food in a particular direction.

Inversion symmetry

The system shown in Figure 2.1 has an additional type of symmetry, as shown in Figure 2.3. This is called *inversion through a point*; the point-of-interest being the center of the circle. After this operation each system point is on the extension of the line between its original location and the inversion point, at a distance equal to its original distance from the center. While the circle of points in Figure 2.3 is indistinguishable after this operation, when students are at the points, they end up up-side-down below the floor after inversion, as shown in Figure 2.4. Thus inversion through a point, while a symmetry operation of the two-dimensional diagram in Figure 2.4, is **not** a symmetry operation of our (idealized) class.

In the mathematical theory of symmetry, called *group theory*, the group of operations that leaves an object indistinguishable defines the *point group*

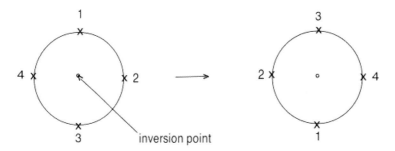

Figure 2.3. Inversion of points through center of circle.

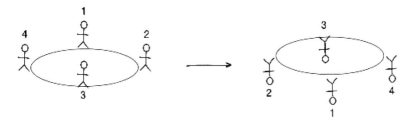

Figure 2.4. Inversion of students through center of circle.

of the object. For example, the point group of our four-student idealized class includes 90° and 180° rotations and the four reflection planes. For mathematical reasons, the identity operation — no change at all — is included in the group. Note that in each of the symmetry operations considered above, the center of the circle did not move.

Translational symmetry

In addition to point groups, there are other symmetry operations that move all points in a system. These operations are called translations and can only leave systems indistinguishable if they are extended. As an example, consider a class seated in a classroom in which there are two rows with two desks each. If we shift (translate) each student to the right by the width of one row, the class will certainly not be indistinguishable. Now we will have one row of empty desks and two students sitting on the floor. We call these edge effects. However, if our four students are sitting in a 2×2 section of a very large lecture hall, surrounded by many other identical classes, edge effects are less important. In the limit, when the lecture hall contains an infinite number of seats, edge effects are totally negligible. (Here we go, idealizing again!) Our infinite hall of identical students is indistinguishable after left-right or front-back translations equal to the distance between rows. These two translations form two sides of little rectangles with which we can fill up the entire infinite lecture hall — or alternatively, two-dimensional space.

The rectangles that fill up the hall are called its ***unit cell***, which defines the ***translation group*** of the infinite system. Unit cells of two-dimensional systems can be thought of as tiles covering a floor. It's possible to tile a floor, leaving no spaces, with tiles that have three sides (triangles), four sides (squares, rectangles or rhombuses) or six sides (octagons). However, tiling with three sided-tiles requires rotating the tiles as well as translating them. Filling up a floor with five-sided tiles is not possible. Thus, only four and six-sided unit cells are unit cells for two-dimensional systems. Floor tiles in these shapes are shown in Figure 2.5.

Molecular crystals are periodic arrays of molecules. Molecules are so small, even a tiny molecular crystal can be thought of as infinite on the molecular scale. Since molecular crystals are three-dimensional objects, movements in three directions are needed to define their unit cells and translation groups. In order for these unit cells to fill up three-dimensional space upon translation, they must have sides that are squares, rectangles, rhombuses or octagons.

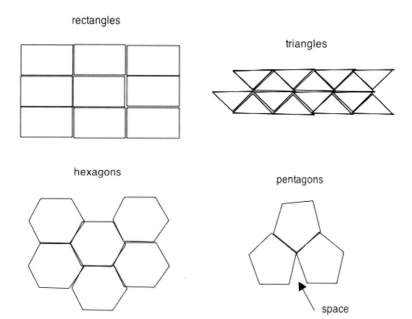

Figure 2.5. Two-dimensional unit cells.

Figure 2.6. Crystal of halite.

In nature, molecules and compounds are often found as macroscopic crystals, with planar faces meeting at angles. The form of the macroscopic crystal is a reflection of the arrangement of the microscopic atoms and molecules in their lattice. Thus, as shown in Figure 2.6, halite crystals, which are sodium chloride — common table salt, have planes that meet at right angles because its sodium and chloride ions are arranged on a cubic lattice.

One difference between floor tiles and atoms or molecules in crystals is that if you want to tile a floor, in addition to buying tiles, you have to hire someone to lay them. Atoms and molecules, on the other hand, self-assemble into their positions in the crystal lattices. Slow growth favors the formation of large crystals, and halite crystals larger than a person have been found in caves. Self-assembly is essential for producing order in the microscopic world and living systems.

Of course, not all naturally occurring substances have macroscopic structures with the straight edges and regular shapes characteristic of crystals. Many show curved, or even spherical, shapes that result from minimizing their surface area. This is because, as will be discussed in Chapter 6, surfaces have higher energy than bulk material.

Symmetry in time

Just like objects can repeat in space, events, such as the swings of a pendulum, repeat in time. The time for repeating is the *period*, **t**, of the phenomenon.

Events can also repeat in space and time, such as occurs when the teacher marches her class in a line to the cafeteria for lunch. If once again we idealize the students as identical, the line repeats with a characteristic distance, l, the distance between the students. When the students are marching, we can also characterize the line by a quantity, f, which is the number of students that pass a given point in a certain time, say a minute. We will make use of these two quantities in Chapter 5 when we consider wave motion. l is called the *wave length* of the line and f its *frequency*. The frequency is the reciprocal of the period.

$$(t = 1/f).$$

We should note that our kindergarten class does not spontaneously achieve the various types of order that we have discussed. Its natural state is disorder, i.e., anarchy. Ordering the class requires a force; in this case the force of the teacher's personality.

Fractals

What sort of symmetry has the object in Figure 2.7? Although it has no obvious rotations, translations, reflections or inversions, instinctively we feel that this computer drawn "fern leaf" has some sort of regularity. The object has what we call *fractal symmetry*, or symmetry of scale.

Figure 2.7. A fractal object.

Figure 2.8. Romanesco broccoli.

What this means is that if you take one branch of the fern leaf, magnify it and rotate it appropriately, it looks just like the original leaf. In fact, the branches themselves are made up of smaller branches with which we can make similar transformations. We can continue this process until we reach some natural limit, which might be the resolution of the computer image or the biological unit of a natural fern leaf.

Fractal order is ubiquitous in Nature. Some examples are snowflakes, clouds, mountain ranges and river branching. Figure 2.8 shows romanesco broccoli, a vegetable in which fractal symmetry is very apparent.

The essence of a fractal object is that

A fractal object appears identical at many scales of observation.

Because of this characteristic, there are properties of fractal objects that are not defined until we know the scale of our observation. Questions such as: what is the length of a river or the coast of Maine, or what is the circumference of the fern in Figure 2.7, are not answerable until we know the length of the ruler with which we are making our measurements. Fractal order is indicative of regularity in the process that is forming the fractal object. For example, the branching of a river results from endless combination of smaller streams into larger ones. Some scientists have invoked fractal ideas to explain structure in the universe at many different scales.

We are familiar with the concept of the dimension of an object. A piece of white paper, for example, is a two-dimensional object (if we neglect its thickness), because we can locate any point on the paper using two perpendicular axes. Likewise, a straight line is a one-dimensional object and a bowl of Jell-O is a three-dimensional object. These three objects also have fractal symmetry, since they look the same at any scale. We generalize the concept of dimension to any fractal object. For example, the coast of Maine is said to have a *fractal dimension* of something between one and two because it fills up space in a manner between that of a straight line and a flat plane. In a similar manner, the surface of a mountain range has a fractal dimension between two and three because it fills up space in a manner between that of a flat plane and a solid block. There are mathematical procedures to determine the exact fractal dimension of an object from the rules required for its construction. However, we will be satisfied with an intuitive feeling for this concept.

Supplemental Reading. Effect of Time and Temperature on Symmetry

Symmetry depends upon the time over which we observe a system. Let's consider, as an example, the monatomic (made up of single atoms) gas argon. If argon is cold enough (below −189°C at one atmosphere pressure), it is a solid, which forms a regular crystal. A two-dimensional slice through this crystal is shown in Figure 2.9a. Figure 2.9b shows the instantaneous position of a number of gaseous argon atoms in a container.

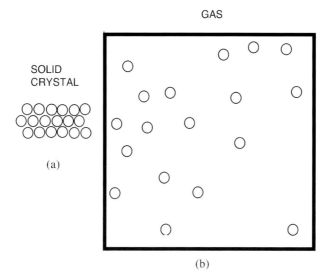

Figure 2.9. Crystalline and gaseous argon.

It would seem that crystalline argon, having a number of translational symmetry operations, is the more symmetric of these two systems. The gaseous argon atoms have no symmetry as drawn. However, we must remember that this is an instantaneous picture of the argon atoms. What would happen if we observed these atoms over a longer time, say by photographing them with a long-exposure film that blurred their instantaneous position? While it is impossible to do such an experiment, we can imagine the result. Argon atoms move really fast. We can assume that the atoms don't interact with each other or the walls of the container, so that they don't prefer any position in the container over any other position. Under such assumptions, our long-term exposure would just show the gas as an equal-intensity blur throughout the container. Such a blur has very high symmetry, being unchanged upon translations, rotations, reflections and inversions. Thus, over a long-time average, a gas has a higher symmetry and higher order than a crystal.

Some time averaging is an essential process for achieving symmetry. Even if the students in our kindergarten class were identical quadruplets, and we clearly marked where in the circle they should stand, the disorder caused by the fidgeting inherent in children would destroy any symmetry, unless we averaged over time. Atoms and molecules fidget more when they are hot than when they are cold. Temperature is a measure of their fidgeting, of their random motion. Thus, when averaged over time, many microscopic systems show higher symmetry when they are hot (gaseous argon) than when they are cold (solid argon). In a similar manner, symmetries existed in the early, very hot, universe, that no longer exist at the present time. We say that these symmetries have been "broken." Although symmetry can be increased by time averaging at higher temperature, such symmetry does not increase information or order. Order generally decreases at higher temperature.

Functional order

For living and life-produced systems probably the most important type of order is *functional order,* the ability of the parts of an individual or the individuals of a society, or the members of an ecosystem to work together to produce an overall effect greater than that possible by a single part or individual. This is what the kindergarten teacher evaluates when she gives a report card grade for "ability to work and play well with others." Anyone who has watched kindergarten students play soccer has probably noted that, while some of them may be amazing athletes, they show little functional order.

Functional order is much more complicated than symmetry or limits, since it must be measured against goals, which it rarely completely achieves. In addition, perfect functional order may be a moving target as external conditions change (e.g., a kindergarten soccer team of ten years ago might not be good enough to compete today, when children in the U. S. start playing earlier.)

Particular note should be made of functional order that is *cyclic*, i.e., that will regenerate itself after a time. Thus, although the kindergarten teacher trains the doctors, engineers and managers that are required for the smooth operation of society, it is also necessary that some of her students become teachers and perpetuate the system. This is the process of *replication*. In fact, it would be best if some of her students grew up to be even better teachers than she is. This is the process of *evolution*. It is generally considered that systems capable of maintaining themselves, replicating and evolving show all of the characteristics of *life*. We will have much more to say about this later in the book.

Summary

The universe shows evidence of order at all scales as well as of disorder resulting from the high temperature of its initial formation. Order results from restriction, disorder from lack of restriction. Order is of three types: limits, symmetry and functional order. Limits can be in position, velocity or time. Symmetry is specified by symmetry operations, such as rotations, reflections or inversions. The group of symmetry operations that characterized a system is called its point group. There may also be a translation group that defines the unit cell of a system. Waves illustrate symmetry in both space and time. Some objects show symmetry of scale, called fractal symmetry. Functional order is required for a system to operate smoothly.

Questions

1. We observe microwave (long wavelength) radiation from the very high temperature Big Bang, because the universe is expanding from its initial very small volume with high velocity. What sort of order did the universe possess in its initial formation?
2. What is the change in order of the water molecules in a small droplet as it evaporates?
3. What would be the symmetry operations of five students equally spaced in a circle?
4. What are the symmetry operations of four points equally spaced on a circle?
5. In the children's game, "ring around a rosie," there is a child at the center of the circle in addition to the children around the circle. Does this change any of the symmetry conclusions of this chapter?
6. What types of symmetry operations has a person?
7. Show that it is impossible to tile a floor with five-sided tiles, and thus pentagons cannot be the unit cell for a two-dimensional system.
8. Discuss the fractal dimensions of: a length of string thrown on the ground, a tightly wound ball of string and clouds.

3 Particles

The beginnings of all things are small.

Cicero

In order to discuss the organization of the universe we need to know what we are organizing — what are the building blocks of the universe? We will take the basic components of the universe to be ***particles***, entities that are located in very small regions of space. These entities are discrete, rather than continuous.

Atoms

What is the world around us composed of? Consider the successive dividing of a bar of gold into halves. For a while it is obvious we still have gold, since the ***intensive properties***, those that are intrinsic to each bit of material and don't depend on amount, remain the same for the smaller pieces. We can even, with some difficulty, continue this process under a microscope. Can this go on forever, or is there a point at which further division is either impossible or no longer leaves us with something we can call gold? Since the time of the Greeks, philosophers such as Leucippus, have suspected that there is indeed a limit to this division, i.e., there is an ultimate particle of gold, called an ***atom*** by Democritus (460–370 BC.). Any division of a gold atom leaves us with something we can no longer call gold. Today there is no doubt about there being ultimate particles of a material, since we have microscopes that can "see" these particles. For example, Figure 3.1 is an image of the surface of a piece of gold, taken with a scanning tunneling microscope (STM), an instrument that is capable of resolving individual atoms on a surface.[9]

Gold is one of the chemical *elements*. There are about 100 such elements, which in various combinations make up the myriad materials observed in the world around us. Atoms cannot be converted into each other with "chemical" amounts of energy. A chemical amount of energy is roughly the amount of energy that we can obtain from a flame or battery.[10] Some examples of chemical elements are hydrogen, oxygen, mercury and iron. Under normal conditions, hydrogen and oxygen are gases, mercury is a liquid, and iron, as well as gold, is a solid. New elements are still being discovered today, but the atoms of these elements live only a small fraction of a second before they spontaneously convert to other particles.

There are also materials that can be converted into each other or into elements by chemical amounts of energy. These are called *compounds*. For example, consider water, the most important compound for life as we know it. We can divide a water droplet down to our observable limit and still have something that keeps the intensive properties of water. However, we can also pass an electric current from a battery through water — a process called

Figure 3.1. STM image of gold surface.[11]

---

[9] A scanning tunneling microscope operates with electrons, rather than light.

[10] Elements can be transmuted (changed into other elements) in particle accelerators, or in stars or by nuclear reactions. These involve energies that are thousands or millions of times greater than those available in chemical systems.

[11] Notice how the gold atoms on the surface arrange themselves in rows.

electrolysis — producing hydrogen and oxygen, two gases with properties very different from liquid water. In doing this we always get eight grams of oxygen for each gram of hydrogen — an example of what is called the *law of definite proportions*. There is also another compound of hydrogen and oxygen, called hydrogen peroxide, which upon electrolysis produces sixteen grams of oxygen for each gram of hydrogen. The existence of more than one compound of two elements with mass ratios that are in the proportion of small whole numbers is called the *law of multiple proportions*.

Observations like these are explained by the *atomic theory* of John Dalton (1766–1844), which specifies that:

> **The atom is the basic building block of matter.**

Elements are made up of a single type of atom, while compounds are composed of several different atoms, in ratios of small whole numbers. Chemical reactions are rearrangements of atoms, rather than their creation or destruction.

Example 1. There are two compounds of carbon and oxygen. One always contains four grams of oxygen for every three grams of carbon and the other eight grams of oxygen for every three grams of carbon. Show that these results are in agreement with the laws of definite and multiple proportions and can be explained by Dalton's atomic theory.

Solution: For compound one, the oxygen carbon ratio is *always* 1.33, while for compound two the ratio is *always* 2.66. Individually these observations satisfy definite proportions, while one being twice the other satisfies multiple proportions. The observations can be explained if an oxygen atom is 1.33 times as heavy as a carbon atom, and in compound one there are equal numbers of carbon and oxygen atoms, while in compound two there are twice as many oxygen atoms as carbon atoms. We designate these compounds by their chemical formulas, which are CO and CO_2.[12]

---

[12] A measurement of molecular mass is required to rule out multiples of these compounds.

Mass

All bodies, have a tendency to resist a change in their motion that is called *inertia*. Some bodies have more inertia than others. For example, it is much more difficult to deflect a bowling ball than a Ping-Pong ball, and even more difficult to deflect an asteroid. The property that measures the resistance of a body to a change in motion is called the ***mass*** of the body. Mass is often determined by measuring weight, the force of gravitational attraction to the Earth, since at a given place, weight is proportional to mass. Weight measurements, however, depend on the particular place at which they are made; a quarter weighs less in Denver than it does in Death Valley, because gravity is lower at Denver's higher elevation, but it has the same mass in both these locations. The difficulty in measuring mass can be overcome by using a ***balance***, which compares the mass of an object with those of objects of known mass.

The atomic theory was supported by measurements, summarized in Antoine Lavoisier's (1743–1794) theory of ***conservation of mass***, which states that mass doesn't change in chemical transformation. Instead of measuring the masses of individual atoms, chemists of Lavoisier's time measured the ratios of masses of the same number of atoms of different elements. 1.00 gram of the lightest element, hydrogen, was called a ***mole*** of hydrogen. The mass of a mole of any other element is its atomic mass (or less precisely, atomic weight).[13] Modern experiments have established that 1.0 gram of hydrogen contains 6.02×10^{23} atoms of hydrogen. This number, which connects the microscopic world of atoms to the macroscopic world of moles, is called ***Avogadro's number***. To get an idea of how large this number is: the height of a stack of Avogadro's number of sheets of copy paper would be four hundred million times the distance between the earth and sun.

Inside the Atom

Within the realm of atomic theory, the atom is the basic building block of matter. Atoms, however, have an internal order; they are made up of

---

[13] Nowadays atomic masses are referred to the mass of 12.00 grams of the carbon isotope carbon-12. Changes from the earlier standard are usually negligible.

smaller particles. The size of a typical atom is a fraction of a nanometer. Atoms are electrically neutral; they have no charge. Experiments indicate, however, that charged particles can be obtained by subjecting atoms and molecules to energetic interactions in flames or electrical discharges, or by dissolving them in certain solvents. This demonstrates that atoms are composed of parts that are positively or negatively charged and these charges cancel in the atom. Starting in the late nineteenth century, scientists began to do experiments to measure the properties of the charged components of atoms. J. J. Thompson (1856–1940, Nobel prize in physics in 1906) studied the negatively charged particles obtained by breaking up atoms. Combined with other experiments by Robert A. Milliken (1868–1953, Nobel prize in physics in 1923), these experiments showed that, regardless of their source, the negative particles have the same charge (which we will call "$-e$")[14] and the same mass. These particles, called *electrons*, have a mass that is only about 1/2000 of that of the lightest atom, hydrogen.

Experiments on the positive parts of the atoms showed that their mass is much greater than that of the electron and depends upon the element. They have a charge of $+e$ (the same magnitude, but the opposite sign of that of the electron), or some multiple of $+e$, corresponding to the removal of different numbers of electrons from the neutral atoms. We call these particles positive *ions*. The lightest positively charged ion is the hydrogen ion, obtained by removing one electron from a hydrogen atom. This particle is called the *proton*. No matter how much energy is used to excite a given atom, there are a maximum number of electrons that can be removed from it. For example, only a single electron can be removed from a hydrogen atom, two from a helium atom and eight from an oxygen atom. In almost all cases, the heavier the atom, the more electrons can be removed from it. Electrons can also be added to atoms, giving rise to negatively charged ions.

Knowing that atoms are made up of light negatively charged particles and heavy positively charged particles tells us nothing about the arrangement of these charged particles in the atom. Rutherford (1871–1937, Nobel prize in chemistry in 1908) and his colleagues performed a key experiment. They directed a beam of alpha particles — heavy

---

[14] The decision of whether to call the charge on the electron positive or negative is arbitrary.

particles with a mass approximately that of a helium atom — from radioactive decay at a piece of very thin gold foil. As shown in Figure 3.2, most of the alpha particles went right through the foil, indicating that most of the atom was populated only by the very light electrons.[15] Some of the alpha particles were deflected just a little going through the foil, but, the key observation was that a few of the alpha particles bounced right back towards their source. This last observation was only consistent with almost all of the mass of the gold atom being concentrated in a positively charged region with a dimension of ca. 10^{-13} meter. Thus, all the positive charge and almost all the mass of the atom is concentrated in a region, called the ***nucleus***, that only comprises ca. a billionth of the volume of the atom.[16]

The charge on the nucleus determines the identity of the atom and is called its ***atomic number***. Electrons are kept from flying out of the atom by their attraction to the positively charged nucleus, which is located toward the center of the atom. However, the chemistry of an atom is determined by the electrons on its outside. It is here that the atom interacts with other atoms.

The properties of atoms do not vary smoothly with their number of electrons; potassium has one more electron than argon, but is very different from argon. It is a metal, not a gas. There are very few things in our experience that behave in such an unusual manner. The observed variation of the properties of atoms on their number of electrons was one of the

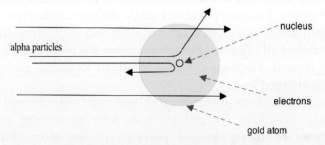

Figure 3.2. Rutherford's experiment.

---

[15] A heavy alpha particle would hardly be deflected in a collision with a very light electron.

[16] One thousandth in linear dimension implies one billionth in volume.

reasons that scientists had to completely change their view of matter at the beginning of the twentieth century.

If you change the number of electrons in an atom to get either a positive or negative ion, you produce something with very different physical and chemical properties from the original atom. A good example of this, as shown in Figure 3.3, is an atom of the element sodium, which has eleven electrons and a nucleus with charge +11. Sodium is nasty stuff; put it in water and it produces sizzling, popping and perhaps even a flame. Since there is a lot of water in your skin and intestines, you certainly wouldn't want to hold or eat a piece of sodium. Common table salt is sodium chloride, which is made up of equal numbers of sodium atoms (11 electrons) and chlorine atoms (17 electrons). In sodium chloride, however, the chlorine atoms each pull one electron from a sodium atom,

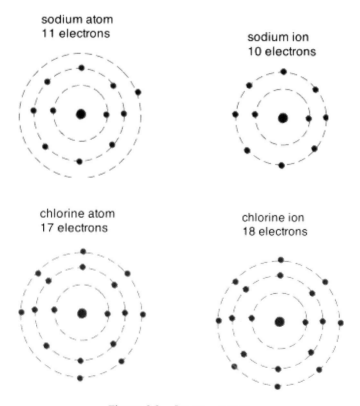

Figure 3.3. Ions vs. atoms.

leaving a positive sodium *cation* with 10 electrons and a net charge of $+e$. In doing this, the chlorine atom gains an electron and becomes a negative chlorine *anion* with 18 electrons and a net charge of $-e$. We consume table salt in many of our foods, but because it is made of ions, we can do so without sizzling, popping and flames. In summary:

> **Atoms are comprised of negative electrons surrounding a small, positive, massive nucleus.**

Inside the Nucleus

We can take the idea of order a step further and ask: what is the order, the structure, of the nucleus. The consideration of isotopes, can tell us something about this order.

In addition to normal hydrogen, which we write as hydrogen-1, indicating that it has one particle (a proton) in its nucleus, there is a different type of hydrogen, called deuterium, hydrogen-2, or "heavy hydrogen," because it has about twice the mass of normal hydrogen. Atoms of the same element, with different mass are called *isotopes*. Heavy hydrogen makes up about .01% of hydrogen found in natural abundance.[17] One hydrogen atom of every ten thousand found in nature is heavy hydrogen. What can we say about the structure of heavy hydrogen? First of all, since heavy hydrogen has all of the chemical properties of normal hydrogen, it too must have a single electron in its outer region, with the difference in mass being due to different nuclear masses of the two species. In order for both to be neutral hydrogen atoms, their nuclei must both have the same nuclear charge, $+e$. Let's zoom in closer on the nucleus in order to discover the difference between hydrogen and deuterium. Two models for the nucleus of heavy hydrogen are shown in Figure 3.4.

One possibility is that the nucleus of heavy hydrogen, called the deuteron, is made up of two protons and one electron. Another is that the nucleus of heavy hydrogen consists of a proton plus a second

---

[17] Heavy hydrogen is an ingredient in hydrogen bombs and fusion power reactors.

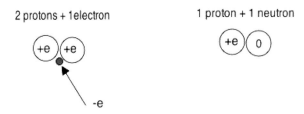

Figure 3.4. Models for the nucleus of heavy hydrogen.

particle with no charge and a mass approximately equal to that of the proton. Although the first hypothesis is simpler — because it doesn't introduce a new particle — the second possibility has been found to hold, and the new particle is called the *neutron*. It is now unquestioned that

> **The nucleus is comprised of protons and neutrons.**

In a later chapter, the manner in which physicists made the choice between these two possibilities is outlined.

The theory that atomic nuclei are made up of protons and neutrons has stood the test of time. Hydrogen-1 and hydrogen-2 are two *isotopes* of hydrogen. The symbols refer to both the neutral atoms and to the nuclei of the atoms, which we call *nuclides*. Protons and neutrons are collectively called *nucleons*, and it is the total number of nucleons that is written after the name of a nuclide or an isotope, as in hydrogen-2. It is sobering to realize that all the matter we see around us is made up of just three different particles, the electron, proton and neutron, in different combinations. How so much variety is obtained from so few components will be discussed in later chapters.

Example 1. How many proton, neutrons and electrons are there in each of the following particles:

(a) An atom of helium-4.
(b) A lithium-7 cation with a single charge.
(c) A neutral atom with atomic number of six and 13 nucleons.

Solution: You can use the periodic table in the appendix to obtain the atomic numbers of helium (He) and lithium (Li).

(a) two protons, two neutrons and two electrons
(b) three protons, four neutrons and two electrons
(c) six protons, seven neutrons and six electrons (This is an isotope of carbon.)

Radioactive Decay

Free neutrons (those existing outside of nuclei) are not stable; they decompose into protons and high-velocity electrons called, for historical reasons, **beta particles**.[18] The decomposition of a free neutron occurs with a half-life of about 15 minutes. That means: if we have 1000 free neutrons at some time, 15 minutes later we will have 500 neutrons and 15 minutes after that (30 minutes after our original time), we will have 250 neutrons.

The stability (how long it lives before it decays) of a neutron depends very much on its environment. For example, the neutron in deuterium is indefinitely stable. However, hydrogen has a third isotope, hydrogen-3 or tritium, which undergoes **beta decay**, with a half-life of 12 years. A very small concentration of tritium is produced in the upper atmosphere due to impact of **cosmic rays**, which are very high-energy particles from outer space, with atmospheric components. Tritium is produced for scientific and weapons[19] use in particle accelerators and nuclear reactors.

Nuclei can decay in other ways besides beta decay. One of the most common of these is **alpha decay**, involving the emission of a high energy particle comprised of two protons and two neutrons — the nucleus of a helium atom — called an alpha particle. This was the projectile used in Rutherford's experiment.

Because of their charge and high velocity, alpha and beta rays are examples of **ionizing radiation**. In the body, blood-producing cells in bone marrow are particularly sensitive to such radiation, and large doses

---

[18] An ephemeral particle, called a neutrino, is also produced in this decomposition. We will discuss neutrinos in Chapter 8.
[19] Tritium is used in fusion reactions, such as occur in hydrogen bombs.

of it will produce anemia and even death. Radiation can also cause changes in cells that lead to cancer. Since radiation damage is cumulative, even low-level exposure to radiation over long times is dangerous. Marie Curie (1867–1934), who won Nobel prizes in both physics (1903) and chemistry (1911) for her discoveries in the field of radioactivity, died of leukemia, most probably due to a lifetime of exposure to ionizing radiation. There are several other modes by which nuclei can decay that will be mentioned in later chapters.

Supplementary Reading: Carbon-14 Dating

In the atmosphere, neutrons initiated by cosmic rays continually convert nitrogen-14 to carbon-14. (The nitrogen atom absorbs a neutron and emits a proton.) Carbon-14 is an unstable isotope of carbon, and decays by emitting beta rays, with a half-life of 5720 years. At the average rate that it is formed and destroyed, atmospheric carbon dioxide is mildly radioactive, producing 15 decays per minute per gram of carbon. This level of radioactivity has been somewhat variable due to factors such as nuclear testing in the atmosphere.

Using photosynthesis, plants grow by incorporating atmospheric carbon dioxide, and they have a level of radioactivity that results from the carbon-14 in the atmosphere during their lifetime. Once a plant is harvested, however, incorporation of atmospheric carbon-14 ceases, the carbon-14 decays, and its material becomes less radioactive. By measuring the number of decays per gram of carbon of a wooden or a cloth artifact, scientists can determine how many half-lives the radioactivity in the artifact has been decaying, and from that, the number of years since its material has been part of the living biosphere. This method of dating materials is called radiocarbon dating. It can be used for items up to about 50,000 years old, beyond which the radioactive emissions become too weak to be measured.

Summary

The atom is the basic building block of matter. Atoms are neutral, but are made up of negatively charged electrons and a positively charged nucleus. In spite of its comprising most of the mass of the atom, the nucleus takes up only about a billionth of its volume. The nucleus contains protons and neutrons. Different numbers of neutrons produce different isotopes of the same element. Some nuclei decay by emitting alpha or beta particles.

Questions

1. What types of scales are you familiar with? Do they measure weight or mass? Are their measurements of mass place independent?
2. If an atom is the size of your classroom, how large would its nucleus be?
3. What is the difference between a nuclide and an isotope?
4. How would a chlorine-37 atom be described assuming that:
    (a) the nucleus is comprised of protons and electrons?
    (b) the nucleus is comprised of protons and neutrons?
5. Potassium bromide (KBr) is similar to common table salt (sodium chloride). Describe the particles of which potassium bromide is comprised.
6. The two major isotopes of naturally occurring uranium are ^{238}U, with 99.3% abundance, and ^{235}U, with 0.7% abundance. Only ^{235}U undergoes fission and can be used in nuclear reactors and weapons. Can these uranium isotopes be separated by chemical means? Do you expect there to be much difference in their physical properties?
7. Show that if the volume of a sphere is $\frac{4}{3}\pi r^3$, and a typical dimension of an atom and a nucleus are 10^{-11} and 10^{-15} m, respectively, the volume of the nucleus comprises ca. a trillionth of the volume of an atom. If an atom has a dimension of a baseball stadium (200 m), what would be the dimension of its nucleus?
8. What are some of the assumptions that are made in the method of radiocarbon dating? Can you think of any way of testing some of these assumptions? Can you think of situations in which the assumptions would probably be invalid?
9. The wood from an ancient wooden tool is found to show a level of carbon-14 beta decay of 0.075 decays per gram of carbon. How old is this tool?
10. Why does the tritium in our nation's stockpile of hydrogen bombs have to be continuously replenished?

11. We call the charge on the electron negative and that on the proton positive. Could we have used the opposite sign convention? What is important about the sign of the charges on the elementary particles?
12. If the formula of water is H_2O, indicating that it has two hydrogen atoms for each oxygen atom, what would be the formula of the other hydrogen-oxygen compound discussed at the beginning of this chapter.
13. How do we know that electrons are on the outside of an atom?
14. For a bar of gold: list two intensive properties (those that are intrinsic to each bit of material) and two extensive properties (those that are proportional to the size of the sample).

4 Forces and Motion

When beggars die, then are no comets seen: The heavens themselves blaze forth the death of princes.

from Shakespeare's "Julius Ceasar"

A *force* is defined as something that can bring about a change. For example, we have talked about the "force of the teacher's personality" in establishing order in a kindergarten class. The order we perceive in the universe is a result of the forces that organize it. There are two types of forces, attractive forces and repulsive forces. Systems with limits on their extent, such as stars or planets, are ordinarily under the influence of attractive forces. Combinations of attractive and repulsive forces produces symmetry, such as exist in crystals. A system solely under the influence of repulsive forces is driven apart without limit, i.e., towards a state of total disorder.

Fundamental Forces

Physicists know

> There are four fundamental forces in nature.

These are:

(1) The *gravitational force*, attractive between all matter.
(2) The *electromagnetic force*, between charged or magnetic particles.
(3) The *strong force*, holding the nucleus together.
(4) The *weak force*, producing processes such as the conversion of a neutron into a proton.

Here we listed these forces in the order in which they were discovered. The relative strength of the forces is: strong force >> electromagnetic force >> weak force >> gravitational force. The strong force, however, is appreciable only over distances the size of the atomic nucleus, and the weak force only over even smaller distances, while the gravitational and electromagnetic force have their effects felt to infinite distance. The differences in the strengths of these forces are so great, that in a case in which one force has an effect, the effect of all weaker forces can be neglected.

The gravitational force

The gravitational force, which was the first force discovered and is the most important force in our daily lives, is by far the weakest of the four fundamental forces. Nevertheless, it is the only force required to accurately describe the motion of the planets and many bodies on Earth. It is told that Isaac Newton (1643-1727) first realized that heavenly and terrestrial bodies both moved under the influence of the same force when he simultaneously observed the motion of the Moon and a falling apple. Newton hypothesized that there is an attractive force, gravitation, between all bodies in the universe. This force is proportional to the product of the masses of the bodies and inversely proportional to the square of the distance (d) between them[20]

$$\left(f_{grav} = \frac{Gm_1 m_2}{d^2}, \text{ where } G \text{ is a universal gravitational constant} \right).$$

If we double the distance between two bodies, the gravitational force between them is reduced by 1/4. The negative sign means that the force

---

[20] For roughly spherical bodies, such as planets, distances between the centers of the bodies can be used.

is attractive, i.e., towards smaller values of d. The gravitational force is an extremely weak force, but since it is always attractive and extends to infinite distance (with decreasing intensity), it is dominant for large masses over long distances. Rather than talk about gravitational force, physicists often discuss the ***gravitational field***,[21] which is the force per unit mass at a point in space due to gravity. Both concepts: gravitational force or gravitational field, do little to answer the fundamental question of how one body "reaches out" to attract another. We will have more to say about this later on.

The electromagnetic force

We are all familiar with static electricity and permanent magnets. These are two manifestations of the interactions of stationary and moving charges that we lump together in what we call the electromagnetic interaction. First we will discuss the ***electrostatic interaction***. *Static electricity* is sometimes produced when we vigorously comb our hair on a dry day. In this process, electrons are transferred from our hair to the comb, leaving the strands of hair with a positive charge. The electrostatic interaction is described by ***Coulomb's law***

$$\left(f = \frac{kq_1q_2}{d^2}, \text{ where } k \text{ is a constant and } q \text{ is the charge on a body}\right)$$

which tells us that there is an attractive force (f negative), if the particles have the opposite charge (+ with −), and a repulsive force (f positive), if the particles have the same sign of their charge (+ with + or − with −). The positively charged strands of hair repel each other and may stand up due to their mutual repulsion.[22] The electrostatic force is proportional to the product of the charges and inversely proportional to the square of the distance between them. It thus has the same distance dependence as the gravitational force. In analogy to the gravitational field we can also talk about the ***electric field***, which is the force per unit positive charge at a point in space due to other charged particles.

---

[21] Analogous to using an interest rate, rather than the total interest paid.
[22] In a similar manner, the electroscope was an early instrument that indicated charge by the mutual repulsion of two pieces of metallic foil.

Lightning is a dramatic example of the buildup of static electricity occurring in the clouds of thunderstorms. Friction between rising water droplets and stationary particles in the cloud results in stripping electrons off the rising particles. The excess negative charge that is produced on the lower parts of the clouds repels electrons on the surface of the Earth, giving the surface an induced positive charge. When the electric field between the negative charge on the bottom of the cloud and the positive charge on the surface of the Earth becomes sufficiently strong, it may ionize some of the intervening air molecules. Normally electrically insulating air then becomes conducting, with a huge release of energy, as the electrons in the cloud flow to Earth.

The electrostatic force is much, much stronger than the gravitational force. Why then, does this force, instead of gravitation, not determine the motion of the planets? The reason for this is that there are equal amounts of positive and negative charge in most regions of the universe and these charged particles arrange themselves so that electrostatic forces cancel over macroscopic distances. For example, a grain of table salt is made up of countless numbers of positively and negatively charged ions. However, to a superb approximation there are equal numbers of these oppositely charged particles and they are arranged in a manner so that outside the grain of salt no electrostatic forces can be detected. On the scale of the heavenly bodies there is even more of a tendency for such cancellation. Even though electrostatic forces balance out over macroscopic distances, in the microscopic world of atoms and molecules, this is the dominant force and the only force that must be considered.

The electromagnetic interaction also includes *magnetic forces*, such as those produced by refrigerator magnets. Such magnets have two poles, which we call North and South, because they point in roughly these directions in a compass.[23] If two North poles or two South poles of magnets are brought close together, the magnets repel each other. However, if the North pole of one magnet is brought close to the South pole of another magnet, the force is attractive. In analogy to the electric field, we can discuss forces between magnets in terms of a *magnetic field*, which is the force on a hypothetical[24] magnetic pole of unit strength. **Magnetic field lines** extend from

---

[23] The North pole of a magnet actually points to the North magnetic pole in the Arctic, rather than the geographic North pole. Magnets are used as the pointers in compasses.

[24] We say hypothetical because, as we will discuss, it is impossible to isolate a single magnetic pole.

North poles to South poles. A good way to visualize the magnetic field of a magnet is, as shown in Figure 4.1, to sprinkle the area around it with iron filings. The filings act as little magnets and line up with the lines of magnetic field; filings close to the magnet are drawn to the poles.

Refrigerator magnets are *permanent magnets*. Magnetic fields are also generated by moving electric charges, called electric currents. This is illustrated by the *electromagnet* shown in Figure 4.2, in which a magnetic field

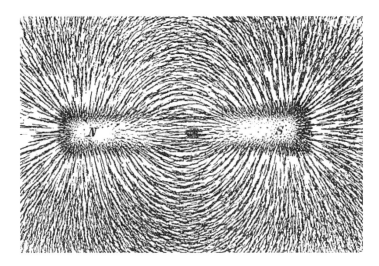

Figure 4.1. Magnetic field lines revealed by iron filings.

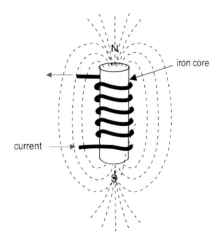

Figure 4.2. An electromagnet.

is produced by a current flowing in a coil of wire. A North pole is produced on one side of the loop and a South pole on the other. Powerful electromagnets operating on this principle can lift automobiles. If the current in an electromagnet is reversed, the poles of the resulting magnetic field are reversed. Although permanent magnets and electromagnets appear very different, they actually work on the same principle, since it is electric currents within atoms that produce the magnetic field of permanent magnets. This explains why it is impossible to isolate a single magnetic pole. If a bar magnet is cut in half, two magnets are obtained, each with a North and South pole.

As shown in Figure 4.3, an electric motor consists of a coil of wire within the poles of a permanent magnet. When a current is passed through the coil, it rotates the coil towards its attractive orientation in the field of the permanent magnet. If the current continued to flow in the same direction through the coil, it would stay in this attractive orientation. However, if the current passing through the coil is alternating current (A.C), its direction and the direction of the coil's magnetic field changes 60 times in a second. The changing direction of the coil's field causes it to continue to rotate in the permanent magnet, and provides us with the rotary motion needed to drive our fans, vacuum cleaners, etc. Motors can also operate on direct current (D.C), which always flows in the same direction, by using a device called a commutator that changes the direction of current flow through the coils as the motor turns.

As pointed out by Michael Faraday (1791–1867), a changing magnetic field produces an electric field. As a result, if a magnet is moved near a

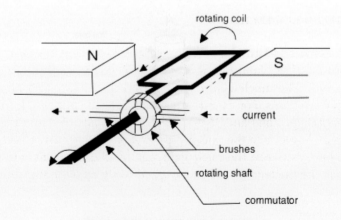

Figure 4.3. A d.c. electric motor.

wire or a wire is moved near a magnet, an electric field will be generated in the wire.[25] The resulting electric force on the electrons in the wire generates a current if the wire is in a closed circuit. Faraday's principle is the basis for producing most of the electricity we use. An electric generator is a motor, like that shown in Figure 4.3, run in reverse. Energy from steam, combustion gases or falling water turns the shaft of the generator, moving the coil through the field of the magnet and producing an electric current. An alternating current is generated as the wires successively pass the two poles of the magnet.

The work of James Clerk Maxwell (1831–1879) provided a complete electromagnetic theory by pointing out that a magnetic field is also produced by a changing electric field. Maxwell succinctly summarized all the laws of electromagnetic theory in four equations. We can summarize *Maxwell's equations* as:

> **Electric fields are produced by charges and changing magnetic fields. Magnetic fields are produced by currents and changing electric fields.**

One of the most exciting predictions of Maxwell's equations is the propagation of self-sustaining electromagnetic waves. These will be discussed in the next chapter.

The strong and weak forces

Electromagnetic and gravitational forces are not sufficient to describe the order produced in Nature — as can be seen by considering the nucleus of an atom. The nucleus contains positively charged protons and uncharged neutrons. For example, an atom of lead consists of a nucleus containing 82 protons and 122–126 neutrons. Since electrostatic forces between the protons in the nucleus are exclusively repulsive, and gravitational force between the nucleons (protons and neutrons) are totally negligible, the nucleus would be expected to fly apart in the absence of

---

[25] Only the relative motion between the magnet and the wire is significant.

other forces. The force that opposes the repulsion is called the strong force. It is always attractive, and over distances the size of the nucleus it can be stronger than the electrostatic force. It falls off very rapidly at larger separations, and does not have to be considered for extra-nuclear structure. The strong force acts on heavy particles, such as protons or neutrons, but not on electrons.

As we add protons to a nucleus the number of electrostatic repulsions rises dramatically. For example, there is one such repulsion in a helium nucleus, which has two protons. However, an oxygen nucleus has eight protons and therefore $(8 \times 7)/2 = 28$ proton-proton repulsions.[26] Heavy atoms need enough neutrons in their nuclei to obtain sufficient strong force to hold the nucleus together against their increased proton-proton repulsion. Typically light atoms have about equal numbers of protons and neutrons, while heavy atoms (such as lead) have about 50% more neutrons than protons.

The weak force dies out faster with distance than the strong force. It usually only has to be considered for processes that occur within individual nucleons for which the strong and electrostatic forces play no role. The decomposition of the neutron, discussed in the last chapter, is of this type.

Motion

One of the most obvious changes in a body that can be produced by a force is a change in its motion. In fact, until the twentieth century, this was the only context in which forces were considered. It was Newton who first realized that, as stated in his first law,

> **In the absence of forces, objects move with constant velocity.**

Velocity (V)[27] is a vector quantity; it has both direction and magnitude; the magnitude of velocity is called *speed* (v). Before Newton, it was commonly thought that an undisturbed body would come to rest. This belief was due to the difficulty of making measurements in the absence of frictional forces,

---

[26] Each proton is repelled by all the other protons. We divide by two to avoid counting the repulsion between two protons twice.
[27] Boldface symbols are used for vector quantities.

which reduce velocity. Only in outer space, away from frictional and gravitational influences, are bodies observed to move with constant velocity.

Momentum is defined as the product of mass times velocity ($\mathbf{p} = m\mathbf{V}$), and has the direction of velocity. Another way of stating Newton's first law is that in the absence of an external force, the momentum of a body is constant; physicists say that momentum is conserved. This is one example of a ***conservation law***. Such laws often provide powerful insight into physical processes.

The law of motion that applies in the presence of a force is Newton's second law,

> **The rate of change of the momentum of a body is equal to the force on the body.**

Thus, if a body, such as an asteroid hurtling towards Earth, has a large momentum, it takes a large force to change its path. For bodies of constant mass, the first law is included in the second, since if the force is zero, the momentum, and thus the velocity of the body, doesn't change.

Forces are exerted by one body on another. Newton's third law tells us that,

> **If body A exerts a force on body B, then body B exerts an equal and opposite force on body A.**

The third law explains the recoil when a gun is fired. The gun exerts a force on the bullet to increase its velocity, and the bullet exerts an equal and opposite force on the gun.

Angular motion

Angular motion is motion around a center. An example is the Earth's orbit in its annual revolution around the Sun.[28] In this motion, the magnitude of

---

[28] To be exact we should say that both the Earth and the Sun revolve around the center of mass of the Earth-Sun system. This point is very close to the Sun.

the Earth's velocity (its speed) remains constant, but the direction of the velocity is continually changing. It is useful to discuss this motion in terms of ***angular momentum***, which is the ***moment*** of the Earth's momentum around the Sun, i.e., the momentum multiplied by its perpendicular distance from the rotation axis.

Angular momentum is a vector quantity and is assigned a direction according to the right-hand rule: if the four fingers of the right hand curl in the direction of the motion, the angular momentum is in the perpendicular direction of its thumb. For the Earth's revolution the angular momentum is perpendicular to the plane in which the Earth moves, towards the northern (defined by the Earth's north pole) direction.

Newton's laws predict that

> **Angular momentum around a center of rotation can only be changed by a *torque* around that center.**

A torque around a center is the moment of a force around that center, i.e., the force multiplied by the perpendicular distance to the axis of rotation. It also is a vector quantity, following the right-hand rule. Because the force of attraction of the Earth to the Sun goes through the center of rotation, it has zero moment and produces no torque. It therefore cannot change the Earth's angular momentum around the Sun. Thus, the angular momentum of the Earth in its orbit around the Sun is constant.

Simultaneous with its revolution around the Sun, the Earth also spins around an axis tilted at a twenty-three degree angle from the perpendicular to the plane of its orbit. In this spinning of the Earth, each small bit of the Earth has an angular momentum equal to the product of its momentum times the distance of the bit from the axis of rotation. The angular momentum for rotation is the sum of the angular momenta of each of these small bits.[29]

For the spin rotation of the Earth, there is also no external torque to change angular momentum, and it remains constant. We will see in Chapter 21 that the revolution of the Earth combined with the tilt of its

---

[29] For revolving around the Sun we don't have to perform a sum, since each piece of the Earth is just about the same distance from the center of the angular motion.

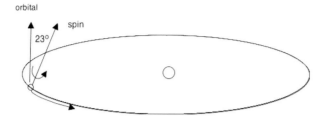

Figure 4.4. Orbital and spin angular momenta of the Earth (not to scale).

rotation axis gives rise to the seasons. The two angular momenta of the Earth are shown in Figure 4.4.

Angular momentum, either orbital or spin or both, is constant in many physical situations, resulting in it being a very important quantity. One example of conservation of angular momentum deals with a spinning ice skater. The angular momentum of the ice skater is the sum of that of all her parts — the mass times the velocity of the part times its distance from the center of rotation. What happens when the ice skater pulls her arms closer to her body? There's almost no *external* torque on the skater, so her spin angular momentum is conserved. Since the distance of her arms from the center of rotation has decreased, the angular momentum of her arms decreases. Her total spin angular momentum remains unchanged because she rotates faster.

Another example involves what happens as a comet comes from far away and travels around the Sun. In this case there is a force on the comet — its gravitational attraction to the Sun — but since this force is directed towards the Sun, there is no torque and its angular momentum around the Sun is conserved. In 1695, Edmond Halley (1656–1742), using Newton's laws, calculated the orbit of a comet that had been observed several times in the past, predicting that it would return in 1758. The comet was observed, but unfortunately Halley did not live long enough to relish his triumph. Halley's comet has been observed since 239 BC, and its return every 75–76 years has in the past been attributed to forces other than gravitation, as indicated by Shakespeare's quotation at the beginning of this chapter.

To a very high accuracy, Newton's laws predict the motion of everything from grains of dust to galaxies. Using only gravitational attraction, they suffice to explain the clockwork regularity of the motions of the planets summarized in the laws of Johannes Kepler (1571–1630). For two centuries, scientists believed that Newton's laws provided the ultimate explanation of motion. It was only in the early twentieth century that these laws were shown to be only approximations. They were useless when considering very light particles (e.g., electrons) or bodies whose velocities approached those of the speed of light or when forces were extremely large (like near a black hole).

Transformations

Another type of change that can be brought about by a force is a change in the nature of a particle. An example of this was the decomposition of a neutron into a proton and an electron, discussed in the last chapter. An important point is that a force can only bring about a transformation, if it interacts with *all* the reactants and products of the transformation. The stronger the force that brings about a transformation, the shorter the time it takes for that transformation to occur. The observation, that the neutron has a half-life of 15 minutes, instead of less than the 10^{-10} seconds typical of most nuclear decompositions, indicates that neutron decomposition results from the weak force, rather than from the strong or electromagnetic forces.[30]

Dark energy

Besides the four forces that are well known by physicists, there is a fifth force that they strongly suspect exists. The evidence for this force comes from the field of cosmology, a subfield of physics that deals with big questions, such as the beginning and end of the universe. As we will see in Chapter 23, there is strong evidence that the universe was formed in a cataclysmic explosion, called the Big Bang, about 14 billion years ago. As the universe rapidly expanded from this explosion, theory predicts that gravitational attraction should slow its rate of expansion. There is observational evidence that such slowing of the expansion did occur, but surprisingly, about 7 billion years ago, the expansion accelerated. This acceleration is attributed by physicists and astronomers to a repulsive

---

[30] The gravitational force is too weak to bring about particle transformations.

force due to an unobserved (and therefore "dark") energy permeating the universe, which becomes dominant at large separations. Dark energy represents a fifth fundamental force in Nature.

States of motion

What types of states of motion do we observe in Nature? Most obvious is the state of no motion, or **rest**, such as our car sitting in the driveway. (Actually it is moving with the speed of the surface of the Earth.) A constant state of no motion implies no change of velocity and therefore zero force, or to be more exact, zero net force; the forces are balanced. The car in the driveway does not move up or down because the downward force of gravity is balanced by the upward force due to the push of the Earth's surface. Likewise, the ions in a crystal of sodium chloride don't move because they occupy positions where the electrostatic forces from the surrounding positive and negative ions just cancel. It is these forces that produce the order in the crystal.

Another type of motion that we are very familiar with is ***damped motion***. This is motion that gradually goes to zero due to frictional forces. If our car suddenly stalls out and we are on level ground, it does not take long before our horizontal motion stops.

When a body is acted on by one or a few forces, we say that the motion is ***predictable***, in that, using Newton's laws and a powerful computer, the time course of the motion can be calculated. Predictability is usually an idealization, in that it may require that we neglect some very small forces or restrict our predictions to a limited period of time.

Periodic motion is repetitive in time. The motion traces out a closed cycle in space. The most obvious example of this is the motion of the Earth around the Sun or the Moon around the Earth. These motions are actually complicated. Not only does the Earth move in a slightly non-circular (elliptical) orbit around the Sun, but the Sun moves under the influence of the Earth as well. In addition, there are subtle influences on the motion of the Earth and Moon from other nearby planets: Venus, Mars and very massive Jupiter. Astronomers can only predict the exact time of occurrence of future astronomical events, such as eclipses, because there are dominant influences in the motion and they limit their predictions to the not very distance future.

It is useful to know whether a periodic motion is **stable**. That is: if the motion is disturbed by some transient external influence, does it return to its periodic behavior. A stable periodic motion is called a *limit cycle*. Limit cycles are very important in Nature. After an asteroid passes through the solar system, the planets return to their periodic motion. After an exertion, our hearts return to their regular pattern of beating. There is often a maximum value of the size of the perturbation that permits restoration of the cycle. Thus an asteroid collision with a planet or a lightning strike on a human will produce a permanent change to the cyclic motion.

The state of motion becomes more complicated when there are many interacting bodies, such as occurs in the belt of asteroids that exists between Mars and Jupiter. While the general extent of the belt and its structure can be predicted, the motion of individual asteroids is more uncertain, occasionally one is even ejected from the belt and flies towards Earth. Even with our best measurements of the current positions and velocities of the larger asteroids, computer predictions of their positions become very inaccurate at modest astronomical time scales. They have inherent disorder. Such motion, exquisitely sensitive to initial conditions, is called *chaos*.

When the number of interacting particles approaches the numbers of atom or molecules that we usually deal with in macroscopic samples (some fraction of Avogadro's number), the prediction of motion of individual particles (even if they followed deterministic Newtonian mechanics — which they don't) becomes impossible. Since we can discern no order in the motion, we treat it as being completely disordered or *random*, which means that we use statistics to determine what probably will happen; knowing that with so many particles, what probably will happen is what we will observe. This is our basis for concluding in Chapter 2 that over appreciable time scales gases have more symmetry than crystalline solids.

Summary
ଔ ଓ ଔ ଓ ଔ ଓ

Physicists know of four basic forces in Nature: the gravitational, electromagnetic, strong and weak forces. While the gravitational force is always attractive, the electromagnetic force may be attractive or repulsive. The strong force holds the nucleus together and the weak force is responsible for slow transformations of elementary particles.

Newton hypothesized that an undisturbed body moves with constant velocity (or momentum) and that the rate of change of momentum is proportional to the force on the body. The angular momentum of a body (either orbital or spin angular momentum) can only be changed by an external torque. Momentum and angular momentum are examples of quantities that are conserved in systems under certain conditions.

States of motion range from those easily predictable because they are under the influence of no or a few forces to those that are completely random because they involve the interaction of a huge number of bodies. Chaotic motion is a state intermediate between these, where some generalization, but not complete prediction, can be made.

Questions

1. In Figure 4.1, the magnetic filings are lined up along curves extending from the north pole to the south pole of the magnet. Can you think of a reason for this alignment? Hint: think of the filings as little magnets.
2. Write the SI unit of force, the Newton, in terms of the basic SI units given in Chapter 1. Hint: use an equation involving force and the idea that both sides of an equation must have the same units.
3. If a meteor is hurtling towards Earth, could we deflect it from its path by exploding a nuclear bomb on the meteor? What would happen if we did this?
4. If a stationary atomic nucleus decayed by emitting a β particle (a high-velocity electron), what does conservation of momentum require for the velocity of the nucleus after the decay?
5. An electricity-generating plant burns coal, producing steam, which rotates a turbine. Show how this produces electricity.
6. How many proton-proton repulsions are there in a sodium atom with 11 protons in its nucleus?
7. Why does a gun have to be much more massive than the bullet it fires?
8. Why doesn't the Earth fly off into space instead of continuing to rotate around the Sun?
9. Why isn't the Earth pulled closer to the Sun by the gravitational attraction between these two bodies?

10. Why is it easier to keep a bicycle upright when it is moving than when it is standing still?
11. Some very small torques produce slow changes in the orbital and spin angular momenta of the Earth. Identify these torques and discuss their consequences.

5 Waves

Are not the Rays of Light very small Bodies emitted from shining Substances?

Do not all fixed bodies, when heated beyond a certain degree, emit light and shine; and is not this emission performed by the vibrating motions of their parts?

Isaac Newton

Everything in nature can be described as being composed of either particles or waves. Particles are localized; waves are spread out and characterized by a repeating behavior in space and/or time. Sound and light are two examples of waves. In this chapter we will discuss the properties of waves. In Chapter 7 we will show that for some things, such as electrons, both wave and particle descriptions are useful.

In physics, a wave is a ***disturbance***, a movement of material, that moves through a ***medium.*** For example, ocean waves move horizontally through water and what is disturbed is the vertical position of the surface of the water. There is practically no movement of water molecules in the direction of the wave. The position of the water at the surface moves up and down, just like the fans in a stadium do when they cheer their team by producing an undulation sweeping through the stands. We call this a traveling wave, and it is a ***transverse wave***, since the disturbance is perpendicular to the direction of its propagation.

Imagine sitting on an offshore oil platform, with the crests of waves on the surrounding ocean separated by ten meters, the *wavelength*, λ, of the wave. Six waves might pass the platform per minute; this is the *frequency*, f, of the wave. The maximum displacement of the surface from the undisturbed position of the water is the *amplitude*, A, of the wave; it is one-half the difference in displacement between crest and trough. The product of the wavelength times the frequency is the wave velocity, $v = \lambda f$. For our ocean waves, the velocity would be 10 m × 6 (1/min) = 60 m/min.[31] The wave velocity is a constant, since it depends only on the medium through which the wave is travelling. Since $v = f\lambda$, or $f = v/\lambda$,

> **Wavelength and frequency of a travelling wave are inversely proportional.**

For example, if in the same medium, the waves are five meters apart, twelve waves would pass our platform in a minute.

A diagram of the waves on water at a given instant is shown in Figure 5.1. The **phase** of a wave is the position of its crest at a particular time. The figure shows two waves with a *phase difference*, ϕ. When two waves are *in phase* their crests are at the same position, and their phase difference is zero.

Note that waves have regions of positive displacement and regions of negative displacement. Positions on the wave where the displacement is

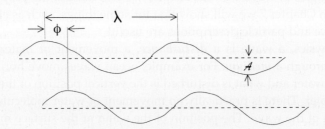

Figure 5.1.   Transverse waves.

---

[31] Frequency is read as "six per minute." Note how the units, as well as the numbers, are multiplied.

zero are called *nodes*. For a travelling wave, the crests, troughs and nodes all move with the wave velocity. Ocean waves are produced by the wind and have many undulations. Tsunamis (tidal waves) are produced by sudden disturbances in or under the ocean (earthquakes, submarine volcanoes or meteor collisions), which produce a wave with just a few crests that moves with moderate amplitude and large velocity through the ocean. As it approaches shore, the wave velocity slows and its amplitude greatly increases, often with disastrous consequences.

The waves on the strings of pianos or violins are of a different type. The ends of these strings are held motionless and the wave is due to the other points of the string moving up and down. Because these waves do not travel, they are called *stationary or standing waves*. Their crests only move up and down. Wavelength and frequency are varied in a piano by having strings of different lengths and materials and in a violin by having the player adjust the lengths of the strings with her fingers.

The strings of a piano or violin produce a disturbance in the air, but how does this disturbance reach and excite our ears? A *sound wave* is a traveling wave comprised of alternating regions of compression (pressure increase) and rarefaction (pressure decrease) of the air (or another medium, such as water). When these pressure waves reach our ear they stimulate our eardrum at a particular frequency, which we recognize as the pitch of a sound. The compression waves move with the speed of sound, which at ambient conditions in air is approximately 330 meters per second. Since in a sound wave, compression and rarefaction are achieved by molecules moving back and forth in the direction of propagation of the wave, they are called *longitudinal waves*.

The wavelength and frequency of a wave are not absolute, but depend on the motion of the wave source with respect to the observer. Have you ever noticed when a train is coming towards you, the pitch (frequency) of its whistle is higher than when the train approaches you than when it is moving away? This is called the *Doppler effect*, and, for a source moving towards the observer, is due to the source partially catching up to waves it has emitted, shortening their wavelength. Since the velocity of the wave is not changed, the frequency of the wave is increased. This is shown in Figure 5.2. At each instant, the moving source emits a spherical wave that is centered on the position of the source *at that instant*. Since the source is moving toward the right, the waves are closer together in that direction. With a shorter wavelength and the same velocity, the frequency is higher in the forward direction.

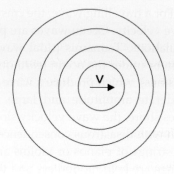

Figure 5.2. The Doppler effect — waves from a moving source.

When a source of sound waves moves as fast as, or faster than, the speed of sound, all disturbance made by it moves in the same wave. Sound waves produced by lightning, called thunder, or by aircraft moving faster than the *speed of sound*, may be so powerful that they cause damage to structures on the ground.

Ocean waves move in a line and are called *planar waves*. When we drop a rock in a quiet pond, *circular waves* emanate from the point of impact. If there is an explosion in the air, such as produced by an anti-aircraft shell, a *spherical wave* moves out in all directions. In all these cases the velocity of the wave depends upon the motion of medium through which it propagates. For example, if the rock is dropped in a moving river, the waves emanating from it will move faster in the downstream direction than in the upstream direction. As shown in Figure 5.3,

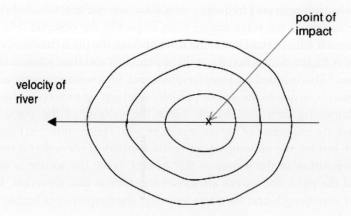

Figure 5.3. Waves in a river.

the wavelength is larger downstream, but the frequency is the same in both directions.

We may also use the term wave to describe phenomena in which there is no motion at all. The waves of someone's hair are of this type. All these different types of waves are important in science.

Diffraction and Interference

Diffraction and interference are two phenomena that are produced by waves. Diffraction, the bending of waves around an edge, is illustrated by the behavior of a planar ocean wave when it encounters a wall with a small opening. By the wave theory of Huygens (1629–1695), each point on the wave can be imagined as the origin of a spherical wave. As shown in Figure 5.4, this wave is circular on the two-dimensional surface of the water.

The wave disturbs all regions downstream of the hole. We say it is *diffracted* around the hole. Diffraction of sound waves is the reason that we can hear around corners.

In a poorly designed concert hall, there may be positions in which the sound is muffled. This results from *interference* of sound waves reaching the position with different path lengths (e.g., a direct and a reflected wave). By wave theory, when two waves arrive at the same point, their amplitudes are added together. Displacements are algebraic quantities; they can be either positive or negative. Thus when two waves overlap, at some points their displacements will reinforce each other — their waves will be in phase, while at other points their displacements will cancel — producing

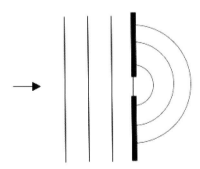

Figure 5.4. Diffraction of ocean waves.

66   Order and Disorder

nodes. As shown in Figure 5.5, interference is produced if water waves impinge on a wall in which there are two closely spaced holes. Due to diffraction, circular waves emanate from each of the holes. In certain directions, where the crests the waves coincide, they reinforce each other and in other directions they cancel.

Diffraction and interference are only appreciable when the wavelength of a wave is comparable to or larger than the size of holes and the distances between them.

Light as a Wave

The results of early experiments on light were explained by using the concept of *light rays*. An important advance in understanding the properties of light resulted from Newton's observation that "white" light from the sun could be separated into light rays of different colors by a glass prism.[32] As indicated by the quotations at the beginning of this chapter, Newton vacillated in his conception of light, sometimes thinking of it as particles and sometimes as waves. (He called them vibrations.)

Christian Huygens (1629–1695) showed how light rays (or beams) could result from interference between many light waves. Due to the very short

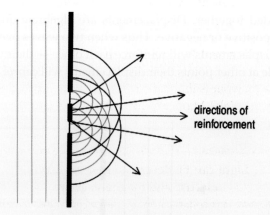

Figure 5.5.   Interference of waves.

---

[32] You see sunlight separated into its constituent colors when you observe a rainbow.

wavelengths of visible light, it is difficult to construct apparatus with dimensions small enough to demonstrate its wave properties. However, Thomas Young (1773–1829) was able to produce an interference pattern by passing light of a single color through two closely spaced, narrow slits. The similarity of the pattern he observed to that produced by water waves passing through two openings, strongly suggested that light had wave properties.[33] From the interference angles of different colors of light, Young was able to determine their wavelength. For visible light, these were found to range from 700 billionths of a meter (which we write as 700 nm, 1 nm = 1 billionth of a meter) for red light to 400 nm for violet light. It is likely that our eyes are sensitive to these wavelengths of light because they have evolved to make use of the wavelengths that are most prominent in sunlight.

Visible light is a type of *electromagnetic radiation*, which also exists at wavelengths both shorter and longer than those to which our eyes are sensitive. For example, Young showed that sunlight also contained wavelengths shorter than 400 nm, called *ultraviolet light*, which could produce chemical effects. We are now familiar with electromagnetic radiation with even shorter wavelengths called *X-rays*, and that with the shortest wavelengths of all called *gamma rays*. Waves with wavelengths longer than those to which our eyes are sensitive, but which we perceive as heat, are called *infrared radiation*, and those with even longer wavelengths are called *microwave radiation* and *radio waves*. The wavelength and corresponding frequency of different parts of the *electromagnetic spectrum* are shown in Figure 5.6.

Nature, Production and Propagation of Electromagnetic Waves

If light is a wave, what is the nature of its disturbance and what medium is it traveling through? Maxwell showed that his famous equations (see Chapter 4) predicted that

> **Light can be described as time-varying electric and magnetic fields.**

---

[33] Although the patterns were similar, their scales were very different. For light waves, the angles of reinforcement were much smaller. This is why it took so long to recognize the wave properties of light.

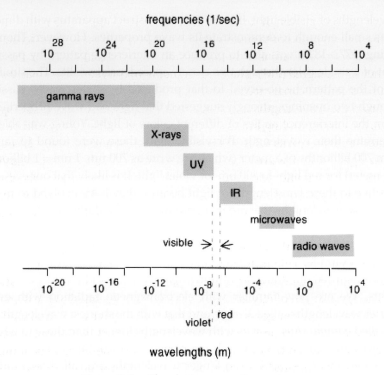

Figure 5.6. The electromagnetic spectrum.

As an example, let's consider a radio station that produces a signal by driving an electric current up and down a wire (an antenna). As the electrons in the current change direction, they accelerate, producing a time-varying magnetic field (**H**). This magnetic field in turn produces a time-varying electric field (**E**). These *electromagnetic waves* propagate in certain directions. Moreover, the waves move away from their source with a velocity determined only by the two fundamental constants in Maxwell's equations. The predicted velocity, c = 300 million meters per second, was exactly that estimated for the velocity of light in a number of experiments.[34]

Of course, if the radio station sent the same number of electrons up and down its antenna at a constant frequency, its waves would have constant amplitude and frequency. Its signal would be a pure tone and carry

---

[34] This is the velocity of light in vacuum; the velocity varies in other media.

very little information. In order to provide a more interesting signal, with all the information contained in music or a baseball game broadcast, the station continually adjusts the amplitude of the signal (AM) or the frequency of the signal (FM). The latter adjustment is over a very small range around a central carrier frequency.

From the wavelengths measured for the extremes of the visible spectrum, the corresponding limiting frequencies for visible light could be calculated from $f = c/\lambda$ as 4.3×10^{14} 1/sec for red light and 7.5×10^{14} 1/sec for violet light.[35] Light waves are transverse waves, with the time varying electric and magnetic fields perpendicular to each other and to the direction of propagation of the light, as shown in Figure 5.7. The intensity of light is proportional to the square of the magnitude of the electric or magnetic field.

The direction of the electric field is called the **direction of polarization** of the electromagnetic radiation. Some sunglasses use lenses that reduce glare by only transmitting light with a particular direction of polarization.

The wavelength of electromagnetic radiation is related to the distance that charge moves in generating it. Thus, radio waves are produced by electrons flowing up and down a wire (an antenna), while microwaves are generated when they move in a small cavity. The vibrations of molecules produce infrared radiation, while the motion of electrons within atoms generates visible and ultraviolet light and X-rays. Gamma rays are created by motions within the nucleus and elementary particles. Radiation that is not in the visible range of wavelengths must be converted to visible light before we can "see it." For example, an X-ray image of our bones is converted to a visible image by X-ray-sensitive film.

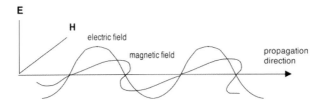

Figure 5.7. Electromagnetic waves.

---

[35] Since the velocity of light in vacuum is a universal constant, we give it the symbol "c," rather than "v." The velocity of light in air is almost identical to its value in vacuum.

Since the wavelengths of X-rays are comparable to the regular spacing of molecular crystals, interference patterns are produced when X-ray beams reflect from crystalline structures. These patterns are used to determine the dimensions of unit cells of the crystals by the technique called X-ray diffraction. Additionally, the intensities of the spots produced in these interference patterns can be used to analyze the contents of the unit cell. X-ray diffraction is a powerful tool for determining the three-dimensional structure of large biological molecules.

A intriguing question is what is the frame of reference for the speed of light. For example, the frame of reference for a swimmer's speed is the water she is swimming in (see Figure 5.3). In a river, she swims faster downstream than upstream. Nineteenth century physicists found the idea of a disturbance propagating through nothing so counter-intuitive that they adopted an idea of a luminiferous ether filling all space from the early Greeks. However, this ether had to have some very contradictory properties. On the one hand, it was known that light traveled through interplanetary and interstellar space. Since the motions of the heavenly bodies were not observed to be slowing, the ether had to be tenuous enough not to provide resistance to their motions. On the other hand the ether had to be dense enough to result in the very high velocity of propagation of light.[36] The final demise of the hypothesis of the *luminiferous ether* occurred when Michelson and Morley in 1887 showed that the velocity of light didn't depend upon whether it was moving parallel or perpendicular to the motion of the Earth around the sun, and thus was independent of its velocity relative to the hypothetical ether.

Another question asked by physicists was whether the speed of electromagnetic radiation depended on the speed of the emitter of the radiation. For example, the reference for the muzzle velocity of a shell is the speed of the gun that fires it. With respect to the Earth, a shell fired by a jet plane moves faster than one fired from a stationary gun. This possibility is ruled out by recent observations that the velocities of gamma rays emitted by decaying elementary particles are unaffected by the velocities of the particles. Light moved with the same velocity with respect to everything. Actually, this was in agreement with Maxwell's ability to calculate the speed of light from his equations, using only the two universal,

---

[36] Sound, for example, travels faster through dense materials than it does through water.

frame-of-reference independent constants in the laws for the electric and magnetic forces. Einstein used the remarkable invariance of the speed of light as one of the basic postulates of his special theory of relativity, which drastically changed the way we think of time and space.

Different types of light are produced by different light sources. For example, an incandescent light bulb contains a filament (typically tungsten) that is hot enough to glow. Usually the temperature of the filament is sufficient to produce approximately white light. (The color of the emissions from hot bodies will be discussed in Chapter 7). As Newton's experiment showed, white light is comprised of different colors, i.e., different wavelengths, which can be separated by a prism. At each wavelength, the waves have all possible phases and all possible directions of polarization. Even light of a single color, such as produced by a glowing neon sign, is made up of all phases and all directions of polarization.

In 1958, light sources with very different wave properties, called *lasers*, were developed by C. H. Townes (1915-, Nobel prize in Physics, 1964) and A. L. Shawlow (1921-1999, Nobel prize in Physics, 1981). In the type of lasers used for laser pointers, the light is emitted continuously and has a very narrow range of wavelengths; it is practically *monochromatic*. In addition, in some of these lasers all the emitted waves have the same phase and the same direction of polarization. Light of this type is called *coherent* and has been employed for many interesting experiments, such as measuring the distance between the earth and moon to within one inch.

Supplemental Reading: Pulsed Lasers

Another type of laser stores up energy and then releases it in a wave that might last for only a millionth of a millionth of a second (10^{-12} sec, called a picosecond) or less. These *pulsed lasers*, which produce the light analogy of a tidal wave, have been used to measure the rates of very rapid processes in Nature, such as molecular rearrangement and ionization. What do the waves produced by lasers that emit very short pulses look like — very short wave trains? To take an extreme example, a green laser pulse emitted for 1/100 of a picosecond would only consist of five waves, as shown in Figure 5.8.

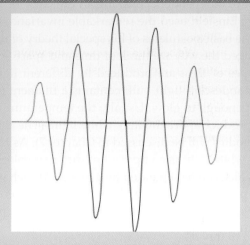

Figure 5.8. Wave train from a pulsed laser.

Although this looks like a fairly clean wave for the duration of the pulse, it goes to zero before and after the pulse. A wave of definite wavelength goes on forever. In order to synthesize pulsed waves we have to combine waves of different wavelengths. You will remember that combining waves is an algebraic process, and by using waves of two different wavelengths we can get interference of a wave at the beginning and the end of the pulse, as shown in Figure 5.9.

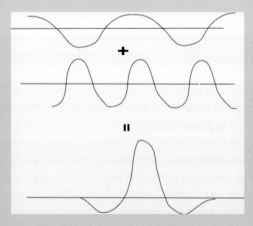

Figure 5.9. Combining waves of different wavelengths.

However, to ensure that the wave remains at zero before and after the pulse, a range of wavelengths must be included in the wave of a short-pulse laser. We see that because of the wave nature of electromagnetic radiation, there is a basic incompatibility between knowing the wavelength (or frequency) of the radiation and locating it in space. This is a basic property of waves and must apply to anything that we describe as a wave phenomenon.

Summary

Waves may be transverse or longitudinal, traveling or stationary, continuous or pulsed. A traveling wave is characterized by a wavelength and frequency; the product of these two quantities is the velocity of the wave. Waves also have amplitude and phase. Waves diffract around corners and two different waves can interfere.

Light is a form of electromagnetic radiation where time-varying electric and magnetic fields propagate through empty space or other media. The velocity of light is predicted by Maxwell's equations and is independent of source velocity. Lasers can produce coherent light, where all the waves have the same wavelength, phase and direction of polarization. Pulsed waves are composed of a superposition of waves of different wavelengths.

Questions

1. If the frequency of a wave is 10 per second, when its wavelength is 1.0 m, what would be the frequency of a wave in the same medium, when its wavelength is 5.0 m?
2. If we see light from a lightning strike almost instantaneously, while the sound from the accompanying thunder travels toward us with the speed of sound in air, how far away is a lightning strike if we perceive a 10 second interval between the lightning and thunder?
3. When we look at a soap bubble, we sometimes see a rainbow of colors. This is due to interference of light reflected from the two sides of the soap film. Explain this phenomenon.

4. Does light exhibit a Doppler effect; i.e., does the color of light depend on the velocity of its source relative to the observer?
5. The police use radar to measure the speed of vehicles. They send out a pulse of (usually microwave) radiation of known frequency and measure its frequency when it bounces off the vehicle and returns to them. Explain how this measurement works.
6. A typical AM station might broadcast at 700 kHz, while an FM station broadcasts at 90 MHz. Which station emits longer wavelength? What do AM and FM stand for?
7. Why is it easier to hear around a corner than see around a corner?
8. What is a "safe harbor"? Are the ships in the harbor perfectly shielded from waves?
9. How could interference adversely affect the acoustics in a concert hall? How could this be avoided?
10. Describe the wave motion of a jump rope, assuming that the ends of the rope can be considered motionless.
11. Sunlight reflected from horizontal surfaces, such as roads, produces glare and is mostly polarized horizontally. How can sunglasses with polarizer lenses reduce this glare?
12. Indicate whether each of the following is in the radio, microwave, infrared, visible, ultraviolet or X-ray region of the electromagnetic spectrum:

    (a) 650 nm
    (b) 250 nm
    (c) 2.0 micron
    (d) 10^{13} 1/s

6 Energy

> *And [energy] is superhuman in the sense that humans cannot create it. They can only refine or convert it.*
>
> Wendell Berry

The organizing principle that produces order can be formulated in terms of *energy*, as well as force. The advantage of dealing with energy is that, unlike force, it is conserved.[37] In addition, energy and energy shortages have become such important themes in modern-day life, that we all need a strong understanding of energy concepts.

We define energy as that which is required to do anything useful. Energy is necessary, but not sufficient, to do useful things. There are types of energy that are not useful, or are hardly useful, for this purpose. For example the background microwave radiation from the formation of the universe that we mentioned in Chapter 2 is of no use to us. The large amount of thermal energy in a lake can be used only with great difficulty. This distinction has to do with the entropy of the energy, which will be discussed in Chapter 13.

Examples of activities that require energy are: moving a train, heating a house and, of course, reading this chapter.

---

[37] We have discussed other quantities, such as charge, momentum and angular momentum that are also conserved.

> **There are three types of energy: potential, kinetic and internal.**

Potential energy is energy of a body's position (where it is) or configuration (the arrangement of its *macroscopic* parts); *kinetic energy* is the energy of its motion as a whole and *internal energy* is the energy of its *microscopic* parts. Energy consumed or generated per unit time is the *power* consumed or generated).

Potential Energy

A boulder on the top of a cliff has considerable gravitational potential energy due to its position in the gravitational field of the earth, as does a charged body due to its position in an electric field. A spring has potential energy when it is stretched or compressed. As shown in Figure 6.1, we can do useful things with potential energy, such as attaching a rope to the boulder, so that as it falls the rope spins an electric generator. This is similar to how the water spilling over Hoover dam generates electricity.

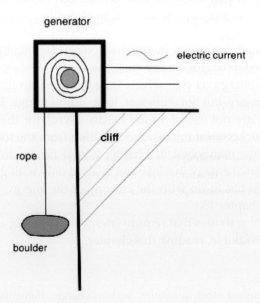

Figure 6.1. Electricity generated by a falling boulder

A boulder on the ground has much less gravitational potential energy than one on top of the cliff. We might think that the potential energy of the boulder on the ground would be zero, but this is not necessarily so. For example, what if there was a well nearby? The boulder on the ground would have a higher potential energy than if it was sitting at the bottom of the well, and more electrical energy could be attained by allowing the boulder to descend into the well. This illustrates an important property of potential energy: the zero of potential energy is arbitrary; it is only changes in potential energy that are important.

In order to increase the potential energy of a boulder by raising it from the ground to the top of a cliff, we must do *work* on it, i.e., we must exert a vertical force on the boulder and move it up a distance. The product of the force times the distance moved *in the direction of the force* is the work done. Note that if we push or pull on the boulder, but do not succeed in moving it, no work is done. Neither is any work done, if the boulder slips to the side (in a direction perpendicular to our force).

Stability and the relationship of potential energy to force

The relationship between potential energy and force is shown in Figure 6.2, using the analogy of the gravitational potential energy of a ball on a one-dimensional surface.

In 6.2a the ground slopes down to the right, and the boulder experiences a force in that direction; in 6.2b the ground rises to the right, and the force is to the left. In 6.2c, the ground is horizontal, so there is no force on the ball. In summary,

> **A force is exerted on a body in the direction in which its potential energy decreases.**

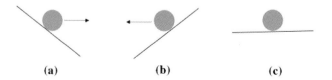

Figure 6.2. Relationship between force and energy.

78 *Order and Disorder*

It is useful to go a step further in discussing the case in which there is no force on the boulder (i.e., the ground is horizontal at the point of contact with the ball). In Figure 6.2c, the boulder is indifferent as to its position on the flat plane. Give it a push to start it moving, and, assuming there is no friction,[38] it will slide in the direction of the push forever. We say that the flat plane is a surface of ***neutral equilibrium***. There are other possibilities for a point where the potential is flat, as indicated by positions a, b and c of Figure 6.3.

If the boulder at position **a** is given a push in either direction, it will experience a force tending to restore it to its initial position. Because position **a** is a ***global minimum*** of the potential energy curve it is a point of ***stable equilibrium***. At such a point there is ***negative feedback***, in that any displacement from the point generates a force that tends to decrease the displacement. Just the opposite happens at position **b** of the curve. Because it is a ***local maximum***, the slightest push of the boulder in either direction from **b** will direct it to either position **a** or **c**. Position b is a point of ***unstable equilibrium***. Alternatively, we can say that there is ***positive feedback*** at point **b**, any displacement from the point tending to grow. Position **c** on the curve is a ***local minimum***; it is stable for small displacements from the minimum, but a displacement to the right, beyond point **d**, will drive the boulder further away from **c**. Point **c** is called a position of ***metastable equilibrium***. In general,

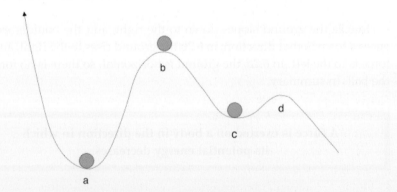

Figure 6.3. Stability of potential energy surfaces.

---

[38] While in practice friction can be made very small, it can never be totally eliminated. Physicists, however, often find it revealing to consider idealized situations, that they call *gedanken* (thought) experiments.

> **Systems are driven towards states of lower energy.**

A good example of this tendency is the ordering of many crystalline materials, which is the arrangement of the lowest electrostatic energy interactions between their component ions.

Energy is usually a function of more than one spatial variable, giving rise to *energy surfaces or landscapes*, rather than energy curves. For example, gravitational potential energy varies with both latitude and longitude on the earth's surface and in three dimensions when traveling in space. The energy landscapes in physical and biological systems are functions of many more dimensions than we can imagine. Computers are very good in extracting information from such systems.

Supplemental Reading: Units of Energy and Power

Since work is converted into potential energy, these quantities must have the same units, which for work is that of force (***Newton***) times distance (m). The SI units for energy, the Newton–meter is called the ***Joule***. A Joule is a small amount of energy. It takes about 10^5 Joules to heat up a cup of coffee and the chemical energy of a gallon of gasoline is about 10^8 Joules.

There are some other commonly used energy units. One of these, the ***electron volt*** (***eV***) derives from the ***electrostatic potential***, measured in ***volts***, which characterizes an electric field. Just like for potential energy, it is only differences in electrostatic energy that are important. A positive charge of one ***Coulomb*** that moves through an electrostatic potential difference of one volt has its energy changed by one Joule. An electron volt (eV) is the potential energy obtained when a proton (with charge 1.6×10^{-19} Coulomb) moves through a potential of one volt. From its definition, $1.0 \text{ eV} = 1.6 \times 10^{-19}$ Joule.

The S.I. unit of power is the Joule/sec, which is called a ***watt***. A typical light bulb requires 60 watts, while a 100 horsepower engine puts out 75,000 watts of power.

Kinetic Energy

Kinetic energy, the energy of macroscopic motion, depends on the mass of an object as well as its velocity through the equation **KE = (1/2)mv^2**. Kinetic energy is thus proportional to the product of mass time the square of the velocity. Because velocity enters into this relation as a squared quantity, even a relatively light meteorite hitting the Earth at high velocity can do a lot of damage.

A moving body can be observed from platforms that are moving with different velocities with respect to the body. For example, a ball thrown from the window of a car can be observed from the car, from the ground, from a passing train or from the moon. Each of these observations will give a different velocity and a different kinetic energy for the ball. Kinetic energy is like potential energy, dependent on the coordinate system from which the body is observed, and only changes in kinetic energy are significant.

Internal Energy

Potential and kinetic energy on a *microscopic* scale contribute to the *internal energy* of a body. For example, *chemical energy* results from the position and motion of electrons within molecules. If atoms are jiggling around (vibrating and rotating) more than the minimum amount, we say that they also have *thermal energy*. There are also attractive forces between molecules, which result in *condensation* (liquid formation) at low temperature. In liquids the attraction of molecules is reduced at the surface, because surface particles are not completely surrounded by other molecules. An energy penalty, the *surface energy*, sometimes called *surface tension*, must be paid for increasing the surface of a bulk material. As a result, an isolated liquid droplet forms a spherical shape,[39] which minimizes its surface area.

Thermal energy is so important that there is a well-known parameter, called *temperature* that determines how much thermal energy resides in each unit mass of a substance. Energy is an *extensive property*; it is proportional to the size of a object, while temperature is an *intensive property*;

---

[39] In contrast, gravity produces deviation from spherical shape.

it characterizes each bit of the object. Thus a body at 80°C (read degrees Centigrade) has more thermal energy than a body at 20°C. One problem with the commonly used Centigrade or Fahrenheit temperature scales is that they are arbitrary. The only significance of 0°C is that it is defined as the melting point of ice. At that temperature, a body has considerably more than its minimum amount of thermal energy. It is much more convenient to add 273 to the Centigrade temperature to get what we call the Kelvin (or absolute) temperature scale. 0 K (read zero Kelvin) is not arbitrary; it is the temperature at which a body has the minimum amount of thermal energy.

For **monatomic** gases (gases composed of single atoms, such as helium, neon or argon.) at low pressure, thermal energy is exclusively *translational*,[40] and this translational energy is proportional to the absolute temperature. For other gases, as well as liquids and solids, thermal energy increases with temperature according to a more complicated relationship derived for the jiggling (vibrations and rotations) of these molecules. Thermal energy is a back and forth motion. At one instant, the velocity of an atom or a molecule is in one direction; a little later, it is in the opposite direction. Also, the jiggling atoms exchange energy, and there is a tendency for the kinetic energies of the atoms to equalize, just as happens when a rapidly moving billiard ball hits a stationary ball. However, this tendency is not complete, and there will be a thermal *energy distribution*, as well as an average thermal energy[41] of the molecules. Some molecules move faster and others slower than the average. It is the more rapidly moving molecules that leave the surface of a liquid when it evaporates, and the more rapidly moving molecules that have sufficient energy for chemical reactions.

Anyone who has lived at the seashore knows waves on water carry energy. Since we are warmed when we stand in sunlight, we also realize electromagnetic waves carry energy.

A boulder on a cliff a certain distance above the ground has a certain potential energy regardless of whether it reached that position by being lowered from a helicopter or pushed up from the ground. Potential energy is a *state function*, depending only on the state of the system at

---

[40] In translational motion, molecules move in a straight line at constant velocity until they undergo collisions.
[41] The average translational energy for a particle at room temperature is ca. 6×10^{-21} Joule/molecule = 0.04 eV/molecule = 3.6 kJ/mole.

the present time. Kinetic energy is also a state function, depending only on a body's current velocity, and thermal energy depends only on its temperature. Thus, the total energy of a body is also a state function.

Energy Conservation

The primary reason that energy is so important is that it is conserved, i.e., it can be neither created nor destroyed. According to Wendell Berry, quoted at the beginning of this chapter, we have been given a certain amount of energy and can only change its form. Going back to our boulder teetering in unstable equilibrium on the edge of the cliff: a slight nudge will make it start to accelerate earthward. As its height decreases, it loses gravitational potential energy, but it accelerates and gains kinetic energy. If the cliff is high enough, the boulder will reach a constant terminal velocity in its fall. The continuing decrease in gravitational potential energy is thereafter converted into thermal energy of the boulder and surrounding air, since it is the frictional force (the rubbing) between these that keeps the boulder's velocity from exceeding terminal velocity.

When the falling boulder hits the ground, it might fragment into a large number of pieces, breaking the bonds that held it together and increasing its chemical energy. The pieces and the surrounding ground can get warm, further increasing thermal energy. In tracing energy flows, we are assured by the *first law of thermodynamics* that

> **Energy can neither be created nor destroyed;
> it is a conserved quantity.**

The first law provides a powerful method of analyzing processes by equating the energy in their initial and final states. Since energy is a state function, it can be calculated from the description of these states. Although for simple processes, conservation of energy follows directly from calculations based on Newton's laws of motion, the general first law of thermodynamics cannot be proved. Its widespread acceptance is based on centuries of experience in which it has never been found to be violated. Not violating the first law has occasionally required the consideration of new types of energy. One of these is *rest energy*, discussed in the next section.

Supplemental Reading: The History of the First Law

It may seem reasonable to us nowadays that potential, kinetic, chemical and thermal energies are all the same type of thing and can be converted into each other. However, before the time of Count Rumford (1753-1814, born Benjamin Thomson), thermal energy was generally thought to be very different from mechanical energy (potential or kinetic). Called heat, thermal energy was considered to be a fluid, *caloric*, that was released from materials. It was measured in calories, one *calorie* being the heat needed to heat one gram of water one degree Centigrade. Only the release of caloric could heat a body.

Rumford showed that in boring cannons, thermal energy could be released without limit and could be used to heat and eventually boil large amounts of water. Since it was unreasonable that the material of the cannon contained unlimited amounts of caloric, he felt that the only possible source of the heating was the motion provided by the horses that turned the cannon on the boring tool. Rumford's idea was the beginning of the end for the caloric theory. In a series of very careful experiments, James Joule (1818-1882) measured the equivalence between thermal and mechanical energy to be 1.0 calorie = 4.18 Joules.

Rest Energy

Albert Einstein (1879-1955) is generally considered to be the greatest physicist, and perhaps the greatest genius, of all time. His novel way of thinking about the universe has led to some amazing predictions, almost all of which have stood the test of time. For example, in his ***special theory of relativity***, Einstein postulated that

> **The laws of physics are independent of the velocity of the observer of manifestations of those laws.**

Maxwell's equations of electromagnetism are a basic law of physics and give a numerical value for the velocity of light and other electromagnetic waves. Thus, according to Einstein, this speed must be independent of the observer's velocity. He reasoned that because of this, no object could move faster than the speed of light (3.0×10^8 m/s), since if it did, it would overtake the light that it emitted. What then happens when we put more and more kinetic energy into an object by having it move through an ever-larger gravitational or electrical potential energy difference? Einstein's conclusion was startling. Potential energy is still converted into kinetic energy. But, since there is an upper limit to velocity, it does this by increasing mass, rather than its velocity. In other words,

> **Mass and energy are equivalent.**

Einstein showed that mass and energy are related by the equation, **$E = mc^2$**, where c is the velocity of light in vacuum. We can think of c^2 as a proportionality factor between mass and energy. However, c is so large, that c^2 is huge. A tiny bit of mass can produce an immense amount of energy. Even when a body is not moving, its mass is equivalent to a certain amount of energy, $m_o c^2$, that we call the *rest energy* of the object (m_o is the mass of the body when it is not moving, its *rest mass*). We include a body's rest energy as part of its internal energy.

Since mass and energy are equivalent, the theories of conservation of mass and conservation of energy are not exactly correct; it is only the sum of mass and energy that is conserved. The reason that this had not been noted until the early twentieth century is that, until that time, the transformations of interest were chemical reactions, in which the energy released or absorbed is relatively small. In these cases, it was a very good approximation to ignore the miniscule changes of mass that accompanied the chemical reaction.[42] In nuclear reactions, the change in mass is much larger, and the resulting energy released is thousands to millions of times as large as in chemical reactions. A few pounds of uranium in an atomic bomb can release as much energy and create as much destruction as exploding a million tons of TNT.

Recognizing that mass and energy are the same sort of thing, physicists usually talk about the rest mass of elementary particles in energy

---

[42] Nowadays with very accurate mass measurements, we can detect mass changes corresponding to the energy released or absorbed in some chemical reactions.

units. In these units, the proton has a rest mass of 938.3 MeV (million electron volts), the neutron has a rest mass of 939.6 MeV and the rest mass of the electron is 0.51 MeV.

Before leaving this section we will say just a few words about Einstein's other theory of relativity, **general relativity**, which he published in 1915. General relativity deals with accelerating coordinate systems and is the theory that physicists currently use to describe gravity. The basic hypothesis of the theory is the **principle of equivalence**, that a gravitational field is indistinguishable from an accelerating reference frame.[43] General relativity leads to a description in which space is warped by the gravity of massive objects and in this warped space, everything, including light, follows curved paths.

Energy Barrier

In spontaneous decay processes of elementary particles, nuclei and molecules, the products have lower energy than the original system. Why then don't these processes occur instantaneously, rather than with a characteristic lifetime, which may range anywhere from 10^{-25} sec to 15 minutes (the lifetime of an isolated neutron) or even longer? Using the analogy of gravitational potential energy, we can diagram the energy variation in such a process as shown in Figure 6.4.

In the gravitational analogy, the system would be a boulder at metastable equilibrium (see Figure 6.3) in the crater of an extinct volcano.

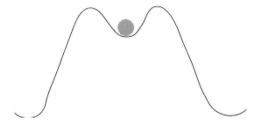

Figure 6.4. Energy variation in a decomposition process.

---

[43] Consider that you are in a spacecraft with no windows. If you feel an upward push on your feet from the floor, you couldn't distinguish whether this was due to a gravitational field or an acceleration of the spacecraft.

The rim of the crater presents an *energy barrier* for the boulder to leave the crater. In order for it to slide down the side of the mountain, it first must find a way to get past the rim.

One possible way for a system to decompose is for part of the system to collect sufficient thermal energy from neighboring parts so that it can surmount its energy barrier. This is what usually occurs in decomposing molecules, where the energy height of the rim is usually less than 100 kJ/mole. (Thermal energies are of the order of 3–5 kJ/mole for each atom, see footnote 41.) For nuclei, however, the energy height of the barrier is much too high to be surmounted by thermal energy. A nuclear boulder would thus be trapped forever in its crater. For nuclei (and some molecules), however, there is another alternative. We will see in Chapter 7, that in such systems, part of the system can *tunnel* through the wall of the crater and then slide down the mountainside. The higher and thicker the wall of the crater, the less efficient is tunneling and the longer is the lifetime of the unstable system.

Energy of Nuclei and Nuclear Processes

Huge amounts of energy are released when small nuclei combine in stars and hydrogen bombs, a process called *fusion*, and when large nuclei decompose in nuclear reactors or atomic bombs, a process called *fission*. In both fusion and fission, energy release results from the formation of more stable nuclei. In a more stable nucleus the *binding energy per nucleon*[44] (proton or neutron) is greater than in a less stable nucleus. Since energy is related to mass, physicists can determine nuclear binding energies by measuring nuclear masses and comparing them to the sum of the masses of their component protons and neutrons.

The binding energy per nucleon reaches a maximum for elements with about 55–60 nucleons. This corresponds to isotopes of iron and nickel, which are fairly common elements in the solar system.[45] In our sun and other stars, as fusion builds up elements to this size, huge amounts of energy are released and stellar temperatures can be maintained. This is what gives rise to the life-sustaining warmth that we obtain from the sun. In the interior of these systems, temperatures are so high that atoms are

---

[44] Binding energy, the energy holding the nucleus together, is negative energy.
[45] The Earth's core, for example, is largely comprised of iron and nickel.

totally ionized, forming a state of matter called *plasma*. Binding energy per nucleon decreases for elements with more than 55–60 nucleons, and their formation in stellar interiors is not favorable. Astronomers believe these elements are synthesized in *supernovae*, the final implosion of some stars, which are among the most violent occurrences in the universe.

The release of energy when a very large nucleus, such as when ^{235}U (uranium-235) or ^{239}Pu (plutonium-239) absorbs a neutron and breaks up into more stable fragments, roughly the size of iron or nickel nuclei, is the basis of nuclear reactors and ordinary, fission-type, atomic bombs. The reactions are self-sustaining, because they produce a number of neutrons, in addition to the nuclear fragments. The increasing number of neutrons produces a branching nuclear chain reaction, which in a reactor has to be controlled by absorbing excess neutrons in graphite control rods. If this is not properly done, the reactor heats up. The Chernobyl nuclear disaster is an example of what can happen when human error or equipment failure negates the safety features built into such a reactor.

A further difficulty with fission reactors is that many of the fragments of nuclear fission are highly radioactive nuclei, which decay with lifetimes of hundreds or thousands of years. A persistent problem is how to securely store these wastes while their radioactivity decays. A typical nuclear transformation releases a MeV (million electron volts) of energy. While this is only 1.6×10^{-13} Joule per nucleus, it is ca. 10^{11} Joules/mole. Uranium fission releases about 100 times this energy. However promising these numbers seem, it is doubtful that nuclear fission reactors will be the answer to the world's energy problems until the problem of radioactive waste disposal can be solved.

Supplemental Reading: The Process of Fusion

Production of helium nuclei ($^4He^{2+}$) from hydrogen nuclei ($^1H^+$) in the interior of the sun goes through the intermediate step of forming heavy hydrogen nuclei ($^2H^+$) called deuterons. As two deuterons approach each other, their energy first rises, due to the long-range electrostatic repulsion, before falling, due to the short-range strong force in the helium nucleus. As shown in Figure 6.5, the potential energy goes through a maximum in this process; there is an *energy barrier* for the fusion reaction.

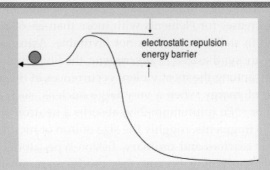

Figure 6.5. Energy barrier for fusion of deuterium nuclei.

In order to surmount this potential energy barrier, the deuterons must have a very high kinetic energy, which is achieved by the very high temperature (ca. 15 million degrees Kelvin) in the interior of stars. Fusion reactions also take place in a hydrogen bomb, in which a fission atomic-bomb trigger provides a temperature sufficient to initiate fusion. A number of programs are in development for fusing deuterium (and tritium) to form helium in a controlled manner. To do this, it is necessary to heat a heavy-hydrogen plasma to temperatures where nuclei can overcome the energy barrier of Figure 6.5. The major challenge of controlled fusion is holding this plasma together, while it is heated to achieve the temperatures and densities required for useful energy generation. Tremendous advantages could accrue to civilization, if the practically unlimited supply of deuterium in the oceans could be used to generate electricity in this manner.

Energy Resources of the Earth

The energy resources of the Earth are of two types: resources that were deposited on the Earth at the time of its formation and the continuous flow of energy from the sun. The former includes the gravitational potential energy that was converted to thermal energy at the time of, and for a while after, the initial formation of the Earth. Because it takes a long time for a rocky body, such as Earth, to cool, much of this energy is still trapped in the inner regions of the Earth. In fact, the center of the Earth is currently about as hot as the surface of the sun. A small fraction of this energy can be tapped as geothermal energy. Long-lived radioactive nuclei, such as uranium-238 (half-life 4.5 billion years) are still contributing thermal energy in inner regions of the Earth.

Nuclei that can be converted into more stable nuclei by either fission or fusion were also deposited on Earth during its formation. Uranium-235 is the fissionable isotope of uranium. While only limited amounts of uranium ore are available to supply uranium at today's market price, this is largely due to the lack of exploration for new highly concentrated ores of uranium. Since fuel cost is a very small part of the cost of nuclear-generated electricity, this industry could probably withstand a large increase in the price of its fuel. At higher prices, new less-concentrated ores could be commercialized to provide fuel for nuclear reactors for many thousands of years.

Deuterium, the isotope of hydrogen used in fusion reactors is available in almost unlimited amounts in the oceans. The problem in using this source of energy is that a good way of carrying out nuclear fusion in a controlled manner has not been developed. Controlled fusion seems to always remain the preferred energy source for the next century.

The amount of solar energy flux reaching the surface of the Earth is huge. In one minute, more solar energy reaches the Earth than fossil-fuel energy is consumed in an entire year. Approximately 0.1% of this energy is captured by plant photosynthesis, both agricultural and in the wild. Solar-produced energy that is used shortly after it is formed is called renewable. We use plant carbohydrates as our grains, fruits and vegetables, and to feed to animals to provide meat for our tables. A recent development is the expansion of the amount of agriculture devoted to producing fuels, both sugar cane and corn for ethanol, and palm oil for bio-diesel fuels. The limiting resources for this production are land and water, and the rapidly increasing price of foodstuffs suggests that these resources are already being taxed. These limitations could be reduced if non-edible forms of plants, such as wood chips or switch grass could be used for this purpose.

Most of the solar energy captured by plants is returned to the atmosphere as carbon dioxide, either annually, or in the case of trees, over decades or centuries. A small fraction of this material decays in an environment where it is exposed to only limited amounts of oxygen before it is buried. Over periods of millions to hundreds of millions of years, these buried materials are converted by heat and pressure to the *fossil fuels* on which we now base most of our society.

Gaseous (natural gas), liquid (petroleum) and solid (coal) fossil fuels are present only in limited amounts. Recently, the expansion of developing economies, such as China and India, strained the supply of petroleum, so that its price rose to near-prohibitive levels. Supply and demand are in better balance with natural gas, and supplies of coal are available for at least a century. The use of fossil fuels, however, deposits large amounts of carbon dioxide in the atmosphere, and this gas retards the escape of thermal energy from the Earth's surface. The expansion of fossil fuel use has produced *global warming*, which threatens the Earth with rising sea-levels, famine, species extinction, and other dire predictions.

Solar energy is also deposited in the atmosphere and oceans, and wind power and *hydroelectric power* provide means of capturing this energy. Since in high-wind environments, wind energy is economically viable, there has recently been an explosion in the use of windmills for capturing wind energy in various parts of the Earth. Hydropower, on the other hand, has been employed for many decades, and its expansion is limited by problems that result from silting and damage to aquatic ecology.

Solar energy can also be directly captured by *photoelectric* methods, which uses semiconductor materials to directly transform solar energy into electrical energy, and by *photo-thermal* methods, which focus the sun's rays to vaporize fluids to drive the turbines that produce electricity. Currently these methods are expensive, but the active research being done in these areas suggests that their prices will decrease over time.

Summary
 octavo

There are three types of energy: potential, kinetic and internal. Potential energy, the energy of position, and kinetic energy, the energy of motion, are relative, while internal energy, the energy of the microscopic parts, is absolute and measured by the Kelvin temperature scale. The energy equivalent of the rest mass of a body is part of its internal energy. Bodies and systems tend towards states of lower potential and kinetic energy. Energy is a conserved quantity. Many chemical and nuclear reactions require surmounting energy barriers, either with thermal energy or by tunneling. Huge amounts of energy are released in nuclear fission or fusion reactions.

The Earth's energy resources were either deposited at the time of its formation or continually flow as electromagnetic radiation from the sun.

Solar energy can be captured by plants, flowing water or wind, or directly by photoelectric or photo-thermal methods, all renewable sources of energy. Fossil fuels result from plants decaying in an oxygen-poor environment with heat and pressure. Currently, use of fossils fuels, stored over millions of years, is depositing large amounts of carbon dioxide in the atmosphere and contributing to global warming.

Questions

1. How does the temperature of the water at the bottom of a waterfall compare to that at the top? What would happen to the temperature at the bottom, if a turbine for generating electricity is installed in the waterfall?
2. Compare the kinetic energy of a 300-pound lineman running at 8 m/sec with a 200-pound fullback running at 10 m/sec. Compare their momentum.
3. Explain why it is impossible to design a racecourse that is downhill throughout, but comes back to the same starting point.
4. Trace the energy flow that occurred when Rumford bored cannons. Start with solar energy, which is the source of almost all the energy that we use in our civilization.
5. What can you deduce from the rest mass of the proton being less than that of the neutron?
6. Where does the energy we expend in pushing on a boulder go, if we do not succeed in moving it, and thus do not change its potential energy?
7. What happens when a little hot water is added to some cold water? Explain this on a microscopic as well as a macroscopic basis.
8. Explain the difference in shape between the diagrams in Figures 6.2 and 6.3.
9. In the biological and social sciences, influences that tend to make entities clump together in time or space are called **positive feedback**, rather than attractive forces. Influences that tend to make entities spread out in time or space are **negative feedback**, rather than repulsive forces. Show that these definitions of positive and negative feedback are the opposite of those used in the physical sciences and discussed in this chapter.

10. Discuss the advantages and disadvantages of satisfying the energy needs of civilization by:
    (a) burning fossil fuels
    (b) burning wood from trees
    (c) generating hydro-electric power with dams
    (d) capturing the energy of the wind with windmills
    (e) directly converting solar energy to electricity
    (f) using solar energy to produce steam that generates electricity.

7 Particle-Waves

On Mondays, Wednesdays and Fridays, I think of light as particles, but on Tuesdays, Thursdays and Saturdays, I think of it as waves.

Lord Rutherford

The first decade of the twentieth century was a period in which the complacency of physicists was sorely shaken. Before that decade they felt they had a pretty good idea about just how to describe Nature: for matter — use forces, masses and Newtonian mechanics; for light — use wave equations. There were, to be honest, a few loose ends that needed straightening out, loose ends that generally involved light interacting with matter. However, physicists didn't realize that tying up these loose ends would require that they completely change their world-view to one that ascribed both particle and wave properties to light and matter.

The Particle Nature of Waves

We discussed particles in Chapter 3 and waves in Chapter 5, so what do we mean by particle-waves, the title of the current chapter? The concept originated with Max Planck (1858–1947, Nobel prize in physics in 1918), who formulated a theory for the electromagnetic radiation emitted from hot objects — such as the light emitted by the heating elements of an electric stove. Planck actually considered a simpler problem, the wavelength distribution of light existing in a cavity within a hot object. The radiation emitted by the object could be considered to be flowing out of a small hole in the cavity. Since the hole absorbs all radiation incident on it and reflects none, it is called a *blackbody*. In order to obtain results that were in agreement with experiments, Planck had to make the outrageous hypothesis that an electromagnetic wave of wavelength, λ, and corresponding frequency $f = c/\lambda$, was not a continuous flow of energy, but rather, was made up of little packages of radiation, each with energy $\varepsilon = hf = hc/\lambda$. h, Planck's constant, has the value 6.62×10^{-34} Joule–second. These packages of radiation energy are called *photons*. According to Planck's equation, the greater the wavelength of light, the less energy each of its photons has

Example 1. Compare the energy of a photon of red light, with $\lambda = 700$ nm with that of a photon of blue light, with $\lambda = 450$ nm.

Solution: for red light

$$\varepsilon = \frac{(6.62 \times 10^{-34} \text{ Joule sec})\left(3 \times 10^8 \frac{\text{m}}{\text{sec}}\right)}{700 \times 10^{-9} \text{m}} = 2.8 \times 10^{-19} \text{Joule}$$

(Note how units cancel in this calculation.)
Similarly, for blue light

$$\varepsilon = 4.4 \times 10^{-19} \text{ Joule}$$

The shorter wavelength light has more energetic photons.

It wasn't obvious to scientists of his time, but Planck's hypothesis holds the key to understanding much of the microscopic world. One important example of its utility concerned the *photoelectron effect*, which is currently used in some of our most sensitive light-detecting scientific equipment. Not long after it became possible to create a good vacuum, Heinrich Hertz (1857–1894) observed that if light (electromagnetic radiation) shines on some metals in a vacuum, electrons are ejected from the metal surface. The light imparts sufficient energy to the electrons to enable them to leave the surface of the metal.[46] However, this effect only occurred if the wavelength of the radiation was shorter than a particular cut-off wavelength. For example, red light impinging on a clean sodium surface in a vacuum does not release electrons, no matter what its intensity — but blue light, even when very faint, ejects electrons. In addition, these electrons are ejected immediately. There is no need to wait for the blue light to deposit some minimum amount of energy on the surface to heat it up.

Einstein explained the photoelectron effect in 1905, using Planck's hypothesis of photons of radiation. Einstein pictured the photons as sufficiently localized, that in the photoelectron effect *only a single photon interacts with an electron*. Essentially he was going back to the particle picture of light discussed by Newton, since, as we stated at the beginning of Chapter 3, localization is the distinguishing characteristic of a particle.

Einstein's hypothesis completely explained the observations made on the photoelectron effect. The reason for the wavelength threshold, above which electrons are not ejected from the metallic surface, is that the electrons are held in the metal with a binding energy, which must be less than the photon energy in order for the electron to leave the surface.

1905 was a very good year for Einstein. In that year he published three papers, one on the photoelectron effect, for which he was awarded the Nobel prize in physics in 1921; one on his special theory of relativity and one on the explanation of Brownian motion — the incessant jumping around of small solid particles suspended in a liquid. All this was done in his free time, while he had a full-time job in the Swiss patent office. It is generally agreed that this was the most productive year for any scientist in history.

How can light be made of particles, if it's an electromagnetic wave as discussed by Maxwell? How can light beams diffract or interfere, if they don't have alternating positive and negative amplitudes? Questions like this led to the type of confusion indicated by Rutherford's quote at the beginning of this chapter. Some light can be thrown on these questions (pardon the pun) by performing Young's two-slit interference experiment with greatly reduced intensity of the light incident on the slits. You might

---

[46] Electrons have lower energy in a metal than in vacuum or in air. (We know this because we move electrons around our house in metal wires, without their escaping into the air.)

we try to avoid influencing the momentum of the electron, we simultaneously reduce our ability to locate its position.

We can either use short wave radiation and limit our knowledge of the momentum (or velocity) of the electron, or use long-wave radiation and limit our knowledge of the position of the electron. Moreover, no one has ever found a way to circumvent this tradeoff. The difficulty in simultaneous position and velocity measurements is enshrined in what is called the **uncertainty principle** of Werner Heisenberg (1901–1976, Nobel prize in physics, 1932), which states that

> **There are limits to which we can simultaneously know the position and velocity of a particle.**

Physicists go a step further and say that an electron cannot simultaneously *have* position and momentum more accurately specified than this limit. As a result, if we apply Newton's laws to the motion of an electron, pretty soon we lose track of where it is. In the world of the very small, physics loses the deterministic causality of Newtonian mechanics. One can no longer exactly predict motion, as one does in predicting the motion of the planets. In discussing atoms and molecules, all we can talk about is the *probability* of particles being at certain positions at a given time!

This new picture of an electron is reminiscent of our discussion in Chapter 5 of the waves that describe a pulsed laser. If a wave has a definite wavelength it goes on forever and we can't locate it at all. If we want to narrowly locate a wave (as for a pulsed laser), we have to construct it by combining waves of lots of different wavelengths. The connection between electrons (particles) and waves was made by Louis de Broglie (1892–1987), a French aristocrat, in his Ph. D. thesis at the Sorbonne, in Paris. Noting some similarities between equations describing the propagation of light and the motion of particles, de Broglie suggested that Planck's formula, $p = h/\lambda$, giving the momentum of light, could also be used to give the wavelength of matter. De Broglie's examiners, not understanding his thesis, asked Einstein for his opinion. It is rumored that Einstein replied that the thesis should not be considered for a Ph. D., but for the Nobel prize. In fact, de Broglie was awarded the Nobel prize for this work in 1929.

By de Broglie's hypothesis, an electron of a definite momentum (or velocity) is a wave with a definite wavelength, and the wave goes on forever; we have no knowledge of the position of such an electron. In order to gain information about the position of an electron, we have to form a localized wave by combining waves of different wavelengths, i.e., waves corresponding to a range of momenta — a range of velocities. We call such a description, the *wave function* of the electron, and say that

> **Matter has both wave and particle properties.**

It's one thing for equations to predict something as strange as wave properties for matter, but is this real? If matter has wave properties, it should be capable of doing things that waves can do, namely, diffraction and interference. However, before you go running full speed at a picket fence, expecting to be diffracted through it (like a diffraction grating does for light), you should be warned that your wavelength is very, very small, and if you try this, you will probably end up impaled on the fence. In order for matter to have an appreciable wavelength, it has to have a very small momentum — a very small product of mass times velocity. Even electrons, with their tiny mass, at the velocities at which they can be prepared, have wavelengths of 0.1–1.0 nm. We cannot build a picket fence with these dimensions. However, they are typical of the regular separation between atoms in a crystalline metal.

Example 2. Calculate wavelength for an electron moving with velocity of 10^6 m/sec and a 100 kg man moving with a velocity of 10 m/sec.

Solution: for the electron,

$$\lambda = \frac{6.62 \times 10^{-34} \text{ Joule-sec}}{(9.1 \times 10^{-31} \text{ kg})(10^6 \text{ m/sec})} = 0.7 \text{ nm}.$$

For the man,

$$\lambda = \frac{6.62 \times 10^{-34} \text{ Joule-sec}}{(100 \text{ kg})(10 \text{ m/sec})} = 7 \times 10^{-31} \text{ m} = 7 \times 10^{-22} \text{ nm}.$$

Note that we can write Joule = kg m²/sec² in these calculations.

100   *Order and Disorder*

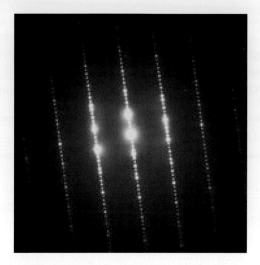

Figure 7.1.   Diffraction pattern of electrons reflected from a nickel crystal.

Not long after de Broglie suggested that electrons had wave properties, Davidson and Germer showed that electrons reflected from a nickel crystal produced a diffraction pattern, as shown in Figure 7.1.

Duality

We've seen that light and matter have *duality*. Since they should both be thought of as particles *and* as waves, we should be able to describe them with similar mathematics. Maxwell showed that light waves are described when his equations are written in the form of a differential wave equation. (Don't worry, you will never see a differential equation in this book.) Erwin Schrodinger (1887–1961) developed a similar differential wave equation for particles in 1926. A solution of Schrodinger's equation is called the *wave function* of an atomic or molecular system.

The interpretations that we give to light and particle waves are also similar. For light, it is the square of the amplitude of the wave (its electric or magnetic field) that is proportional to the intensity of the light beam — or to probability of the light arriving at a certain point. For particles like an electron, it is also the square of the *wave function* that is proportional to the probability of the particle being at a certain place at a certain time. There is greater probability of the electron being where its wave function is large[49] than where it is small. According to the current interpretation of Schrodinger's theory, this probabilistic interpretation is as much as can be known about an atomic or molecular system. It was this interpretation of particle waves that many scientists, including Einstein, objected to. Einstein said "God does not throw dice." Nevertheless, today scientists overwhelmingly accept the probabilistic interpretation.

Another type of uncertainty results from considering the time variation of a wave. A wave of a single wavelength has a definite frequency and goes on for all time. Since frequency is proportional to energy, this means the energy of a system can only be accurately known if the system can be observed for a long time. From a practical point of view: very accurate measurements of energy take a long time. If a particle is only observed for a short time, or only lives for a short time before it decays, its wave must be a combination of waves with a range of wavelengths, i.e., its frequency and energy must be very uncertain.

The uncertainty in energy implied by the Heisenberg principle implies that conservation of energy — or more exactly, mass-energy — does not hold over very short times. Even in a vacuum, devoid of particles and thus of mass-energy, transient particles and their antiparticles can appear for short times before they annihilate each other. This phenomenon is called *quantum fluctuations*.

Quantum mechanics does not change our ideas of the particles or forces that exist in Nature. It just provides us with a way to predict (probable) arrangements and motion that works on the microscopic scale. When we deal with macroscopic objects, the predictions of quantum mechanics are the same as those of Newtonian mechanics. We say that quantum mechanics **reduces** to Newtonian mechanics in the macroscopic limit of heavy particles.

Quantization

Describing matter with Schrodinger's equation is one form of *quantum mechanics*, the physics that describes everything in the microscopic world — all atoms and molecules and many solid-state devices. The

---

[49] Actually, where the *magnitude* of the wave function is large, since the wave function can be either positive or negative.

name derives from the observation that, when describing an unchanging atomic or molecular system — called a *time-independent* system — certain numbers that appear in the equation cannot take on arbitrary values, but are limited to particular discrete allowed values. For example, one of these quantities, which we call a ***quantum number***, can be 0, 1, 2, 3, etc., but nothing in between these values. Each set of permissible quantum numbers produces a solution of the Schrodinger equation — a wave function. The wave function determines an allowable distribution of the particle in space and the values of all its observable properties. Energy is thus also quantized; it can take on particular values and nothing in between. Contrast this with the behavior of the macroscopic world, where energy can be smoothly increased. We would be very surprised if the speed of our car jumped from 40 mph to 60 mph, without ever taking on intermediate values. Energy is just one property that is quantized in quantum mechanics. In general, each quantity that is conserved in classical systems will result in a quantized property in quantum mechanics.

We can see how quantization comes about by exploring the solution of the Schrodinger wave equation for a simple system, the ***particle in a box***. This is a particle completely confined to a region in which it is not subject to any force. The region can be one, two or three-dimensional. In Newtonian mechanics, the solution to this problem is simple: the particle can be stationary (sitting in the box and not moving) or bouncing around in the box with any amount of kinetic energy.

In quantum mechanics, we must take account of the wave nature of the particle, and since it is totally confined to the box, the particle wave function must go to zero at the walls of the box. Just like for a violin string, where there are only certain waves allowed by the constraint that the ends of the string are fixed, there are only certain particle waves allowed in the box. Each allowed wave results in an allowed energy of the system. In particular, a particle in a box cannot have zero energy. This would be a violation of the Heisenberg uncertainty principle, since we would know the velocity of the particle exactly (it would be zero) and still have some knowledge of its position (it would be in the box). The state with the least permissible amount of energy is called the ***ground state*** of the system, and states with more energy are called ***excited states***. Only integer values of the n quantum number are allowed. In Figure 7.2, the energy and wave function of the ground state (n = 1) and first excited state (n = 2) of a one-dimensional particle in box are shown.

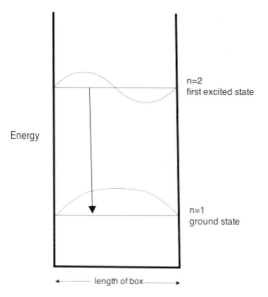

Figure 7.2. Energy and wave functions of states of a particle in a box.[50]

The length of the arrow in Figure 7.2 is proportional to the energy difference between the ground and excited state. The smaller the length of the box, the larger this energy difference and the shorter the wavelength of light emitted in going from the excited to the ground state.

As esoteric as the particle-in-the-box model seems, it helps to describe devices that are finding increased use in our society. In these devices, called *quantum dots*, electrons are confined to small semiconducting regions of materials that are surrounded by insulating regions. An electric current is used to "pump" electrons into excited states of the dots. When the electrons "decay" back to the ground state, they lose the amount of energy corresponding to the quantized energy difference between the ground and excited state of the dot. This energy goes into an emitted photon. The photons can only have frequencies equal to $\Delta E/h$, where ΔE is the difference between the two states and h is Planck's constant. Niels Bohr (1885–1962, Nobel Prize in Physics in 1922) was the first to relate the

---

[50] In the first excited state the wave function as drawn is negative in the right half of the box. However, since probabilities are related to the square of the wave, the probability is positive in both halves of the box.

Figure 7.3. Fluorescence of quantum dots.

frequency of emitted or absorbed light to quantized energy differences between the states of microscopic systems. The particle-in-the-box model explains why the light emitted from quantum dots shifts to shorter wavelengths as the size of the dots decrease. Some of the pure colors emitted by quantum dots of different dimensions suspended in a solvent and excited by ultraviolet light — a process called *fluorescence* — are shown in Figure 7.3. By attaching quantum dots to biological molecules, scientists can observe the movements of these molecules within living cells.

Probably the most important quantized property is the energy of a bound[51] atomic or molecular system. We will see in Chapter 9, that much of chemistry depends on understanding this energy. Even though the wave for an electron in an atom does not have to go to zero at a particular place — as it does for the particle in a box — the wave must die out far from the atom. Otherwise the electron would quickly escape from the atom. As a result, the energy of an atom or a molecule can only have certain allowed values, and photons emitted or absorbed by the atom or molecule can only have energies corresponding to the differences in the energy between these states. Only particular frequencies of light (of energy hf) are absorbed or emitted in these transitions, producing what is called a **line spectrum**. This differs from the emission from hot bodies, such as the sun or a tungsten incandescent bulb, which is a **continuous spectrum**. The use of absorbed or emitted light to probe the quantum nature of an atom or molecule is called **spectroscopy**. The existence of

---

[51] By bound, we mean a system that does not have enough energy to fall apart.

helium, the second lightest element, was first predicted by Pierre Janssen (1824–1907), based on its line spectrum in the solar corona during a solar eclipse.

You or a member of your family may have had a medical diagnostic procedure called **MRI, *magnetic resonance imaging*.** In this procedure, a patient is placed in a large magnet and exposed to radio-frequency electromagnetic radiation. MRI is based on another important quantized property, angular momentum. While in Newtonian mechanics angular momentum is continuous, in quantum mechanics, it can only have integral (0, 1, 2, ...) or half integral (1/2, 3/2, ,,,) multiple of $h/2\pi$.[52] The half-integral values are only allowed for **internal** angular momenta, called *spin*. The spin angular momentum of elementary particles (electrons, protons and neutrons) can be thought of as due to circulating electric charge within the particle. Just as a current circulating in a coil of wire produces the magnetic field of an electromagnet, the circulating charge due to the spin of atoms, nuclei and elementary particles produces magnetic fields. We say that they have ***magnetic moments***. Since spin is quantized, so is magnetic moment.

In the same way that a magnetized compass needle lines up in the magnetic field of the Earth, the magnetic moments tend to line up when a magnetic field is applied to an atom or a nucleus. However, in the case of atoms and nuclei, their magnetic moment can only have particular orientations in a magnetic field — those corresponding to a component of the spin in the direction of the magnetic field that is an integral or half-integral multiple of $h/2\pi$.[53] In quantum mechanics the direction, as well as the magnitude, of angular momentum are quantized.

The possible orientations of a magnetic moment in a magnetic field are related to the spin that produces that magnetic moment. As examples, a measurement of the orientation of a spin of 1/2 can only give one of two orientations; we say such a spin can only point "up" or "down." A spin of 1 can only be measured to have one of three orientations (up, down or sideways) in a magnetic field. Observing the effect of a magnetic field on

---

[52] This is not exactly correct. The exact quantized values of angular momentum are $\sqrt{l(l+1)}\,h/2\pi$, where l is the angular momentum quantum number.

[53] The positions of the compass needle are also quantized, but since it is a macroscopic magnet, these positions are so closely spaced that we perceive them as continuous.

the properties of a particle is a good way to determine the value of its spin. Electrons and protons have both been found to have spin 1/2 and thus two orientations of their magnetic moments in a magnetic field.

Magnetic resonance imaging (MRI) usually makes use of the two possible orientations of the nuclei of hydrogen atoms (protons) in a magnetic field. Our bodies are filled with hydrogen atoms in different environments. When taking an MRI, the patient is placed in a magnetic field, and energy is supplied to "flip" the spins of the protons in her tissues between their two possible orientation states in the field. In different tissues (e.g., skin, bone, cartilage and tumors) the protons are in different environments and have different abilities to absorb this radiation. The tissues therefore show up with different intensities on the MRI image. MRI is non-invasive, and the applied radio-frequency radiation, does not cause tissue damage. It has become a very important diagnostic technique in medicine. The Nobel Prize in Medicine and Physiology was awarded in this area to Paul Lauterbur and Peter Mansfield in 2003.

Example 3. *Are the following quantized or continuous?*

(a) The change in your pocket.
(b) Your weight.

Solution:

(a) Your change is quantized; it can be 37 cents, but not 37.461 cents.
(b) Your weight is continuous. With a sufficiently accurate scale, it can be determined to any accuracy, e.g., 147.615 pounds. Note that your weight is continuous on the macroscopic scale, but on the microscopic scale, it can only change by intervals corresponding to the weights of individual atoms.

Fermions and Bosons

Philosophically, the major difference between the Newtonian approach, called classical mechanics, and the Heisenberg-Schrodinger approach, called quantum mechanics, for describing systems is that quantum mechanics brings the observer into the process. A ***perfect observer*** is limited only by the laws of physics. Quantum mechanics institutionalizes the observer in the wave function, which gives the most intimate description that can be provided by a perfect observer, and not a bit more.

For consistency in quantum mechanics, we must also consider our perfect observer when we assign names to particles (e.g., the different electrons in an atom). In the macroscopic world, we have no problem in naming objects. In our solar system, we call one planet Mars and one Venus, with no chance of mixing these up. Even when we have identical macroscopic objects, we can either mark them or just keep track of which is which, by concentrating on their movements. However, if a perfect observer wants to distinguish between microscopic particles, say electrons in an atom, how would she go about doing this? Electrons are identical and too small to be marked. They are also too light to watch by bouncing photons off them, since the momenta of the photons produces sudden changes in their motion. These considerations lead to the conclusion that in the microscopic world nothing real can depend upon the names that we use to distinguish identical particles. In other words, nothing real can change, if we exchange the names of identical particles.

In quantum mechanics, it is the **square** of the wave function that is proportional to probability, which is something that can be observed and is therefore real. If we switch the names that we give to two particles, there are two ways that the square of the wave function can avoid changing. Either the wave function can be unchanged when we perform this operation, or it can change sign, since $(-1)^2 = 1$. Amazingly, we find microscopic particles with both these properties in Nature. Electrons, protons and neutrons, the building blocks of atoms and molecules, are of the second type; their wave functions change sign when we interchange any two such identical particles. We call these particles *fermions*. Particles whose wave functions do not change on interchanging the particles, are called *bosons*. Particles of light (photons) behave this way, as do any particles that are made up of an even number of fermions.[54] Thus, a hydrogen atom (a proton plus an electron), a deuteron (a proton plus a neutron) and an alpha particle (two protons and two neutrons) are bosons. Fermions and boson can be distinguished by their spin angular momentum. All bosons have integer values of spin (0, 1, 2, ...), while all fermions have half-integer values (1/2, 3/2, ...).

---

[54] Since (-1) to an even power is 1.

If two identical particles in a wave function have the same name, the wave function will not change if we exchange these particles. Since the wave function for fermions must change sign upon such an interchange, we have a contradiction: A wave function using the same name for two identical fermions must change sign but be unchanged when the fermions are exchanged. The only quantity that can change sign, but not be changed, is zero. Zero is not a meaningful wave function; it tells us that the particles are nowhere. Thus, wave functions using the same name for two fermions are not permissible. Since in quantum mechanics we name electrons by their quantum numbers, we obtain the Pauli exclusion principle, which states: in forming a wave function for an atom or a molecule, we cannot use the same set of quantum numbers for more than one electron. The *Pauli exclusion principle* requires that, since electrons are fermions,

> **We must use a different set of quantum numbers for each electron in an atom or molecule.**

This is an important key in understanding chemistry. Since the quantum numbers for each electron determines its distribution in space, the Pauli principle tends to keep the electrons apart and, as we will see, has many important consequences. The Pauli principle applies to all fermions, and it is sometimes considered an additional fundamental force, the *exclusion force* — one that is always repulsive. We will see in Chapter 23 that it is this force that halts the compression by gravity of many stars after they have exhausted most of their nuclear fuel — resulting in stars that are called white dwarfs. Some stars, however, are so massive that gravity overwhelms the exclusion force, resulting in further compression to a neutron star or even a black hole.

Unlike fermions, bosons have no restrictions on their use of quantum numbers. Wave functions for bosons can be much more compact in space. Scientists have recently been able to collect groups of bosons (such as hydrogen atoms) in very small regions of space and cooled to very low temperatures, so that their momenta are very small and their wave lengths are very long. Under these conditions, all the boson waves overlap, giving them a single wave function, and they act coherently, like light in a laser. This unusual state of matter is called a *Bose-Einstein condensate*.

Supplementary Reading: Mixed Wave Functions

You may have noticed we have been very careful in our language when describing quantum systems. For example, we have said that the orientation of an electron spin can be *measured* to be up or down. We have not said anything about what this spin is *before we measure* its orientation. Quantum mechanics allows that, before we make a measurement on the electron, it can be in a mixed state, with both up and down orientation, each with certain probabilities.[55] It is not until we make a measurement that the electron decides in which direction its spin is pointing.

If you are having trouble accepting this, you are in good company; Einstein thought that such mixed states would lead to some ridiculous results. Along with Podlosky and Rosen he considered a "thought experiment," where an atom with no angular momentum simultaneously emits two photons. To conserve ordinary momentum, these photons travel in opposite directions, each at the speed of light. To conserve spin angular momentum, the spins of these photons have to point in opposite directions. Until the spin of one of the photons is measured, both photons are in mixed states, with equal probability of having spin up or spin down. As soon as one photon's spin is measured, the spin of the other is determined by spin conservation, even if at the time the photons are huge distance apart. Einstein reasoned that this result would be impossible, since information cannot be transmitted faster than the speed of light.

In this matter, however, Einstein has been shown to be wrong by experimental measurements. Somehow, in this two-photon experiment, the wave functions of the two photons are *entangled*, so that when one changes the other instantaneously changes. This entanglement provides useful possibilities for data encryption, where the key for decoding a message could be transmitted in a way in which it could not be intercepted. Unlike data encrypted with more-or-less random numbers, a quantum-encrypted code would be unbreakable.

---

[55] These two probabilities must sum to unity, since there are no other possibilities for the orientation of an electron spin in a magnetic field.

> Scientists are very excited about using mixed states for quantum computers, where data is stored and manipulated as **qbits**. A qbit could, for example, be an electron in a state before we have determined its spin. Such an electron is simultaneously in the spin-up and spin-down states. While it would require $2^{10} = 1024$ magnetic bits to represent all settings of 10 on-off switches, these settings could be represented by just 10 qbits, and calculations could be performed on all these settings simultaneously. Unlike a magnetic bit which can only take on one of two values or a biological information bit (DNA), which we will see can take on only one of four values, a qbit can take on a huge number of values depending on the precision with which the probability of the two states is specified. Quantum computing (manipulating qbits) holds out the possibility of huge increases in computation power, which as of the present are far from being realized.

Summary

Phenomena, such as radiation of hot bodies and the photoelectron effect, indicate that light has particle properties as well as wave properties. Light of frequency, f, is comprised of photons each having an energy hf and a momentum h/λ.

Because we cannot know both the position and momentum of a microscopic particle, it must have wave properties and be described by a wave function, i.e., a solution of the Schrodinger wave equation. A particle wave has a wavelength of $h/momentum$. The wave equation of an unchanging bound system contains certain quantum numbers, which label the wave functions and can only assume discrete values. As a result, energies and angular momenta of atomic and molecular systems are quantized.

Particles are either fermions or bosons. Because electrons are fermions, in an atom or molecule, no two electrons can be labeled by the same set of quantum numbers.

Questions

1. Combine Planck's equation for the energy of a photon with Einstein's equation relating mass and energy, to show that the momentum (mass x velocity) of a photon is h/λ.

2. A problem somewhat similar to those of quantum mechanics is figuring out how many different ways that the students in a class can be arranged in the classroom. What are quantized quantities if
   a) the students can be anywhere in the classroom?
   b) the students are confined to seats?
3. For students sitting in seats in a classroom, the "wave function" would indicate which student is sitting in which seat. Students are macroscopic objects, so there is no problem in having two Marys in the class; we can distinguish these by physical characteristics. What would happen if, instead of students, we had electrons in the seats? Could two of them be called Mary? What if they were photons?
4. Identify whether each of these particles is a fermion or a boson:
   (a) a helium atom
   (b) an atom of ^{14}N (7 electrons)
   (c) an atom of ^{85}Rb (37 electrons)
5. Which of these quantities are quantized and which can take on continuous values:
   (a) the number of siblings that you have.
   (b) the amplitude of the fundamental note of a violin.
   (c) your potential energy when you are at rest on a flight of stairs.
6. At the time Einstein explained the photoelectron effect, light sources were available only up to moderate intensities. Nowadays, pulsed lasers can provide light of such huge intensity that the electrons in a metal can interact almost simultaneously with more than one electron. How do you think that this might change the observations on the photoelectron effect?
7. You are to be executed by a firing squad using a powerful monochromatic (single wavelength) laser. In order not to have to look at his victim, the executioner will fire the laser through a slit in a wall. Do you have any recourse? Do you have any recourse if the laser is fired through two parallel slits in a wall?
8. In the first excited state of the particle in a box, shown in Figure 7.2, the wave function has a node (zero amplitude) at the center of the box. Since a particle cannot be found where its wave function is zero, how can a particle in this state spend some of its time in the left side of the box and some of its time in the right side of the box? How can it get from one side of the box to the other?

8 The New Physics

Three quarks for Muster Mark!/Sure he hasn't got much of a bark. And sure any he has it's all beside the mark.

from James Joyce's "Finnegans Wake"

The previous chapters have covered physics through the early part of the twentieth century. In this chapter you will be given a glimpse of some of the discoveries of physics since then.

Conservation Laws

Processes that occur in Nature always conserve a number of physical properties, including charge, momentum, angular momentum and energy (actually mass-energy), which we have discussed in previous chapters. Other conserved quantities will be discussed below. A process that violates a conservation law is called *"**forbidden**,"* and physicists will go to great lengths, including assuming the existence of previously unobserved particles, in order to avoid invoking such processes. On the other hand, it is generally believed that an *"**allowed**"* process, one that does not violate a conservation law, will occur, although it might be very infrequent. If an "allowed" process does not occur, physicists will suspect that it actually is forbidden and violates a conservation law that has not yet been recognized. Many of the recent advances of physics resulted from consideration of conservation laws.

Emily Nother proved in 1925 that there is a connection between conservation laws and the basic symmetries of the universe. Nother's theorem is

> **Every conservation law implies a continuous symmetry of the universe and vice versa.**

As an example, consider conservation of energy, which implies that the laws of physics are symmetrical (do not vary) over time. Consider what would happen if a law of physics, say a fundamental constant such as appears in Newton's law of gravitational attraction or Coulomb's law of electrostatic attraction, varied with time. In this case, by moving attracting particles away from each other when the constant is low and returning them to their original position when the constant is high, a net increase in energy could be achieved, violating conservation of energy (actually mass-energy). Similar considerations connect conservation of momentum with the requirement that the laws of physics are independent of position in the universe, while conservation of angular momentum requires that the laws of physics are the same in all directions. Since there is no evidence that conservation of mass-energy, momentum and angular momentum are ever violated, Nother's theorem indicates there is nothing special about the time or place that we occupy in the universe.

The Neutrino

One example of the confidence physicists have in their conservation laws is the discovery of the neutrino. In Chapter 3, we discussed beta decay, which is the emission of a high-energy electron when a neutron changes to a proton in a nucleus. For example, carbon-14 undergoes beta decay with a half-life of 5720 years. When an electron is ejected at high velocity from a nucleus, conservation of momentum requires that the nucleus hardly move — similar to when a battleship fires a shell. Since the nucleus acquires negligible kinetic energy, by conservation of energy, the mass difference between the initial and final nuclei must go into kinetic energy of the emitted electron. This is what the theory predicts; what was observed was quite different. In beta decay, very few of the detected electrons had the requisite amount of energy; most of them had much less energy. Where was the extra energy going?

Rather than admit failure of conservation of energy in beta decay, Wolfgang Pauli hypothesized in 1930 that in this process, another particle was emitted. Pauli called this particle a neutrino (little neutron), since for conservation of charge, it had to be uncharged. The mass of the neutrino had to be very small, since in a fraction of the decays the electron did account for almost all the required energy (and thus mass, by conservation of mass-energy). The zero charge and near zero mass would explain why the neutrino had never been observed; it just doesn't interact appreciably with any particle. An average neutrino can pass through 3,500 light-years[56] of lead before being absorbed.

Nowadays, we know that

> **Neutrinos, particles with no charge and very little mass, are emitted in nuclear reactions and particle transformations.**

You might think that with neutrinos interacting with other matter so weakly, there would be practically no chance of ever witnessing such an interaction. However, there are trillions of neutrinos streaming out of nuclear reactors and the sun. By building detectors consisting of huge vats of liquids, and placing these in abandoned mines, where they are protected from cosmic radiation, scientists are now able to observe a few interactions of neutrinos with matter. Neutrinos travel at close to the speed of light, and their detection can be the first indication that a violent event has occurred far away in the universe.

Neutrinos interact so weakly with matter because neither the strong force nor the electromagnetic force influences them. Since their mass is so small, their gravitational interaction is also negligible. Thus, neutrino interactions result from the short range and very feeble weak force.

---

[56] A light year is a unit of length, equal to the distance that light travels in a year. In SI units it is 9.5×10^{12} km.

That neutrinos interact only by means of the weak force explains why a free neutron has such a long lifetime (15 minutes). Fifteen minutes might not seem a long time in our lives, but it is eons on the scale of particle decompositions, since most particle decay processes occur in 10^{-10} to 10^{-20} seconds. We have indicated that for a transformation, all the particles in the transformation, both reactants and products, must interact with the same force. Since the strongest force that interacts with all the particles in neutron decay is the weak force, the interaction is very feeble and the decomposition takes a very long time.

The production of a neutrino in beta decay is also needed to conserve angular momentum. When a neutron decays the only angular momenta that have to be considered are internal angular momenta, i.e., the spin angular momentum of the individual particles. (Orbital angular momentum, mvr, terms are negligible, since r is essentially zero; all the particles are emitted from the same point.) The neutron, proton and electron all have spin of 1/2. The initial state of the decay (the neutron) therefore has an angular momentum of 1/2. If the final state consisted only of a proton and an electron, it could only have an angular momentum of 0 (if the two spins pointed in opposite directions) or 1 (if they pointed in the same direction). A third particle with spin 1/2, i.e., the neutrino, is therefore necessary to conserve angular momentum in this process. Since it has spin 1/2, a neutrino, like the proton, neutron and electron, is a fermion.

Besides the decay of the free neutron, other reactions that are very slow because they rely on the weak force are:

(1) the reverse of neutron decay — the reaction of a proton and an electron to form a neutron.
(2) the reaction that combines two protons to form a deuteron (hydrogen-2), a positron and a neutrino. This reaction is the first step of the main pathway for helium formation in the sun. Since it relies on the weak force, it is unusually slow, resulting in the sun releasing its fusion energy in a controlled manner over a long period of time, rather than very rapidly.

Classification of Elementary Particles

Historically, elementary particles were classified on the basis of their mass, using the terms: **_leptons_**, **_baryons_** and **_mesons_** to indicate light, heavy and intermediate mass particles, respectively. With this classification, electrons and neutrinos are leptons and protons and neutrons are baryons. The term meson encompasses a variety of short-lived particles that have been detected as products of interactions between particles in accelerators or between cosmic rays and atmospheric particles. *Cosmic ray* is a term used to describe a variety of super high-energy particles that impinge on the Earth's atmosphere from outside the solar system. As more particles were discovered, the particle classification was changed to recognize that leptons and baryons are fermions (half-integral spin), and mesons are bosons (integral spin), and only mesons and baryons interact by means of the strong force. The mesons and baryons were called *hadrons* collectively.

As particle accelerators advanced, so that collisions at higher and higher energies could be studied, more and more hadrons were discovered, until their number reached almost 100. Physicists began to despair of ever making sense of this "hadron zoo." Additional complexity was introduced by the discovery of *antimatter*.

Antimatter

The *positron* was predicted by P. A. M. Dirac (1902–1984) when he combined Einstein's special relativity with quantum mechanics. A positron is a form of antimatter.

> **Antimatter is identical to ordinary matter, except that the charges of the particles are reversed.**

The positron was first detected in 1933 by Carl Anderson (Nobel Prize in physics, 1936) using a cloud chamber. The first recorded track of a positron is shown in Figure 8.1.

In a cloud chamber, water droplets are formed around the charges produced along the path of high-energy particles that travel through water vapor in the chamber. The chamber is placed in a magnetic field in which the paths of charged particles are bent. In Figure 8.1, a high-energy particle produced by a cosmic ray comes from the bottom and loses some energy passing through lead, before entering the upper part of the chamber.

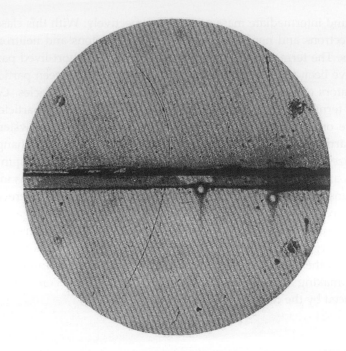

Figure 8.1. Cloud chamber track of a positron.

From the bending of the path and the energy lost in passing through the lead, the particle was found to have the mass to charge ratio of the electron. However, the particle is bent in the opposite direction expected for an electron, indicating that it has a positive charge.

Positrons are difficult to detect, because as soon as they come near an electron, the two particles *annihilate* each other and change their combined rest mass into energy, usually two 0.5011 MeV gamma rays (photons), which travel in opposite directions. The annihilation process does not violate any conservation law: charge is conserved and so is spin (the resulting photons each have spin of 1); mass-energy is the same before and after the process. *Pair production* is the reverse process, conversion of a gamma ray with sufficient energy into an electron-positron pair.

The proton also has its antiparticle, the antiproton, which is just like a proton except it has a negative charge. In 1959 a particle accelerator achieved sufficient energy to create proton-antiproton pairs, which were separated from each other in a magnetic field.

With positrons (anti-electrons) and anti-protons detected, only anti-neutrons were needed for the complete creation of anti-matter.[57] However, the neutron has a charge of zero; reversing this would still leave us with a neutron. How could an anti-neutron be different from an ordinary neutron? The key to the existence of the anti-neutron involves other charge-like quantities that can be assigned to particles in addition to electrical charge. The theory of anti-matter also requires that in an anti-particle these other charge-like quantities reverse sign in the antiparticle.

We will discuss only one charge-like quantity, called the baryon number. Protons and neutrons both have a baryon number +1, and the baryon number of a nucleus is just the sum of the baryon numbers of its constituent particles (the number of nucleons in the nucleus). Anti-protons and anti-neutrons have baryon number −1. Thus, an anti-nucleus has a baryon number equal to the negative of the number of nucleons in the nucleus. Like charge, baryon number is also conserved in transformations of elementary particles.

So much energy is released in the annihilation of particles with anti-particles, that anti-matter is of interest for military and space missions. The big problem is that, even if useful amounts of anti-matter can be created, how could it be stored until it is used?

Example: Approximately how much energy is released if one gram of antihydrogen annihilates one gram of hydrogen?

Solution: Two grams of matter would be annihilated, and since $E = mc^2$,

$$E = .002 \; kg \times \left(\frac{3 \times 10^8 \, m}{\sec} \right)^2 = 1.8 \times 10^{14} \; Joule$$

This is the energy released in detonating 43,000 tons of TNT. (Note that mass must be in the SI unit, the kilogram, for this calculation.)

---

[57] The photon is its own anti-particle.

Positron-Emission Tomography (PET)

You might think that antimatter is so esoteric that you would never come in contact with it. However, this is not the case; positrons, one form of antimatter, are increasingly used in medicine.

Because the two gamma ray photons produced in annihilation of a positron by an electron travel in opposite direction, two detectors that coincidentally (at the same time) record the arrival of these photons can be used to pinpoint their point of origin in a sample. This provides the basis of positron-emission tomography (PET) scans, a powerful medical diagnostic technique, which exposes the functioning, as well as the structure of bodily tissues. A number of isotopes that decay by the conversion of a proton into a neutron, such as carbon-11 (half-life 20 min), oxygen-15 (half-life 2.0 min) and fluorine-18 (half-life 110 min), are used for PET scans. Almost immediately after emission, the positron is annihilated by an electron, producing two gamma rays. Collinear, coincident detection of these gamma rays permits tissues in which the radioactive material is collecting to be imaged with very high resolution.

Although the diagnostic procedure involves ingesting or inhaling radioactive isotopes, the extremely high sensitivity of the detection method permits very small amounts of these isotopes to be employed. In addition, the isotopes used usually decay with half-lives of minutes, and the gamma rays formed are not strongly absorbed in the body. One disadvantage of the method is that the short half-lives of the emitters require they be produced in the vicinity of the diagnostic center.

Structure of the Hadrons

How were physicists to make sense of the multitude of hadrons — mesons and baryons, including the proton and neutron — and their antiparticles? The situation was similar to what existed during the days when new elements were being discovered and arranged in the periodic table — before the structure of the elements was revealed to derive from just three particles, the proton, neutron and electron.

The hadrons, however, were not broken up by collisions in high energy accelerators, so there was no experimental information on their internal structure. The search for the constituents of the hadrons was therefore led by theoreticians, using ideas of symmetry. A satisfying theory of the structure of the hadrons was presented in 1964 by Murray Gell-Mann (Nobel prize in physics in 1969). Gell-Mann proposed that

> **Hadrons, such as protons, neutrons and mesons, are composed of particles called *quarks*.**[58]

According to the ***standard model***, proposed by Gell-Mann and accepted by almost all physicists, there are two quarks, an up quark (u), with charge +2/3, and a down (d) quark, with charge −1/3. Since quarks have spins of 1/2, they are fermions.

The leptons of the standard model of physics are the electron (*e*) and its neutrino (v_e, called the electron neutrino). These fermions are fundamental particles of matter. For each particle in the standard model, there is a corresponding antiparticle. According to the standard model, these quarks and leptons comprise the first generation of particles.

The standard model also encompasses two matching heavier generations of particles and antiparticles. The second generation contains the strange (s) quark, with charge +2/3 and the charm (c) quark, with charge −1/3, with their corresponding leptons, the muon (*μ*) and muon neutrino (v_μ). The third generation contains the top (t) quark, with charge +2/3 and the bottom (b) quark, with charge −1/3) quarks, with their corresponding leptons, the tau (*τ*) and tau neutrino (v_τ). With these fundamental particles and a few rules, physicists can explain the entire hadron zoo. A diagram of the standard model is shown in Figure 8.2.

For each of these particles there is a corresponding antiparticle.

In the standard model, baryons are composed of three quarks, for example, a proton is composed of two up quarks and one down quark (uud), giving it a charge of 2/3 + 2/3 −1/3 = 1, while a neutron is composed of two down quarks and one up quark (udd), giving it a charge of 2/3 − 1/3 − 1/3 = 0. Because quarks have spin of 1/2, baryons have half-integral spins and are fermions. Mesons are composed of two quarks, and are bosons.

---

[58] Taken from the quote from Finnegan's Wake, by James Joyce, given at the beginning of this chapter.

122 *Order and Disorder*

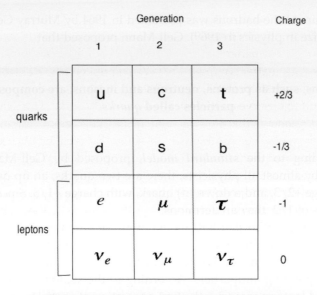

Figure 8.2. Particles of the standard model.

Are quarks (and leptons) the ultimate particles of matter? You may have heard that some physicists are working on a theory of the composition of quarks (as well as of everything else). In this theory, the ultimate components of matter are called strings. How can a string not be composed of even smaller particles? We won't even try to go into this, except to say that, once again, quantum mechanics comes to the rescue by pointing out that there are certain questions like this that cannot be answered.

Dark Matter

We have discussed the various types of particles with which physicists are familiar. Might there be other types of particles that have not yet been discovered? It seems that this is very likely. Evidence for new types of particles comes from astronomers. These scientists have estimated the amount of visible matter (such as stars, gas and dust particles) in the large structures of the universe, such as galaxies and clusters of galaxies, and compared this with the amount of material that would be necessary to keep these structures together by gravitational attraction. Their estimates have come up very short. The mass of the visible matter is only about 20%

of what is needed to produce the large-scale order of the universe. The additional mass required must be in forms not accessible to telescopes, and is called "dark matter."

What is dark matter composed of? We will not go into a detailed evaluation of this question in this book. Intergalactic gas and neutrinos are very difficult to observe and might be considered "dark." However, there are strong arguments that rule them out as the major contributors to dark matter. There is also very strong evidence for what is called "black holes," objects so massive, that not even light can escape from their gravitational attraction. However, black holes are also insufficient to explain dark matter. A very popular proposal for the composition of dark matter is **WIMP**s (weakly interacting massive particles), particles heavier than the proton or neutron, which interact with themselves and other particles only by gravity and perhaps the weak force. To date, such particles have not been observed, although a number of experiments have been proposed for their detection.

Virtual Particles

In our discussion of the fundamental forces of physics, we invoked fields (e.g., the gravitational field or the electric field) that transmit forces between particles. However, since fields are defined as *the force per unit mass at a point* or *the force per unit positive charge at a point*; invoking them really does nothing to explain how one particle reaches out and influences another. The means by which this is achieved was particularly difficult to understand once the idea of the luminiferous ether as a means of propagating electromagnetic radiation had been dispensed with.

A hint of how two stationary electric charges interact can be obtained from notions about what happens when a charged particle is accelerated. An accelerating electric charge radiates electromagnetic radiation, which can be thought of either as a wave or as a beam of photons. By sending electric currents up and down an antenna, your local radio station uses this method to generate the radio waves that are picked up by your receiver.

Physicists believe that the electrostatic force between two charged particles is also transmitted by emission of photons by each particle. However, since these particles do not lose energy, they cannot give up these photons. Unless they interact with another charged particle, the photons return to their source; i.e., they are "***virtual photons***."

This idea of forces being "carried" by virtual particles has been extended to the other fundamental forces as well. We visualize that

> **All the fundamental forces are transmitted by virtual particles.**

For the gravitational force, which also extends to infinity, the carrying particle, like the photon, has no mass, and is called the *graviton*. The graviton has never been detected. For the short range strong and weak forces, these particles, can have mass. The particle carrying the strong force between quarks is called the *gluon* and those carrying the weak force are the *W and Z bosons*. These particles have been detected in high-energy accelerators.

Summary

Physicists believe so strongly in conservation of quantities, such as charge, mass-energy and angular momentum, that they will infer the existence of particles that have never been observed to avoid violating these conservation laws. The neutrino was first inferred in this matter. Neutrinos have zero charge and very little mass and interact with other particles extremely weakly.

For every particle there is an antiparticle, identical in every way, except that its charges are the negative of the particle's. When any particle interacts with its antiparticle (such as an electron interacting with a positron) the particle and antiparticle annihilate each other and form gamma rays. Hadrons, such as the proton and neutron and mesons, are built up of quarks, which have fractional charge. The existence of not-yet-observed dark matter is necessary to explain gravitational clumping in the universe. The current explanation of forces is that they are transmitted by virtual particles, rather than by action-at-a-distance.

Questions

1. Explain how conservation of momentum requires that when carbon-14 undergoes beta decay, the resulting nitrogen-14 nucleus is produced with almost zero velocity.

2. Show that the reaction of two protons to form a deuteron (hydrogen-2 nucleus with spin 1), a positron (positive electron) and a neutrino conserves charge and angular momentum. Why is this reaction very slow, keeping our sun from burning up in a flash?
3. Show that conservation of angular momentum (spin) requires that the deuteron (the nucleus of heavy hydrogen, with spin of 1) cannot be composed of two protons and one electron, but can be composed of a proton and a neutron.
4. Why would it be difficult to run a PET scan with isotopes of half-lives of a millisecond or ten thousand years?
5. Show that with quarks having spin ½, protons and neutrons are fermions and mesons are bosons.
6. What is a possible explanation for our visible universe being composed almost entirely of matter, with practically no antimatter?
7. Would it be conceivable to communicate with an antimatter individual by email? Could you ever meet in person and shake hands — or more?
8. Nother's theorem indicates that all positions and times in the universe are equivalent. How does this differ from other (e.g., some religious) views of the universe?
9. Gas at low temperature does not give off light. Why is it not considered dark matter (matter not observed by a telescope)?

9 Atoms

Atoms are completely impossible from the classical point of view ...

Richard Feynman

Three basic particles: the proton, neutron and electron, produce all the complexity of our world by forming a hierarchy of structures, beginning with the formation of atoms. We have seen in Chapter 3, how protons and neutrons combine to form nuclei, positively charged structures held together by the strong force. Rutherford showed that the nuclei contain almost all of the mass, but practically none of the volume of atoms. However, it is the electrons, which are on the outside of atoms that determine their properties.

In this chapter, we will first consider two different models, those of Bohr and Schrodinger, for explaining the structure of hydrogen, the simplest atom. Both models are based on the ideas of quantum mechanics, but only that of Schrodinger can be used to describe more complicated atoms.

The Hydrogen Atom

The quote by Richard Feynman (1918–1988, Nobel prize in physics, 1965) at the top of this chapter indicates the impossibility of describing atoms using Newtonian mechanics. With Rutherford's very small nucleus, stationary electrons would be pulled into the nucleus. A planetary model is also not stable, since the charged electron is in circular motion around the nucleus. In circular motion the direction of the velocity of the electron continually changes, i.e., it accelerates. By Maxwell's laws, an accelerating electron radiates energy. In losing its energy, the electron would very rapidly spiral into the nucleus.

In his theory of the hydrogen atom, Neils Bohr (1885-1962, Nobel prize in physics in 1922) approached this problem by ignoring it.

> **Bohr proposed that there are stable orbits for the electrons in atoms.**

In these stable orbits, according to Bohr and in contradiction to Maxwell's laws, the electron does not radiate as it travels around the nucleus. The stable orbits are characterized by a quantum number "n" which can only have integer values (1, 2, 3,...), leading to quantized values of the radius of the circular orbit and the energy of the atom in these stable states. When the state of the hydrogen atom has $n = 1$, it is called the ground state of the atom. Higher values of n corresponded to excited states, which can be accessed from the ground states by collisions with photons or electrons.

In the ground state the hydrogen atom is stable indefinitely, unless it interacts with light or another particle, which can put it into an excited

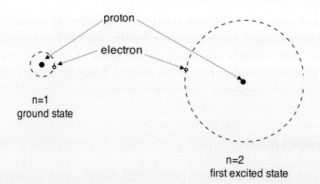

Figure 9.1. Hydrogen atoms in the ground and first excited state according to Bohr.

Figure 9.2. Hydrogen line spectrum in the visible region.

state. Excited states can decay to lower energy states by spontaneously emitting light. Planck's constant times the frequency of the emitted light is the energy difference between the two states ($\Delta \varepsilon = h\nu$). This condition insures conservation of energy in the emission process. Bohr's theory explained the hydrogen line spectrum, the visible region of which is shown in Figure 9.2, to within the very high accuracy with which it was measured.

An alternative description of the hydrogen atom was proposed by Erwin Schrodinger (1887–1961, Nobel prize in physics, 1933). In Schrodinger's theory, each state of the hydrogen atom corresponds to a wave function — a solution of a wave equation. The wave functions of the hydrogen atom are used as the basis to explain all the properties of matter.

Schrodinger's wave equation will not be given here, but its solutions show that

> **Each hydrogen atom state is a wave characterized by three quantum numbers, n, l and m_l.**

Schrodinger's theory predicts exactly the same results for the energy levels (and the spectrum) of the hydrogen atom as the Bohr theory. In Schrodinger's theory, however, the electron wave is characterized by three quantum numbers, not just the single quantum number of the Bohr theory. In addition to the n quantum number of the Bohr theory, which takes on integral values (1, 2, 3, ...), Schrodinger's theory predicts two additional quantum numbers, l and m_l.

In the Schrodinger theory, the n quantum number is a measure of how close, *on average*, the electron is to the nucleus. An electron with $n = 1$ is closer to the nucleus than one with $n = 2$.

The second quantum number, l, which may take on any integer value from 0 to $n - 1$, characterizes the shape of the electron wave function. Wave functions with $l = 0$ have spherical shape, meaning the probability of finding

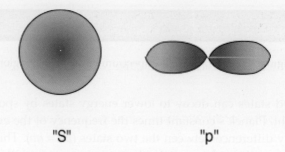

Figure 9.3. Shapes of s and p wave functions.

the electron is the same in all directions. Wave functions with $l = 1$ have dumbbell shapes, as shown in Figure 9.3. Wave functions with higher values of l have more complicated shapes that will not be given here. For historical reasons, states with $l = 0$ are called "s" states, those with $l = 1$ are called "p" states and those with $l = 2$ are called "d" states. The l quantum number also tells us about the angular momentum the electron has in its motion around the nucleus. This is called its *orbital angular momentum*. The greater l, the larger the orbital angular momentum.

Example 1. What is the lowest value of n that can give rise to a "d" state?

Solution: A d state means that $l = 2$. Since the largest value of l is $n − 1$, n must be greater than or equal to 3 for a d state.

The third quantum number characterizing the wave function is called "m_l" (spoken as m-sub-l). It specifies the orientation of the wave function, say with respect to an external electric or magnetic field. Values of m range from $-l$ to $+l$ in integer steps. There is just a single m_l value for an s state ($m_l = 0$) and three values of m_l for a p state ($m_l = -1, 0, +1$). In general, there are $2l + 1$ values of m_l for a given value of l.

Example 2. What are the allowed values of m_l when l = 2?

Solution: Since m_l goes from $-l$ to $+l$ in integer steps, its allowed values are $-2, -1, 0, +1, +2$, i.e., there are five allowed values of the m_l quantum number.

States of the hydrogen atom with a given value of n and different values of l and m_l all have the same energy. These are called **degenerate**. The number of states at a given energy level is called the **degeneracy** of that level and is the key to understanding the periodic table of the elements.

Example 3. What is the degeneracy of the n = 2 energy level?

Solution: When $n = 2$, l can be 0 or 1. For $l = 0$, m_l can only be 0. However, when $l = 1$, m_l can be -1, 0 or $+1$. Thus, there are four states with this energy; the degeneracy is four.

In Figure 9.4, we show the probability of the electron being at different positions — called its **probability density** — for the hydrogen atom in its 1s, 2s and one of its three 2p wave functions.

The lowest energy state (the ground state) of the hydrogen atom should be nondegenerate, with unique set of quantum numbers: $n = 1$, $l = 0$, $m_l = 0$. However, if a beam of ground-state hydrogen atoms passes through a non-uniform magnetic field, it is found to split up into two beams of equal intensity. This is due to the internal angular momentum of the electron, called its **spin** (s),[59] which provides a fourth quantum number for the hydrogen atom.

The electron has spin of 1/2. Like orbital angular momentum, a spin of s can have $2s + 1$ orientations. For a spin of 1/2 this gives two orientations — we

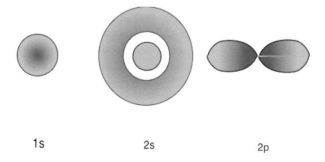

1s   2s   2p

Figure 9.4. Probability densities in lowest energy hydrogen-atom states.

---

[59] The proton in the hydrogen atom also has two possible orientations of its spin. However, the magnetic moment of the proton in the hydrogen atom is much smaller than that of the electron and has negligible effect in this experiment.

say that the spin can be "up" or "down," corresponding to two possible values, +1/2 or −1/2 of a spin quantum number called m_s. The lowest energy level of the hydrogen atom is therefore doubly degenerate, with quantum numbers: $n = 1$, $l = 0$, $m_l = 0$, $m_s = +1/2$, or n = 1, $l = 0$, $m_l = 0$, $m_s = −1/2$. The degeneracies of the first three energy levels of the hydrogen atom, corresponding to $n = 1$–4, are given in Table 9.1.

Multi-Electron Atoms

All atoms other than hydrogen have more than one electron, with the negative charge of the electrons balanced by the positive charge of the protons in the nucleus. The number of protons in the nucleus determines the identity of an element. Different isotopes of the same element have different numbers of neutrons in the nucleus. Each element has a name that is represented by a one or two-letter symbol. Most of the symbols are recognizable from the names, but others, being derived from non-English names, are not. In Table 9.2, names and symbols are listed for the elements that will enter into discussions in this book. Symbols not derived from the English name for the element are marked with a star. The symbols of all

Table 9.1. Degeneracies of hydrogen-atom energy levels

n	l and m_l	m_s	degeneracy
1	0,0	+1/2 or −1/2	2
2	0,0; 1,−1; 1,0; 1,+1	+1/2 or −1/2	8
3	0,0; 1,−1; 1,0; 1,+1; 2,−2; 2,−1; 2,0; 2,+1; 2,+2	+1/2 or −1/2	18

Table 9.2. Names and symbols for some elements

hydrogen - H	helium - He	lithium - Li	beryllium - Be
boron - B	carbon - C	nitrogen - N	oxygen - O
fluorine - F	neon - Ne	sodium - Na*	silicon - Si
phosphorous - P	sulfur - S	chlorine - Cl	argon - Ar
potassium - K*	calcium - Ca	iron - Fe*	bromine - Br
krypton - Kr	rubidium - Rb	iodine - I	xenon - Xe
cesium - Cs	tungsten - W*	mercury - Hg*	uranium - U
plutonium - Pu			

Figure 9.5. Line spectrum of iron in the visible region.

elements are given in the *periodic table*, a systematic arrangement of the elements that reveals many of their properties, which can be found in the appendix of this book. The periodic table will be discussed in detail in the next chapter.

As indicated by the line spectrum of iron, shown in Figure 9.5, the electronic structure of multi-electron atoms is complicated.

The Bohr theory cannot describe multi-electron atoms. The Schrodinger approach is based on quantum mechanics, and solving the wave equation becomes increasing difficult as the number of electrons in the atom increases. Nevertheless, using powerful computers, it has been shown that quantum mechanics is capable of predicting the properties of atoms to experimental accuracy.[60] Moreover, by considering approximate solutions to these equations, chemists have been able to obtain a very useful "intuitive" feel for the properties of atoms. In this approach,

> **Wave functions for multi-electron atoms are constructed from one-electron wave functions (orbitals).**

The one-electron wave functions that are used to construct multi-electron wave functions are called "hydrogen-like," because the charge on the nucleus has to be adjusted to that in the multi-electron atom. For example, for the helium ion, we use a nuclear charge of +2e (where e is the charge on the proton), which pulls the electron closer and binds it more strongly to the nucleus than in the hydrogen atom. The lowest approximation to the ground state of a helium atom has two electrons with 1s hydrogen-like wave functions. We say that, in its ground state, helium has two electrons in the 1s orbitals, an *orbital* being a one-electron wave function. The *electron configuration* of the ground state of the helium

---

[60] For some properties, relativistic quantum mechanics, a combination of special relativity and quantum mechanics must be used.

atom — the designation of the orbitals employed — is written $(1s)^2$. Since the Pauli exclusion principle requires the use of a different wave function for each electron, one of these electrons has $m_s = +1/2$ and the other $m_s = -1/2$. When we combine the m_s quantum number of the electron with its spatial wave function we have a *spin-orbital*. An alternative way of stating the Pauli principle is

> **In multi-electron systems, each electron must be in a different spin-orbital.**

Hydrogen-like orbitals can also be used to describe excited states of atoms. For example, 1s2s and 1s2p would be electron configurations of excited states of the helium atom.

In lithium, with three electrons, a new problem arises. The first two electrons can go into the two $n = 1$ spin-orbitals, but where will we put the third electron? An obvious choice, if we are looking for the ground (lowest-energy) state of the lithium atom, is to use the next lowest energy orbital, one of the $n = 2$ orbitals. Because the n quantum number indicates the size of orbitals, we call these orbitals the $n = 2$ **shell** of lithium. However, there are two different types orbitals in the $n = 2$ shell, those with $l = 0$, called the 2s **subshell**, and those with $l = 1$, called the 2p subshell. Which one will we use? Which one has the lowest energy in the Li atom? The answer to this question depends on some subtle differences in the probability distributions of these wave functions. Fortunately, the result is easy to remember.

> **For a given value of n, orbitals with lower l values have lower energy.**

Thus, the 2s subshell in lithium has lower energy than the 2p subshell, and the electron configuration of lithium is $(1s)^2$ 2s.

Figure 9.6 is a schematic representation of the ground states of hydrogen, helium and lithium. In these representations the circles indicating the positions of the various electrons are placed at radii that are the average position of their Schrodinger electron densities.

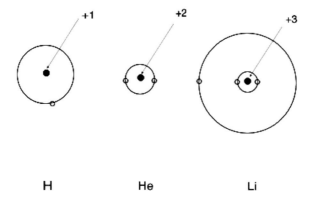

Figure 9.6. Ground states of H, He and Li atoms.

Example 4. How many electrons can be placed in a 2p subshell?

Solution: Since a p subshell ($l = 1$) can have $m_l = -1$, 0 or +1, each of which can have $m_s = +1/2$ or $-1/2$, it can hold up to six electrons.

Example 5. How many electrons can be placed in a 3d subshell?

Solution: Since a d subshell ($l = 2$) can have $m_l = -2, -1, 0, +1$ or +2, each of which can have $m_s = +1/2$ or $-1/2$, it can hold up to ten electrons.

The approximate order of filling the subshells of atoms with up to 56 electrons, along with the number of electrons that can be put in each subshell, is given in Table 9.3. We build up the electron configurations of the ground states of atoms by successively adding electrons to these subshells (2 to an s subshell, 6 to a p subshell, 10 to a d subshell).

Note that d subshells ($l = 2$) are filled between s and p subshells with the next higher value of n; e.g., 3d is filled between 4s and 4p.

Table 9.3. Order of filling electron subshells

1s (2), 2s (2), 2p (6), 3s (2), 3p (6), 4s (2), 3d (10), 4p (6), 5s (2), 4d (10), 5p (6), 6s (2)

Beryllium, with four electrons, completely fills the 2s subshell and in boron the fifth electron goes into the 2p subshell. We will be particularly interested in the chemistry of carbon, whose electron configuration is $(1s)^2 (2s)^2 (2p)^2$. The 2p subshell is completely filled at neon, which we say has a filled $n = 2$ shell, since its 2s subshell is also filled. The electrons in the outer shell of an atom are its valence electrons, which determine the atom's chemical and physical properties. Thus lithium has a valence of one, beryllium of two, boron of three and carbon of four.

In sodium, the Pauli principle requires that one electron be placed in the 3s subshell. The $n = 3$ shell is complete at argon, and in potassium (symbol K in the periodic table), the $n = 4$ shell begins to be filled. We will continue discussing the electron configurations of atoms in the next chapter, when we take a tour through the periodic table of the elements.

Example 6. Write out the electron configuration of iron (Fe) with 26 electrons.

Solution: $1s^2\, 2s^2\, 2p^6\, 3s^2\, 3p^6\, 4s^2\, 3d^6$

Note that the 3d subshell is not completely filled; it contains only six of its maximum of ten electrons. This is one reason for the complexity of the spectrum shown in Figure 9.5.

Example 7. What is the electron configuration of the sodium cation?

Solution: The sodium cation has a charge of +1, and thus one fewer electron than the 11 protons in its nucleus; it has 10 electrons. Its electron configuration is $1s^2\, 2s^2\, 2p^6$, i.e., the same as the neon atom.

Using quantum mechanics, we have learned how electrons, protons and neutrons are ordered to produce atoms and atomic ions. At higher levels of approximation, quantum mechanics can provide microscopic properties of atoms, such as electron distributions and energetics to very high accuracy. However, just knowing the electron configuration of an atom allows chemists to estimate these properties. As we will see in the next few chapters, from the properties of atoms we can understand the macroscopic physical and chemical properties of the elements and the compounds formed from them.

Summary

∽ ∾ ∽ ∾ ∽ ∾

In Bohr's model of the hydrogen atom a single quantum number describes stable orbits of the electron and their associated energies. In Schrodinger's model, three quantum numbers describe the size, shape and orientation of an electron's wave. A fourth quantum number gives the spin orientation of the electron.

The Schrodinger description can be extended to atoms containing more than one electron, by using the Pauli exclusion principle, which requires a unique set of quantum numbers for each electron. Electrons in atoms are arranged in shells, with those in the outer shell being the valence electrons, which determine the atom's chemical and physical properties.

Questions

1. For the hydrogen atom, what is the degeneracy of the first excited level (n = 2)?
2. Why is a helium atom with two electrons so much smaller than a hydrogen atom with one electron?
3. In a neon sign, electrons at high velocity pass through neon gas to produce red light of a very "pure color" (monochromatic wavelength). Explain what is happening in such a device.
4. Discuss similarities and differences between the planetary model for the hydrogen atom and the Earth-Moon system.
5. What is the electron configuration of an oxygen atom?
6. Write out the electron configuration of bromine with 35 electrons.
7. Put the following in order of increasing size: neon atom, sodium atom, sodium cation. Explain your choice.
8. Isoelectronic species have the same number of electrons. Which species in question 7 are isoelectronic? What determines the relative size of two isoelectronic species?

10 The Elements

"There's antimony, arsenic, aluminum, selenium, And hydrogen and oxygen and nitrogen and rhenium…"

Tom Lehrer song

Elements are chemical substances that cannot be broken down into simpler substances by chemical means.[61] Each element is made up of a single type of atom. In this chapter, we will see how the macroscopic properties of elements are determined by the microscopic structure of their constituent atoms.

The approximately 100 elements are the fundamental building blocks of the chemical world, each of which element has a unique set of properties. In modern society, some use has been found for practically every element. Most of the elements are metals. They conduct heat and electricity and are ***malleable***, meaning they are distorted by forces without breaking, the opposite of ***brittle***. Among these, melting points range from –39°C for mercury to 3410°C for tungsten. Almost all the other elements are gases. These range from the almost completely non-reactive gases: helium, neon, argon, krypton and xenon to violently reactive fluorine. Mercury and bromine are the only two elements that are liquids. Bromine liquid exists in equilibrium with a considerable amount of gas.

---

[61] The energies available to chemists are those typically found in flames and electrical discharges and visible and UV light. Elements can be transformed by the much larger energies available to physicists, for example, in particle accelerators.

The Periodic Table

Until the middle of the nineteenth century, scientists had to remember the properties of each of the more than 50 elements known at the time. They searched for regularity in the properties that might help them organize this information.

Dimitri Mendeleev (1834–1907) took an early step in this direction. While preparing a list of the elements for the chemistry textbook he published in 1869, Mendeleev noted certain elements had similar properties. He was arranging elements in a row by increasing atomic mass (atomic numbers were not known at that time), and decided to fold over the row, so that similar elements were in the same column. This put potassium in the same column as sodium and bromine in the same column as chlorine.

> **Mendeleev's periodic table demonstrates regularities in the properties of the elements.**

In the *periodic table*, elements in the same column are called a *group* (a chemical family), and each row is called a *period*. In the first five periods there are 2, 8, 8, 18 and 18 elements respectively. A recent periodic table of the elements, including many that are only produced in particle accelerators and decay in very short times, is given in the appendix of this book.

In the periodic table, elements fall into three types of groups. Elements in groups 1A through 8A are called *main group elements*. These elements are present in all the periods of the table. Starting in the fourth period, ten new groups, called groups 1B through 8B, are placed between groups 2A and 3A. The elements in these groups are called the *transition elements*. Finally, 14 new groups of elements, called *inner transition elements* are placed at the bottom of the periodic table. More exactly, these elements should be placed in the sixth and seventh periods, after the elements lanthanum (La) and actinium (Ac). However, doing so would make the table inconveniently wide. These elements are often called the *lanthanides* and *actinides*, respectively. We will see how the structure of the periodic table results from the electronic structure of the elements, as indicated by their electron configurations.

Mendeleev was so convinced of the correctness of his arrangement that in several places he deviated from listing elements in terms of increasing mass.[62] In two places in the table, he had to leave empty spaces in his table. He was vindicated in his decisions, when elements for the empty spaces were later discovered and when Henry Moseley's (1887–1915) work on X-rays showed the elements were properly ordered in terms of increasing atomic number (number of protons in the nucleus), rather than increasing atomic mass.[63] This is because atomic number, the number of protons in the nucleus, is equal to the number of electrons in the atom, and it is the electrons that interact with the external world and produce the properties of the element.

Mendeleev's table originally did not include column 8A for the non-reactive gases: helium, neon, argon, krypton and xenon (called the *noble gases*), which had not been discovered in his time. In spite of this limitation, the periodic table has remained the supreme guide for chemists in their everyday work for almost 150 years.

Explanation for the Periodic Properties of the Elements

Although Mendeleev's arrangement of the elements helped to systematize their properties, it was not a theory in the sense of explaining these properties at a deeper level. In fact, Mendeleev didn't even believe in the electrons that determine this periodicity. An explanation of the properties of the elements awaited the development of quantum mechanics, with its description of atoms, in the early twentieth century. In the previous chapter we saw how the electronic structures of multi-electron atoms can be described in terms of hydrogen-like wave functions (orbitals), using the electron configuration of the atom.

> **The properties of an element can often be predicted from the electron configuration of its atoms.**

---

[62] For example, iodine, with atomic number 53, has a slightly lower atomic weight than tellurium, with atomic number 52.

[63] Moseley's untimely death at 28 at the battle of Gallipoli made him ineligible for the Nobel prize, which can only be awarded to a living scientist.

142 *Order and Disorder*

In this chapter we take a more extensive tour through the periodic table of the elements. We will begin our discussion by focusing on the first three rows of the periodic table, which only contain elements in groups 1A to 8A. The electron configurations of these elements, plus those of potassium and calcium, the first two elements in the fourth row, are given in Table 10.1.

Hydrogen and helium

In the first period (row), the $n = 1$ shell is being filled. Since this shell only has an s subshell ($l = 0$), the first row contains just two elements, hydrogen (H) and helium (He).

The electromagnetic force is the only force important in chemistry.[64] In fact, the structure of molecules is completely determined by the electrostatic

Table 10.1. Electron configurations

Hydrogen (H, 1A) - 1s
Helium (He, 8A) - $(1s)^2$
Lithium (Li, 1A) - $(1s)^2$ 2s
Beryllium (Be, 2A) - $(1s)^2 (2s)^2$
Boron (B, 3A) - $(1s)^2 (2s)^2$ 2p
Carbon (C, 4A) - $(1s)^2 (2s)^2 (2p)^2$
Nitrogen (N, 5A) - $(1s)^2 (2s)^2 (2p)^3$
Oxygen (O, 6A) - $(1s)^2 (2s)^2 (2p)^4$
Fluorine (F, 7A) - $(1s)^2 (2s)^2 (2p)^5$
Neon (Ne, 8A) - $(1s)^2 (2s)^2 (2p)^6$
Sodium (Na, 1A) - $(1s)^2 (2s)^2 (2p)^6$ 3s
Magnesium (Mg, 2A) - $(1s)^2 (2s)^2 (2p)^6 (3s)^2$
Aluminum (Al, 3A) - $(1s)^2 (2s)^2 (2p)^6 (3s)^2$ 3p
Silicon (Si, 4A) - $(1s)^2 (2s)^2 (2p)^6 (3s)^2 (3p)^2$
Phosphorous (P, 5A) - $(1s)^2 (2s)^2 (2p)^6 (3s)^2 (3p)^3$
Sulfur (S, 6A) - $(1s)^2 (2s)^2 (2p)^6 (3s)^2 (3p)^4$
Chlorine (Cl, 7A) - $(1s)^2 (2s)^2 (2p)^6 (3s)^2 (3p)^5$
Argon (Ar, 8A) - $(1s)^2 (2s)^2 (2p)^6 (3s)^2 (3p)^6$
Potassium (K, 1A) - $(1s)^2 (2s)^2 (2p)^6 (3s)^2 (3p)^6$ 4s
Calcium (Ca, 8A) - $(1s)^2 (2s)^2 (2p)^6 (3s)^2 (3p)^6 (4s)^2$

---

[64] As discussed in Chapter 4, the electrostatic force is much stronger than the gravitational force, while the strong and weak force only act within the nucleus or elementary particles, respectively.

H :H   H—H

Figure 10.1. Covalent bond in H_2.

interaction. For example, in elemental form, hydrogen forms a gaseous diatomic molecule, H_2, in which the two hydrogen atoms are held together by shifting electron density into the region between the two atoms. The negative electrons serve as the "glue" that holds the two positively charged protons together. The build-up of electron density between the atoms is considered to result from their *sharing* electrons in order to fill the 1s shell of each atom. This is the essence of a *covalent bond*.

> **A covalent bond between two atoms results from their sharing electrons.**

Chemists represent covalent bonds between two atoms either by two dots or by a line between the atomic symbols as shown in Figure 10.1.

Hydrogen, which is the most abundant element in the universe, makes up only about one percent of the Earth's crust, mostly in the form of water. This is due to the ease with which very light hydrogen atoms escape from the atmosphere when they are not combined with other elements. H_2 is an ideal fuel, producing only water when it burns in oxygen. Alternatively, hydrogen can be combined with oxygen in a fuel cell to directly produce electricity. The problem in using hydrogen as a fuel is that it must be compressed to high pressure to have appreciable energy density (Joules per unit volume of fuel). In addition, due to its reactivity, reservoirs of uncombined hydrogen do not exist on earth, so it must be produced from other energy sources.

We are most familiar with hydrogen as a gas. At temperatures below 20 K it becomes a liquid, and below 14 K, a solid (both at 1.0 atmosphere pressure). All these phases are insulators, meaning they do not conduct an electric current. However, under extreme pressures, hydrogen becomes an electrically conducting metal, either a solid metal when it's cold or a liquid metal when it's hot. This latter phase of hydrogen exists in the inner regions of the giant planet, Jupiter.

The physical and chemical properties of materials usually depend only on the electrons that are on the outside of their constituent atoms.

The only appreciable exception to this rule is the case of hydrogen, for which the deuterium isotope has twice the mass and the tritium isotope three times the mass of normal hydrogen. As a result, the properties of isotopic variations of H_2 and other compounds of hydrogen can differ noticeably.

Helium, a **monatomic gas** (comprised of individual atoms), is the second most abundant element in the universe. The sun, for example, is 71% hydrogen and 27% helium by mass. Because of the low mass of helium atoms, they are not held strongly by Earth's gravity and are continually escaping from the upper atmosphere. However, since the helium nucleus is the α particle produced in radioactive decay of nuclei, there is a continual source of helium on Earth. Nevertheless, the abundance of helium is less than a millionth of a percent of the Earth's crust. If helium were not concentrated in natural gas reservoirs (sometimes up to 7% abundance), we would not be able to use it to fill our party balloons or obtain the very low temperatures needed to operate our magnetic resonance diagnostic instruments.[65]

Other main group elements

The electrons in the outermost shell of an atom, the one with the highest value of n, are called its *valence electrons*. The valence electrons are on the outside of the atom, and they play the dominant role in determining its physical and chemical (reaction) properties, since only they can interact with other atoms or the outside world. Electrons in the inner shells (having n less than its highest value) are called the *core electrons*, and have very little effect on chemical and physical properties. Since an s subshell can hold two electrons and a p subshell six electrons, an atom has a maximum of eight valence electrons. For all the atoms in Table 10.1, except helium, the number of valence electrons is just equal to the element's group number. Since elements in the same group have similar properties, we conclude that

> **The properties of elements are primarily determined by the number of their valence electrons.**

---

[65] These operate with liquid helium, which exists below 4.2 K.

The noble gases

Of particular interest are the elements in group 8A: helium (He), neon (Ne), argon (Ar), krypton (Kr) and xenon (Xe). The outer shell of helium contains two electrons, while those of the others contain eight electrons. Since the valence shells of these atoms are completely filled, they have practically no tendency to interact with other atoms. No compound of helium, neon and argon, stable at room temperature, has ever been prepared, and krypton and xenon only form compounds with very reactive elements.

The elements of group 8A are all gases because their thermal energy is larger than the energy of interaction between their atoms. If the temperature of a noble gas is lowered sufficiently, its thermal energy will eventually become smaller than the interaction energy, and it will liquefy or even solidify.[66] The structure of one layer of a solid noble gas is shown in Figure 10.2.

In the layer, each atom has six *nearest neighbors,* atoms with which it is in contact. Around the atom there are also six voids — three above and three below the paper. Thus, each atom can have twelve nearest neighbors in this *close-packed structure*.[67] When atoms weakly interact through non-directional forces, they are ordered just like oranges in the supermarket. This isn't surprising.

Figure 10.2. Arrangement of atoms in one layer of a solid noble gas.

---

[66] Helium can only be solidified at low temperature, if its pressure is increased.
[67] There are two different close-packed structures, depending on whether the same or alternate three voids are filled in the layers above and below the one shown.

Other main group elements

The electron configurations of the noble gases are particularly stable, and as a result

> **Elements tend to achieve noble-gas electron configurations by losing, gaining or sharing electrons.**

Alkali, alkaline-earth and aluminum group elements

Elements in groups 1A (the *alkalis*) and group 2A (the *alkaline earths*) have a tendency to lose their valence electrons. These *electron donors* form +1 or +2 cations respectively, having noble-gas electron configurations. The elements in these groups are metals; in the solid they form a lattice of cations surrounded by a sea of electrons. This description is called the *free-electron model* of metals. The electrons that are lost have considerable mobility, giving the metals their characteristic physical properties of high electrical and thermal conductivity and metallic luster (reflection of visible light). The use of the alkalis in metallic form is limited, since they rapidly react with oxygen in the air. Cesium is the most reactive of these; it reacts explosively with water. This is due to the valence electron in its 6s subshell being further from the nucleus and thus more weakly bound than in the other alkalis. Alkali compounds, such as sodium chloride or potassium oxide, are *ionic* — with the positively-charged cations (e.g., Na^+) stabilized in a lattice by electrostatic attraction to nearest-neighbor negatively-charged anions (e.g., Cl^-). Sodium and potassium comprise 2.8 and 2.6% of the Earth's crust, respectively, and are essential for many aspects of life.

The alkaline earths (group IIA), magnesium and calcium comprise 2.1 and 3.6% of the Earth's crust, where they are components of many minerals. They are major constituents of our bones and teeth and also play other roles in our bodies. Alkaline earths are less reactive than alkalis, and beryllium and magnesium are used as lightweight structural metals. However, this is only possible due to an impervious film of oxide, which forms on the surface. Oxygen cannot penetrate this layer and thus oxidation ceases after a thin layer is formed. Once ignited, the layer is breached. As a result, magnesium will burn in air and even in nitrogen. Magnesium plays an important role in chlorophyll, the light gathering pigment of plants; it is provided in our diets by green vegetables. Almost all the compounds of the alkaline earths are ionic.

All the elements in group 3A are metallic, except for boron, whose valence electrons, being close to the nucleus in the $n = 2$ shell, are difficult to remove. Aluminum is the most abundant metal in the Earth's crust. Being light and strong, it is commonly used for structural purposes. Aluminum is also protected by an oxide film, which makes it stable in air.

The oxidizing elements in groups 6A and 7

The elements towards the right side of the periodic table are called *non-metals*. The atoms of the elements in groups 6A (the oxygen family) and 7A (the halogens) can achieve noble gas configurations by adding electrons. They do this by removing electrons from other atoms — a process called *oxidation*. These elements are called *oxidizing agents*.

The *halogens* add a single electron to become the fluoride (F⁻), chloride (Cl⁻), bromide (Br⁻) and iodide (I⁻) anions. In the pure elements, since there are no electron donors available to provide electrons, the halogen atoms share a pair of electrons to form diatomic molecules. At ambient conditions, the lightest halogens (F_2 and Cl_2) are gases, Br_2 is a liquid and the heaviest halogen (I_2) is a solid. In these diatomic molecules, six electrons, in addition to the shared bonding pair of electrons, surround each atom. Since electrons pair up their spins, chemists call these six electrons three *lone pairs*. This gives a total of eight valence electrons around each atom. In Figure 10.3, the structure of F_2 is shown in two different ways, using a line or two dots for the bonding pair of electrons. In the latter structure the lone pairs are also shown. Molecular structures that place eight valence electrons around each atom (except for hydrogen, which being a first-period element, only has two valence electrons around it) are called Lewis *octet* structures after G. N. Lewis (1875–1946).

The oxygen atom needs two additional electrons to achieve a filled shell. In the presence of electron donors, it forms the oxide anion, O^{2-}. In the elemental form, it shares electrons in two different ways, forming two

Figure 10.3. Lewis structures of F_2.

gaseous *allotropes* of the element oxygen. The most common form of elemental oxygen, the form in which it is mostly found in our atmosphere, is as a diatomic molecule. In this molecule four electrons (two pairs) are shared between the two atoms, to produce a *double bond*, as shown in Figure 10.4. One of these electrons pairs lies along the line joining the two oxygen nuclei and is called a *sigma bond* (s), while the other electron pair lies above and below this line and is called a *pi bond* (p). The additional electron density in the double bond provides extra electron "glue" to produce a shorter and stronger bond than for the pair of electrons in a single bond. The second allotrope of oxygen is ozone, which is comprised of three oxygen atoms. Ozone is a very reactive molecule that can be formed in the atmosphere by lightning or photochemical air pollution. Its presence in the atmospheres of cities, such as Los Angeles, Denver and Atlanta, presents a very serious health hazard. The Lewis octet structure of ozone has a single bond and a double bond, as shown in Figure 10.4.

The Lewis structure drawn for the ozone molecule in Figure 10.4. shows a single bond between the left oxygen atom and the central oxygen atom and a double bond between the right atom and the central oxygen atom. Such a structure would have a short, strong O=O bond on the right and a long, weak O–O bond on the left. However, there is another structure in which the double bond is on the left and the single bond is on the right. How does the ozone molecule know which of these equivalent structures to adopt? The answer to this question is that it doesn't know, and all experimental measurements on the ozone molecule show the two bonds to be identical. These observations are explained as a result of *resonance*, which predicts that when a molecule has two or more Lewis structures, it adopts a geometry that is some average of those predicted by the different structures.[68] Resonance is indicated by

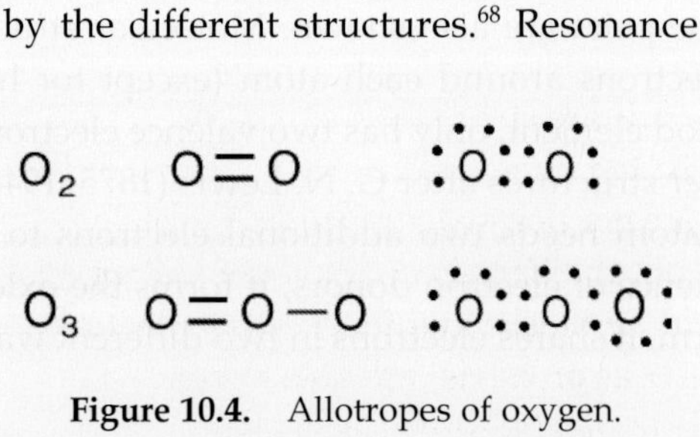

O_2 $O=O$ $:\ddot{O}::\ddot{O}:$

O_3 $O=O-O$ $:\ddot{O}::\ddot{O}:\ddot{O}:$

Figure 10.4. Allotropes of oxygen.

---

[68] Resonance is different from rotation of the molecule, which interchanges the two bonds and thus averages them over the time of the rotational period. Resonance provides an instantaneous average of the two structures.

$$O=O-O \longleftrightarrow O-O=O$$

Figure 10.5. Resonance structures for ozone.

chemists by a double-sided arrow between the Lewis structures and is shown in Figure 10.5 for ozone. Since in ozone the two resonance structures are equivalent, they are equally weighted.

It is predicted and found that a molecule with multiple resonance forms will generally be more stable than a molecule for which only a single Lewis structure can be drawn.

Sulfur and selenium, two other group 6A elements, are both essential for life. Keshan disease, endemic in parts of China in the early twentieth century, was caused by a lack of selenium in certain soils. Both elements are solids, with sulfur forming a variety of allotropes, the most important being orthorhombic sulfur, comprised of 8-membered rings of sulfur atoms.

The nitrogen group

In group 5A, nitrogen needs to gain three electrons in order to form the nitride ion, N^{-3}, which has a filled shell. It does this with many metals. In the elemental form, it completes its shell by forming a triple bond with another nitrogen atom, as shown in Figure 10.6.

$$N\equiv N \quad :N::N:$$

Figure 10.6. Structure of N_2.

Since triple bonds are stronger than double bonds, which are stronger than single bonds, the bond in N_2 is very strong indeed. Because of this strong bond, N_2 is very non-reactive, which explains why it is the largest component (78%) of the Earth's atmosphere. Nitrogen is a component of proteins, the building blocks and tools of life. Bacteria in root nodules of legumes (such as peas or beans) are able to convert atmospheric nitrogen to forms needed by plants, (*nitrogen fixation*). Nitrogen fixation is also carried out on a massive scale by the chemical industry by reacting N_2 with hydrogen to form ammonia NH_3, in a process developed by Fritz Haber (1868–1934, Nobel prize in chemistry, 1918), that is the basis of most of the fertilizer and explosive industries.

Phosphorus and arsenic, both solids, are comprised of tetrahedral molecules, P_4 and As_4. Both elements occur at moderate concentrations in the Earth's crust (0.1% and 2 ppm, respectively). Phosphorus is very important for the energy and information providing molecules of life, while arsenic is poisonous for humans in all but the lowest concentrations. In some parts of the world, natural contamination of ground water with arsenic is an important health issue.

The carbon group

In the elemental form, the atoms in group 4A need four electrons in order to complete their shell. Since four pairs of electrons would be very crowded between two atoms forming a diatomic molecule, these elements prefer to form extended structures with single bonds (combined with double and triple bonds in the case of carbon). In elemental silicon and in the diamond form of carbon, each atom is surrounded by four singly bonded atoms arranged in a **tetrahedron**. A small piece of the diamond structure is shown in Figure 10.7.

The tetrahedral structure is chosen by nature because it is the arrangement that puts the electrons in the four bonds the furthest apart, as shown in Figure 10.8.

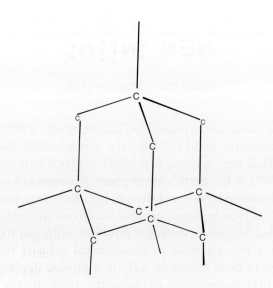

Figure 10.7. The structure of diamond.

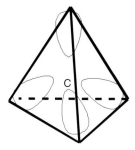

Figure 10.8. Tetrahedral arrangement of electrons around carbon in diamond.

The electrostatic repulsion is thus minimized. The C-C single bonds in this structure are very strong, making diamond the hardest material known and resistant to chemical attack. Unlike the situation in metals, the electrons in diamond are held strongly in place, resulting in diamond being an electrical *insulator*.

Crystalline silicon has the tetrahedral structure of diamond, but since the silicon nuclei are not as closely spaced, they do not attract the electrons as strongly. Under the influence of an electric field, a few electrons leave the bonding region and travel through the crystal. As a result, silicon is a *semiconductor*, which is used in computer chips, making it the most important element of the modern information society.

Diamond is only one of the allotropes of elemental carbon. In graphite, a slightly more stable form of carbon, each carbon atom is surrounded by three other carbon atoms, forming planar hexagons. In order to share its four electrons, one of these bonds is a double bond and the other two are single bonds. However, resonance makes the three bonds equivalent. This structure is shown in Figure 10.9.

In contrast to diamond, graphite, the black material that pencil "lead" is made of, is very soft. Why is this, when the average bond in Figure 10.9, being between a single and double bond, is stronger than the average bond

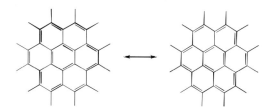

Figure 10.9. The structure of one layer of graphite.

in diamond? The reason is that Figure 10.9 only describes a two-dimensional plane in graphite (called a graphene sheet). In solid graphite these planes have to be held together. Since the carbon atoms in Figure 10.9 have used all their electrons to form covalent bonds in the plane, only much weaker *intermolecular forces* hold the planes together. These forces derive from the interaction of the π electrons above and below each of the planes, when the planes are stacked. It is the breaking of these weak *intermolecular bonds* that allows graphite to be deposited on paper when you are writing. Although graphite is slightly more stable than diamond at ambient conditions, the diamond of an engagement ring will not spontaneously turn into graphite. In order for this to occur, all the C-C bonds would have to rearrange simultaneously. There is a huge energy barrier for such a rearrangement.

For centuries it was thought that diamond and graphite were the only allotropes of elemental carbon. It came as a great surprise therefore, when in 1985, two new classes of stable elemental carbon compounds were discovered. In one of these, graphene sheets are rolled up to form *carbon nanotubes*. The tubes may be either single-walled or nested (tubes within tubes). They are amazingly strong and some of them conduct electricity. In the other class of compounds, hexagons are combined with pentagons to give closed curved surfaces. Probably the most beautiful of these is a molecule called *buckminsterfullerene*,[69] which is shaped like a soccer ball. The Nobel Prize in chemistry was awarded in 1996 to R. F. Curl, H. W. Kroto and R. E. Smalley for the discovery of this molecule, whose symmetry inspired awe in chemists. Currently, a very active field of research is exploring uses for these newly discovered allotropes of carbon. The structures of buckminsterfullerene and one example of a carbon nanotube are shown in Figure 10.10.

 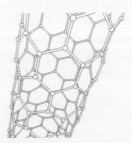

Figure 10.10. Buckminster fullerene and a carbon nanotube.

---

[69] Named after Buckminster Fuller, the inventor of the geodesic dome. Molecules of this compound are sometimes called "buckeyballs."

Supplemental Reading: The Transition and Inner Transition Elements

In the hydrogen atom the 3d state is lower in energy than the 4s state. However, you may have noticed in Table 10.1, that in potassium and calcium electrons are added to the 4s subshell before the 3d subshell. This is due to interactions between electrons in multi-electron atoms. It is only after calcium that we begin to put electrons into the 3d subshell. The elements in the B groups, in which we are adding electrons to d subshells, are said to be *transition elements*. The transition elements are all metals, since their properties are mainly determined by the s electrons in the shell outside the one to which d electrons are being added.

When transition elements are ionized, they may lose d electrons in addition to their outer s electrons. For example, iron (Fe), with electron configuration $(1s)^2 (2s)^2 (2p)^6 (3s)^2 (3p)^6 (4s)^2(3d)^6$, not only forms a +2 cation by losing its two 4s electrons, but in addition forms a +3 cation by additionally losing one of its 3d electrons. We say that iron can have +2 or +3 *oxidation states* and indicate iron in these two states as Fe(II) and Fe(III), respectively. When a substance loses electrons and its positive charge increases, we say that it has been *oxidized*. When the opposite occurs and it gains electron, we say that it has been *reduced*. Iron is easily oxidized from the metallic state or from the Fe(II) state to the Fe(III) state. The agency for doing this, the species that takes up the electrons that the iron loses, is called an *oxidizing agent*. Iron readily reacts (especially when wet) with molecular oxygen, O_2 to form rust, iron (III) oxide. Because of this reaction, no free O_2 could have existed in the Earth's primitive atmosphere until almost all exposed iron had been oxidized.

The elements in the fourth period, from scandium (Sc) to zinc (Zn) are called the first transition series. A second transition series is obtained when the 4d subshell is filled from yittrium (Y) to cadmium (Cd) and a third transition series with the 5d electrons from lanthanum (La) to mercury (Hg). After the element lanthanum, we fill the 4f subshell for elements that are collectively called the "lanthanides." The 4f electrons are way, way inside the outer electrons, and addition of such electrons has very little effect on the properties of these elements. For many years these elements could not be separated, because their properties were so similar.

Another group of f-shell atoms are those of the very heavy elements starting with thorium. These elements, collectively called the actinides, are all radioactive. However, several uranium and thorium isotopes have such long half-lives (of the order of billions of years) that atoms incorporated at the formation of the Earth, 4.5 billion years ago, are still undergoing decay. The lanthanides and actinides, the elements in which the inner f subshell is being filled, are collectively called the inner transition elements.

Crystal structures of metals

Due to electrostatic repulsion, metallic cations generally arrange themselves with fewer than the twelve nearest neighbors characteristic of a close-packed structure. Many metallic lattices are based on cubic structures, such as the simple cube and the body-centered cube shown in Figure 10.11.

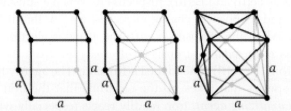

Figure 10.11. Cubic crystal lattices.

In the simple cube, each ion has six nearest neighbors, and in the body-centered cube, it has eight nearest neighbors. Also shown in Figure 10.11, is a face-centered cube, in which each atom has only four nearest neighbors.

Summary

ଔ ଓ ଔ ଓ ଔ ଓ

The properties of the elements, as illustrated in the periodic table, are determined by their electron configuration, in particular the number of electrons in their outer shell (valence electrons). The noble gases, with a filled outer shell, form monatomic gases and rarely react. Elements with one or two electrons in their outer shells (except for hydrogen) form metals, in which electrons are free to move in a lattice of cations. Other elements generally share electrons to form covalent bonds among two or a few atoms.

Questions

1. What are other uses of helium besides that of filling party balloons?
2. Close your eyes and point to the periodic table. Look up one use of the element that you first touch.
3. Hydrogen gas is lighter than helium. Why don't we generally fill party balloons with hydrogen?
4. In electric-arc welding, metal pieces are joined when they are melted by a large electric current. Such welding is usually done under a blanket of argon gas. Why is argon uniquely suited for this purpose?
5. Phosphorus is usually found as tetrahedral P_4 molecules. Draw this molecule, showing its three-dimensional shape. In a Lewis structure of P_4 how many lone pairs does each phosphorus atom have in this molecule?
6. At ambient conditions sulfur is a solid comprised of S_8 rings. Draw a Lewis structure of S_8.
7. Locate the six nearest neighbors of an atom in the simple-cubic lattice shown in Figure 10.11.
8. What are the differences between the following types of bonds: covalent bonds, ionic bonds, metallic bonds.
9. Why do atoms tend to have an octet of electrons around them?
10. Photosynthesis by plants adds huge amounts of oxygen to the atmosphere. Additions of nitrogen are much less. Why then is nitrogen the largest component (78%) of our atmosphere?

156   Order and Disorder

11. Which of the forms of carbon: diamond or graphite, would be a better lubricant? Why?
12. Metallic sodium reacts with gaseous chlorine to form sodium chloride (NaCl), which is made up of sodium cations and chlorine anions. In this oxidation-reduction reaction: What is oxidized? What is reduced? What is the oxidizing agent?

11 Compounds

Double, double toil and trouble; Fire burn and cauldron bubble

from Shakespeare's "Macbeth"

Dalton recognized that compounds are composed of several different atoms in ratios that are usually small whole numbers.[70] For example, water contains two hydrogen atoms for each oxygen atom and table salt contains equal numbers of sodium and chlorine atoms. By combining different elements in a single moiety, a whole new range of chemical and physical properties of substances becomes possible. There are no elements that exhibit properties anything like those of water, H_2O (a liquid at ambient conditions that is a powerful solvent and is necessary for life as we know it) or table salt, NaCl (a non-conducting crystalline solid that becomes an electrical conductor when melted or dissolved in water). We say that new properties have *emerged* when elements form compounds — when atoms form molecules. However, even in compounds, individual atoms can be recognized; chemical combination affects only their valence electrons; their nuclei and core electrons remain unchanged.

In this chapter *inorganic chemistry*, which deals with compounds of all elements except carbon, will be discussed. Most inorganic compounds are ionic; we think of them as resulting from the complete transfer of

---

[70] This is called the Law of Definite Proportions.

electrons between atoms. In the next chapter, **organic chemistry**, the chemistry of the compounds of carbon,[71] will be discussed. Most organic compounds are covalent, we think of them as resulting from the sharing of electrons. No matter how we decide to think of a molecule, it is important to remember that

> **The forces holding molecules together are electrostatic.**

Chemical Formulas

There are three ways of referring to chemical compounds: by a *common name*, by a *systematic name* and by a *chemical formula*. However, not all compounds have common names. For example, hydrazine (common name), a rocket fuel, is dinitrogen tetrahydride (systematic name), which is N_2H_4 (formula).

> **Different types of chemical formulas provide different types of information.**

A formula such as NaCl indicates that this compound contains a 1:1 ratio of sodium atoms and chlorine atoms. We call this an *empirical formula*, since it just gives the ratio of the atoms in the compound. There are no NaCl molecules in solid table salt, which exists as an almost infinite lattice of Na^+ and Cl^- ions.

In other cases, individual molecules of a compound do exist. This holds for N_2H_4 discussed above and for water, H_2O. Distinct molecules of water, consisting of one oxygen atom and two hydrogen atoms exist in its gaseous (steam), liquid and solid (ice) forms. Since the formula H_2O gives information about a discrete molecular entity, it is called a *molecular formula*. There are other formulas, which provide even more detail about the

---

[71] Simple compounds of carbon with oxygen, sulfur or nitrogen are often considered inorganic.

water molecule. For example, HOH tells us that both hydrogen atoms are attached to the oxygen atom and not to each other, while

H H
 \ /
  O

indicates that the three nuclei in the molecule do not lie on a straight line. Formulas of the latter type are called **structural formulas**, and are increasingly important for complicated molecules, for which there are a number of possible arrangements of the atoms, forming different *isomers* of the molecule. An example of isomers is HCN (hydrogen cyanide) and HNC (hydrogen isocyanide).

Even more information is obtained from formulas that give information about the electron distribution in a molecule. For example, the formula in Figure 11.1, shows that in the covalent bonds of the water molecule, electrons are not shared equally; there is a small amount (δ) of electron transfer from the hydrogen atoms to the oxygen atom.

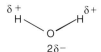

Figure 11.1. Structural formula of water, showing the dipole moment

Such bonds are called **polar bonds** and produce a **dipole moment**, a charge separation, in the molecule. Chemists use a property called *electronegativity* to predict whether bonds will be polar or not. Electronegativity is a measure of how well an atom holds on to its electrons and how well it attracts electrons from other atoms. In general, elements on the right side of the periodic table, such as fluorine, chlorine, oxygen and nitrogen have high electronegativity, while elements on the left side of the periodic table, such as hydrogen and the metals, have very low electronegativity. In a bond, electrons are shifted from atoms with low electronegativity to atoms with high electronegativity, as shown in Figure 11.1.

The most information that can be obtained about a chemical compound is its molecular wave function, obtained as the solution of Schrodinger's equation. This provides the electron density at each point in the molecule. With improvements in computers, accurate wave functions are becoming available for larger and larger molecules.

Ionic Bonds

Ionic compounds are those whose properties can be best explained by attraction of oppositely charged ions. They result from combination of atoms with very different electronegativities. Almost all compounds comprised of a metal with a non-metal are ionic. Sodium chloride (NaCl), calcium sulfide (CaS), potassium oxide (K_2O), magnesium bromide ($MgBr_2$) and aluminum fluoride (AlF_3) are examples of such compounds. Notice, that while the empirical formulas for ionic compounds provides the relative numbers of the combining atoms (their stoichiometry), their systematic names do not. Thus to write down the formula for magnesium bromide, one must know that bromine, being in group VIIA, forms –1 ions, while magnesium, being in group IIA, forms +2 ions. Since magnesium bromide is electrically neutral, it must contain two Br^- ions for each Mg^{2+} ion.

Since transition metals can form cations with different charges, the systematic names for ionic compounds involving transition metals must indicate the oxidation state of the transition metal ion. Thus FeO is named iron (II) oxide, while Fe_2O_3 is iron (III) oxide.[72]

Ionic compounds can also be formed from metals combined with polyatomic anions. The most important polyatomic anions are those in which negative charge is stabilized by adding electronegative oxygen to another element. Some of these important oxy-anions are given in Table 11.1. Also listed in the table is the ammonium ion, a polyatomic cation, which forms ionic compounds by reacting with anions.

In writing formulas for ionic compounds involving these polyatomic ions, the ion formula (without its charge) is enclosed in parentheses

Table 11.1. Important polyatomic anions and the ammonium cation

Hydroxide	OH^-
Nitrate	NO_3^-
Sulfate	SO_4^{2-}
Carbonate	CO_3^{2-}
Phosphate	PO_4^{3-}
Ammonium	NH_4^+

---

[72] These are often called ferrous oxide and ferric oxide.

when there is more than one of the ion in the formula. For example, aluminum sulfate is written as $Al_2(SO_4)_3$. Note how, with +3 aluminum ions and −2 sulfate ions, this empirical formula provides an electrically neutral compound.

Crystals

Ionic compounds form ordered crystals in the solid phase. Their bulk crystalline material is characterized by flat planar surfaces intersecting at definite angles, giving them a beauty to which have occasionally been ascribed mystical properties. The macroscopic planar structure results from a microscopic unit cell that repeats in a periodic crystal lattice. The unit cell of one such compound, cesium chloride, is shown in Figure 11.2.[73] The cell is a simple cube with side of length 0.41 nm that has a Cs^+ ion at its center and one-eighth of a Cl^- ion at each of its corners. (The Cl^- ions are shared between eight cubes.) This maintains the one-to-one stoichiometry of the compound. A macroscopic crystal of cesium chloride is generated by repeating the unit cell in three dimensions.

An important parameter of an ionic crystal structure is the number of nearest-neighbors of each ion, called its **coordination number**. From Figure 11.2, we can see that each Cs^+ ion has eight Cl^- nearest-neighbors.[74]

We can explain the crystal structures of ionic compounds by considering ions to be hard spheres with definite radii. Ionic compounds adopt crystal structures that are most stable, provide the lowest energy. In this case the energy is the electrostatic interaction energy of the

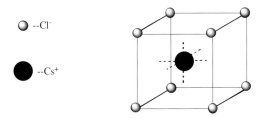

Figure 11.2. Unit cell of cesium chloride.

---

[73] The ions are not drawn to the proper relative size in this structure.
[74] It is less obvious, but each Cl^- ion also has eight nearest-neighbor Cs^+ ions.

charged ions — called the *lattice energy*. Minimization of this energy is achieved by having the ions *in contact* with as many hard sphere ions of opposite charge as possible. In Figure 11.2, the large Cs^+ can be in contact with eight (its coordination number) hard-sphere Cl^- ions. Because Na^+ ions are considerably smaller than Cs^+ ions, in NaCl it is not possible to put the Na^+ in contact with eight Cl^- ions. If this were done, the Cl^- ions would overlap — their repulsive energy would be very large. As a result NaCl adopts a different crystal structure, shown in Figure 11.3, where each ion has a coordination number of only six. In a similar manner

> **Most crystals structures are explained by simple electrostatic energy considerations.**

Even though ionic solids are made up of charged particles, they do not conduct electricity, because in the crystal these particles are held in place in the lattice and are not free to move under the influence of an electric field. If we heat the crystal to a temperature high enough that it melts (e.g., 800°C for NaCl), the ions become free to move. Thus, molten NaCl is a good conductor of electricity.

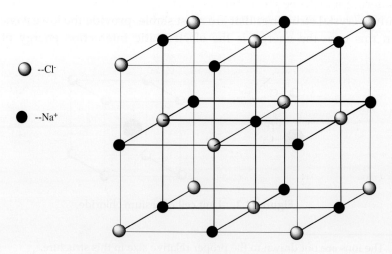

Figure 11.3. Crystal structure of NaCl.

Solutions

Pure water is a very poor conductor of electricity, because it contains very few ions. If we add a small amount of crystalline NaCl to water, the crystals disappear and we get an *aqueous solution* with a much larger electrical conductivity, because NaCl *dissolves* in water to form Na^+ and Cl^- ions. In order for this dissolution process to occur, the attractive energy of the ions in the lattice (the lattice energy) must be overcome. This is achieved by attraction of the ions in solution to the molecules of the water *solvent*. The nature of this attraction is shown in Figure 11.4.

In solution the positive Na^+ ion is attracted to and forms transient combinations with the more negatively-charged end (the oxygen side) of the water molecule, while the Cl^- ions form transient combinations with the more positively-charged end (the hydrogen side) of the water molecule. This energy of *hydration* just about compensates for the loss of the lattice energy of NaCl. Since the dissolution process is close to energy neutral, it is driven by the tendency of the ions to achieve a nearly disordered arrangement in solution. This entropy increase will be discussed in more detail in Chapter 13.

There is a maximum amount of a substance that will dissolve in solution at a given temperature, called its *solubility* at that temperature. Solutions containing that amount of solute are *saturated* with the substance. Solubility generally rises with increasing temperature. Thus, if a hot, saturated solution is cooled, it usually will exceed its solubility at lower temperature, and crystals will be formed. This

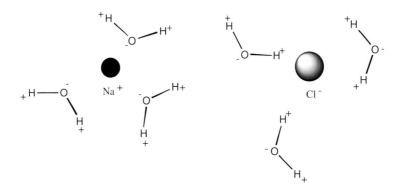

Figure 11.4. Sodium and chlorine ions in aqueous solution.

process, called *crystallization*, may be used to purify substances, since impurities will often not be included in the newly formed crystals. In some cases, solutions can be cooled to temperatures below that at which they are saturated, without crystallization occurring. This situation is called *super saturation*, and results from the absence of small crystals, which act as templates or *nucleation centers* on which subsequent deposition can occur.[75] Addition of a few such crystals to a supersaturated solution, will usually result in sudden massive deposition of material.

The attraction between ions and water is so strong that ionic compounds may incorporate water into their crystal lattices, forming *hydrates*. For example, calcium chloride, $CaCl_2$, when exposed to water, forms $CaCl_2 \cdot (6H_2O)$, calcium chloride hexahydrate. The water in hydrates is not held very strongly and can easily be driven off by heating. $CaCl_2 \cdot (6H_2O)$, for example, loses four waters at 30°C and the other two waters at 200°C. Ionic substances that form hydrates are sometimes used as drying agents.

Covalent Bonds

Non-metallic elements usually bond by sharing electrons in covalent bonds. The sharing of the valence electrons of two atoms results in an increase of negative charge in the region between the positively charged cores of the atoms, holding them together by electrostatic attraction. Covalent molecules are typically comprised of just a few atoms. For covalent compounds we use molecular formulas that specify the number of atoms of each type in the molecule. Since they do not involve ions, the relative numbers of the atoms in the compound is not determined by electrical neutrality, as it is with ionic compounds. The systematic name of covalent compounds specifies the molecular formula by using prefixes such as: *di* (2), *tri* (3), *tetra* (4), *penta* (5) and *hexa* (6). The prefix *mono* (1) is usually omitted. A few examples of the systematic names and molecular formulas of covalent compounds are given in Table 11.2. Common names are given in parentheses in some cases.

---

[75] The supersaturated solution that you are probably most familiar with is that of sweet iced tea, made by adding sugar to hot tea and allowing it to cool.

Table 11.2. Some important covalent compounds

Hydrogen chloride	HCl
Carbon monoxide	CO
Carbon dioxide	CO_2
Carbon tetrachloride	CCl_4
Nitrogen dioxide	NO_2
Nitrogen oxide	NO (nitric oxide)
Dinitrogen tetroxide	N_2O_4
Nitrogen trihydride	NH_3 (ammonia)
Sulfur dioxide	SO_2
Sulfur hexafluoride	SF_6

Structural formulas for covalent compounds can be drawn using the **Lewis octet rule**, which requires that eight valence electrons (an octet) be arranged around each atom (except for hydrogen, which being a first row element, only requires two valence electrons). Although the Lewis rule predates the application of quantum mechanics to molecules it can be rationalized by there being two s and six p orbitals in each row of the periodic table. A pair of electrons between two atoms represents a covalent bond between them and counts for filling the octet of both of the bonded atoms. The Lewis octet structures of several covalent molecules are shown in Figure 11.5. The structures are drawn two ways: with two dots representing a shared electron pair (bond) between two atoms, and with a single line representing the bond.

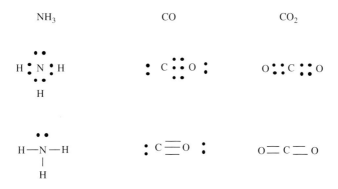

Figure 11.5. Lewis structures of covalent compounds.

Lewis formulas predict some important properties of molecules. For example, in NH_3 nitrogen contributes five valence electrons and each hydrogen one valence electron. The eight valence electrons are arranged in four pairs around the nitrogen atom. Three of these electron pairs form *single bonds* to hydrogen, leaving one *lone pair* of electrons on the nitrogen, which can interact with positively charged regions in molecules. The most important example of this behavior is when ammonia interacts with a H^+ cation, forming the ammonium cation, NH_4^+. All four bonds in this species are equivalent covalent bonds. A covalent bond, in which one of the bonding atoms provides both of the electrons, is called a *coordinate covalent bond*.

Carbon monoxide has $4 + 6 = 10$ valence electrons. In order to have an octet around both carbon and oxygen, six electrons must be shared between these atoms. Six electrons correspond to three covalent bonds, or a *triple bond*, between these two atoms. A triple bond is very strong and pulls the bonded atoms very close; both of these predictions have been verified in measurements on carbon monoxide. Carbon dioxide has $4 + 6 + 6 = 16$ valence electrons. It is known that the arrangement of atoms is OCO. In order to obtain an octet around each of these three atoms, we place four electrons between carbon and each of the oxygen atoms, bonding these pairs of atoms with *double bonds*. Double bonds are intermediate in strength and separation between single and triple bonds. Thus, one predicts that carbon dioxide has longer and weaker CO bonds than does carbon monoxide, which is observed. Collectively, double and triple bonds are called *multiple bonds*.

While Lewis formulas characterize the bonds in molecules, they do not accurately represent their geometry. When we report the *geometry* of a molecule, we specify the positions of its atomic nuclei, but not of its electrons. For example, the geometry of ammonia is a pyramid, with the nitrogen above the plane of the three hydrogens, not the planar structure shown in Figure 11.4. The four groups of electrons around the nitrogen in ammonia, however, have the tetrahedral arrangement shown in Figure 10.8 of Chapter 10. Carbon dioxide has the linear geometry shown in Figure 10.4.

There are some covalently bonded compounds for which Lewis octet structures cannot be drawn. For example, AlH_3, having only six valence electrons, cannot have an octet around aluminum, while ClF_5 has too many valence electrons to arrange them all in octets. In addition, compounds, such as OH are defined as *free radicals*, since they have an odd

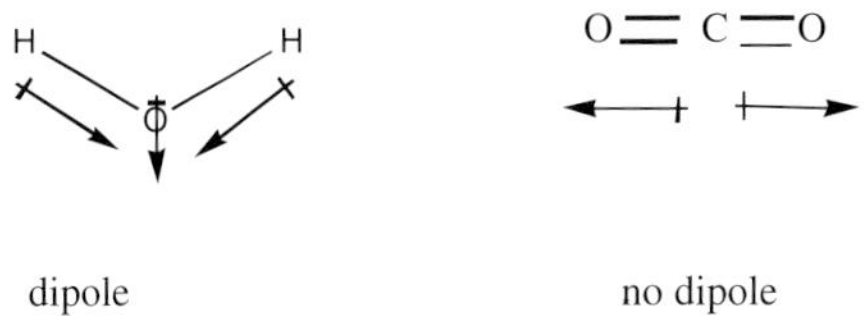

Figure 11.6. Polar bonds in water and carbon dioxide

number of electrons, which precludes them having octets or electron pairs around each atom. Free radicals are very reactive and have been implicated in diseases in the human body.[76]

We mentioned that in the water molecule, the electrons in the bonds between hydrogen and oxygen are not shared equally between the atoms, but are *polarized* to give *polar bonds*, with a partial negative charge on oxygen and a partial positive charge on hydrogen. As a result of its polar bonds, a water molecule has a *dipole moment*, i.e., a positive end and a negative end. A molecular dipole moment is represented by an arrow pointing from the positive end of the molecule towards its negative end. It should be noted, however, that having polar bonds is not a sufficient condition for a molecule to have a dipole moment. Carbon dioxide, for example, has polar bonds, but no dipole moment, because, as shown in Figure 11.6, in this linear molecule the charge separations in the two bonds cancel. In water, which is a bent molecule, the right-left charge separations in the two bonds cancel, but the up-down separations reinforce each other.

Hydrogen Bonds

A very important interaction in biologically important molecules is the hydrogen bond. In a *hydrogen bond*, an electronegative element, such as oxygen or nitrogen, pulls most of the electron density away from a hydrogen atom, exposing the concentrated positive charge of its nucleus, the proton. This positive charge is then attracted to the negative charge of the lone pair electrons on a nearby oxygen or nitrogen, not bonded to the hydrogen. Hydrogen bonds may be within a molecule (*intramolecular*), helping to establish its three-dimensional structure, or between molecules

---

[76] Some nutritionists have suggested that antioxidants, such as vitamins C and E, can protect our body from damages caused by free radicals.

168 Order and Disorder

H—O···H----:O—H
                  |
                  H

Figure 11.7. Hydrogen bonds in water.

(*intermolecular*), holding two molecules in a particular orientation. Hydrogen bonds are usually indicated by dashed lines, as indicated for water in Figure 11.7.

While not as strong as covalent or ionic bonds, hydrogen bonds are considerably stronger than the local interactions that exist between non-bonded atoms. Biological systems are characterized by very large molecules, which present many opportunities for hydrogen bonding. Often the sum of all these hydrogen bonds is the dominant interaction producing the three-dimensional structure of these molecules and allowing them to perform their biological function.

Acids and Bases

There are a number of definitions of acids and bases. We will use that of Bronsted who defined an *acid* as a substance that releases H^+ ions in solution and a *base* as a substance that combines with H^+ ions. Some acids, such as hydrochloric acid (HCl), are *binary*, containing only hydrogen and one other element. Other acids are *oxyacids*, containing oxygen in addition to another element, such as sulfuric acid, H_2SO_4; nitric acid, HNO_3, and phosphoric acid, H_3PO_4. The structures of these oxyacids are shown in Figure 11.8.

Figure 11.8. Structures of oxyacids.

The acids mentioned above are all *strong acids*, meaning that in aqueous solution, one hydrogen is completely ionized. Other acids are *weak acids*, in that they only partially ionize in water. HF is of this type, as is acetic acid, an organic acid that is the major component of vinegar.

The hydroxide ion, OH^- is the prototype base. It is obtained from ionic compounds, such as NaOH, $Ca(OH)_2$ and $Al(OH)_3$. Of these, only those with metals from groups 1A and 2A are called *strong bases*, because they release all their OH^- ions in aqueous solution. Some weak bases do not contain OH^- ions, but produce such ions when added to water. Ammonia, NH_3, is a base, since in aqueous solution it reacts with water to form the ammonium cation, NH_4^+ and the hydroxide anion. Since only a small fraction of the NH_3 molecules in solution react in this way, ammonia is a *weak base*.

The concept of pH measures the acidity or basicity of a solution. An aqueous solution of pH = 7 is said to be neutral,[77] because it has equal concentrations of H^+ and OH^- ions. Aqueous solutions with pH less than 7 are acidic, and those with pH greater than 7 are basic. Each unit of pH corresponds to a factor of 10 in H^+ ion concentration. The pH of a number of familiar aqueous solutions are given in Table 11.3.

Solutions that are highly acid or basic are very corrosive, and their contact with your skin or eyes should be avoided.

Table 11.3. pH of some common aqueous solutions

Household ammonia - 11.5
Blood - 7.4
Tomatoes - 4.5
Cola drinks - 2.5
Gastric juices - 1.8

Chemical Reactions

Dalton's atomic theory tells us that chemical reactions are just the rearrangements of atoms. Thus, there must be the same number of atoms

---

[77] For those of you who are mathematically inclined: pH is formally defined as the negative of the logarithm of the hydrogen ion concentration.

Table 11.4. Some balanced chemical reactions

$2CO + O_2 \rightarrow 2CO_2$
$H_2SO_4 + 2NaOH \rightarrow 2H_2O + Na_2SO_4$
$N_2O_4 \Leftrightarrow 2NO_2$

of each element in the *reactants* and the *products*. In writing *balanced chemical reactions* this is achieved by placing numbers in front of the formulas of compounds. It is not permitted to change the formulas, since these provide the qualitative description of what is occurring. Several examples of balanced chemical reactions are shown in Table 11.4.

The compounds to the left of the arrow are called the *reactants* of the reaction, and those to the right are the *products*. In the first reaction, deadly carbon monoxide is oxidized to carbon dioxide. In the second, sulfuric acid is *neutralized* by the base sodium hydroxide. This reaction probably occurs in *aqueous* (water) solution, where the ionic reactants are ionized. These first two reactions proceed to *completion*, i.e., until one of the reactants is completely consumed. The third reaction uses a double arrow to indicate that the reaction is *reversible*. It proceeds to a state where nitrogen dioxide and dinitrogen tetroxide coexist. At this *equilibrium state* N_2O_4 is breaking up at the same rate at which the NO_2 molecules are combining.

In addition to knowing the *stoichiometry* of a chemical reaction as indicated by its balanced chemical equation and whether it goes to completion or an equilibrium state, we also want to know the speed of reactions.

The kinetics of chemical reactions

Kinetics deals with the speed of processes. We discussed the kinetics of nuclear reactions when we talked about the half-lives of radioactive decay. Such reactions slow down as the radioactive substance decays and are unaffected by their environment, since the nucleus is "tucked away" at the center of an atom.

Chemical reactions show much more variability in their kinetic behavior than nuclear reactions. They are influenced by temperature and pressure and by *catalysts*, which are substances that change the speed of reactions without being consumed. Large biological molecules called *enzymes* are extraordinary catalysts, sometimes accelerating reactions more than a million-fold. Living systems use these molecules to control their reactions.

Supplementary Reading: Catalytic Destruction of Stratospheric Ozone

An important example of catalysis is found in an upper part (15–50 km) of the Earth's atmosphere called the stratosphere. Ozone, O_3, in the stratosphere absorbs solar ultraviolet radiation. Because of its short wavelength ($\varepsilon = hf = hc/\lambda$), this radiation is dangerous to life residing on the surface of the Earth. It causes sunburns, skin cancers and plant damage. We all have a vested interest in maintaining the "ozone shield" in the stratosphere.

Ozone is continually produced by photodecomposition of O_2 in the stratosphere. The resulting oxygen atoms combine with O_2 to produce ozone. In an atmosphere containing only the major atmospheric components (N_2, O_2, Ar and CO_2) the primary destruction reaction for ozone is $O + O_3 \rightarrow 2O_2$. This reaction has a large enough energy barrier, so that it is not too fast and permits adequate concentrations of ozone to exist in the stratosphere. A number of human activities,[78] however, introduce nitric oxide, NO, into the stratosphere. Nitric oxide introduces a catalytic-mechanism, shown in Figure 11.9, for removing ozone from the stratosphere.

The sum of reactions 1 and 2 is reaction 3, the same as the uncatalyzed reaction between ozone and oxygen atoms. NO is consumed in reaction 1, but regenerated in reaction 2. This is the hallmark of a catalyst; it changes the speed of a reaction without being used up. Because of this, a small amount of catalyst can affect a very large amount of reaction. The concentration of NO in the stratosphere is a thousand times less than that of ozone. Catalytic removal of ozone in the stratosphere is a property of NO and NO_2 (collectively called

1) $O_3 + NO \rightarrow O_2 + NO_2$
2) $O + NO_2 \rightarrow O_2 + NO$
3) $O_3 + O \rightarrow 2O_2$

Figure 11.9. Catalytic destruction of ozone by nitric oxide.

---

[78] Such as flying supersonic aircraft in the stratosphere, or using excessive amounts of nitrogenous fertilizers in agriculture.

> NOX); it is not a property of the elements oxygen and nitrogen. Only when these are combined as NOX does catalysis result. Chlorine or bromine containing-compounds can also catalyze ozone removal in the stratosphere. An international agreement, called the Montreal protocol, has been successful in minimizing the amounts of such compounds introduced into the stratosphere.

Reactions can be accelerated by interaction of reactants with surfaces, called *heterogeneous catalysis*, as well as by their interaction with molecules in solution or gas phase, called *homogeneous catalysis*. The chemical industry favors surface catalysts, since they do not have to be removed from product mixtures in order to be reused. One surface catalyst that you may be familiar with is the exhaust catalyst in your automobile, which brings oxidation of fuels to completion. This catalyst is usually made of a thin layer of expensive metals, such as platinum or palladium, deposited on a solid support.

The series of steps that give rise to an overall chemical reaction, such as those for catalytic destruction of stratospheric ozone given above, is called the *mechanism* of the overall reaction. Just as for formulas, mechanisms can be given with different levels of detail up to quantum mechanical descriptions of the various interactions that are occurring.

Autocatalysis and inhibition are interesting chemical mechanisms. In *autocatalysis*, a product of a reaction acts as a catalyst to speed up the reaction. In *inhibition*, a compound slows down a reaction, and this product may be a product of the reaction. When a chemical mechanism allows both autocatalysis and inhibition it can produce oscillations in time. In combination with diffusion of reagents, spatial waves can also be produced, as will be shown in Chapter 14.

Summary

Although bonding in molecules is always due to electrostatic forces, chemists think in terms of ionic bonds, involving complete transfer of electrons, and covalent bonds, involving sharing of electrons. Since ionic compounds are comprised of repeating arrangements of ions, their empirical formulas provide only stoichiometric information. Covalent compounds are composed of individual molecules, whose atomic content is represented by a molecular formula that can be drawn with the aid of the Lewis octet rule. Covalent bonds between atoms may be single, double or triple. Polar bonds between atoms with different electronegativity can produce a molecular dipole moment. Acids are compounds that release hydrogen ions in solution, while bases remove such ions. Catalysts increase the speed of chemical reactions without being consumed.

Questions

1. Complete the table below.

Common name	Systematic name	Chemical formula
Epson salt	Magnesium sulfate	
ammonia	dihydrogen oxide	
		O_3

2. Write the empirical formulas of: magnesium oxide, sodium sulfide and aluminum hydroxide.
3. Show that in the CsCl crystal structure shown in Figure 11.2, each Cl^- ion also has eight Cs^+ nearest neighbors? Note that the crystal must be extended in three-dimensional space, so that each corner Cl^- ion is shared between eight unit cells.
4. In an aqueous NaCl solution is there perfect order, i.e., do all the water molecules around a Na^+ ion have their negative ends pointing to the ion? If not, explain why.
5. Draw Lewis structures for the isomers hydrogen cyanide (HCN) and hydrogen isocyanide (HNC). Characterize the bonds as single, double or triple in each of these molecules.
6. List the quantities that are conserved in a chemical reaction, considering that the mass equivalents of their energy changes are negligible and that they do not change nuclei.
7. If HSO_4^- is the hydrogen sulfate anion, what is the hydrogen phosphate anion? What is the dihydrogen phosphate anion?
8. The bonds in ammonia are polar. Does ammonia have a dipole moment?
9. The bonds in carbon tetrachloride are polar and the molecule has a tetrahedral geometry. Does it have a dipole moment?
10. For the reaction, $CH_4 + O_2 \rightarrow CO_2 + H_2O$: What are the reactants? What are the products? Is it balanced? Is it reversible?

12 Organic Compounds

Plastics!

From the movie, The Graduate

Organic compounds are almost all those that contain the element carbon.[79] They comprise the fuel you put in your car, the food you eat, the beer you drink, the clothes on your back and almost all of life (except for water and a few ions). Because of their prevalence in life, at one time chemists thought organic compounds were infused with a "vital force." However, since Frederick Wohler showed in 1828 that urea (an organic compound) was produced when ammonium cyanate (an inorganic compound) decomposed, we have known that the chemistry of carbon follows the same basic principles as that of all the other elements.

In higher education, the huge number of organic compounds and their myriad uses are usually treated in a separate organic chemistry course. The enormous variety of organic compounds results from a number of factors, the most important of which are:

1. The stability of bonds between carbon atoms and between carbon and hydrogen atoms,

---

[79] Simple ionic compounds or acids containing groups like carbonates, cyanides and cyanates (OCN^-) are generally considered inorganic.

2. The ability of carbon to form double and triple bonds, and
3. The valence of carbon, which is four.

Organic chemistry can be simplified by realizing that

> **Many organic compounds can be described as hydrocarbons with functional groups replacing hydrogen.**

The *functional groups* determine the reactivity and the hydrocarbon (containing only carbon and hydrogen) determines the arrangement of the groups in space. We will first discuss the hydrocarbon "backbones" of organic compounds.

Hydrocarbons (Compounds of Carbon and Hydrogen)

Hydrocarbons are divided into classes that depend upon whether or not they contain multiple bonds. Hydrocarbons with only single bonds are called *alkanes*. The simplest alkane is methane, CH_4, whose structural formula and three-dimensional structure are shown Figure 12.1a and b.

In the structural formula (a), the molecule is drawn in the plane of the paper, and it appears as if the HCH bonds are either 90° or 180°. Actually, the three-dimensional structure of methane is the *tetrahedron* shown in the three-dimensional formula in (b). In this formula, the solid triangular

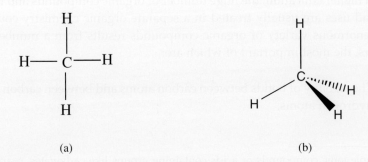

Figure 12.1. Structural and three-dimensional formulas of methane.

bond points out of the paper towards you and the dashed triangular bond points behind the paper. Both solid bonds are in the plane of the paper. All the bond angles in a tetrahedron are 109.5°, which is the preferred angle between two single bonds on the same carbon.

Structural formulas of some higher alkanes are shown in Figure 12.2. Their general molecular formula is C_nH_{2n+2}, where n is an integer.

Alkanes are also called *saturated hydrocarbons*, indicating that they contain as much hydrogen as possible. Alkanes can be named systematically with a stem that indicates the number of carbons in the molecule, followed by the suffix "ane." Except for ethane, propane and butane, the stems are the same as the prefixes given in Chapter 10. The structural formulas of alkanes are generated by the insertion of CH_2 (methylene) groups into the CC and CH bonds of the ethane structure. The three-dimensional structures of these molecules maintain the tetrahedral arrangement of atoms around each carbon. The molecules can rotate around the CC bonds, with the transient structure at different rotational angles called different *conformers* of the molecule. For ethane, the lowest energy conformer has the hydrogens on the two ends of the molecule staggered, as shown in Figure 12.3, in order to minimize the repulsive interaction between hydrogens on the two carbons.

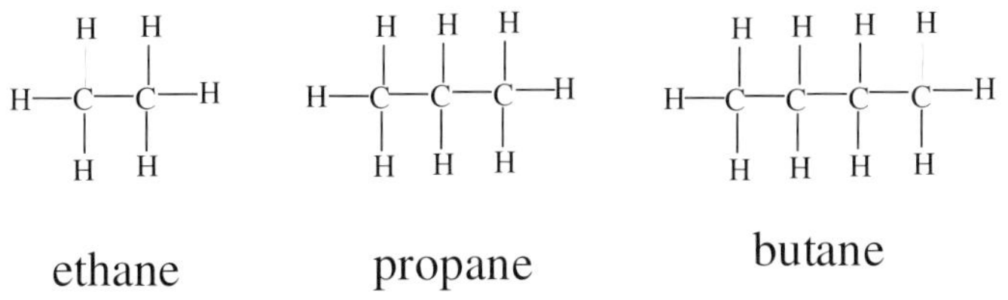

Figure 12.2. Structural formulas of alkanes.

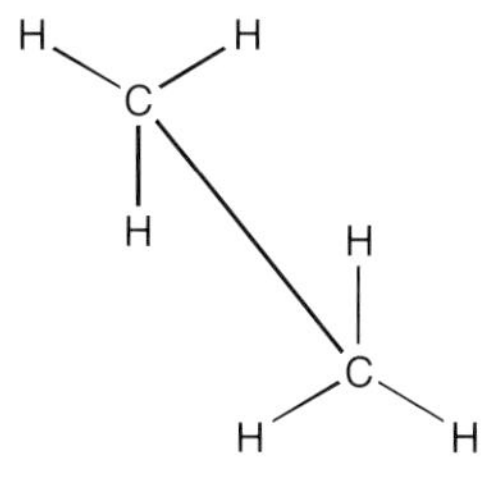

Figure 12.3. Lowest energy conformer of ethane.

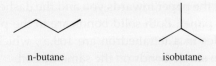

n-butane   isobutane

Figure 12.4. Butane isomers.

Structural isomers are different molecules with the same molecular formula that can be isolated. Since at room temperature an ethane molecule has enough thermal energy that it rapidly visits all conformations, ethane has no isomers. There are also no structural isomers of propane, because the molecule can freely rotate around both CC single bonds. Note that for butane, there are two different isomers, with common names n-butane and iso-butane, which are shown in Figure 12.4. These structural isomers cannot be converted into each other by rotation around CC single bonds.

In these *skeletal formulas*, only the carbon backbone is shown; all hydrogens are omitted. N-butane is called a *linear* alkane and isobutane a *branched* alkane. However, as shown in Figure 12.4, n-butane is not actually linear, since tetrahedral bond angles are maintained around each carbon.

Supplemental Reading: Cycloalkanes

By removing the hydrogens from the ends of an alkane chain and connecting the terminal carbons, a *cycloalkane*, with molecular formula C_nH_{2n}, is obtained. Skeletal formulas for some cycloalkanes are shown in Figure 12.5.

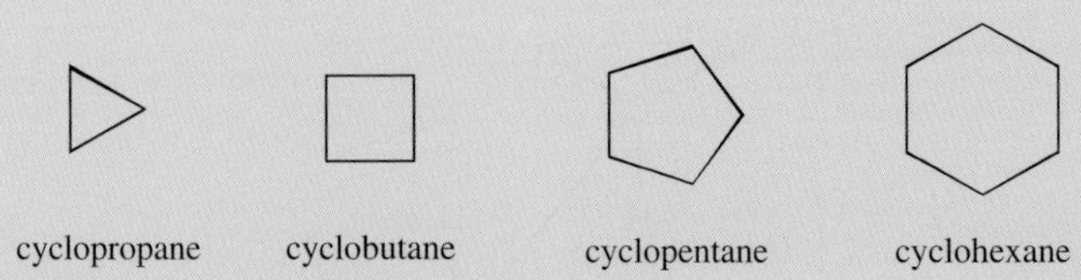

cyclopropane   cyclobutane   cyclopentane   cyclohexane

Figure 12.5. Cycloalkanes.

Except for cyclopropane, the carbon ring in cycloalkanes is not planar, as might be thought from the structural formulas. In order to approximate the tetrahedral geometry around carbon and lower the interaction energy of hydrogen atoms, these rings "pucker" away from the planar arrangement. In the puckered arrangement for cyclohexane, the CCC bond angle is, of all the cycloalkanes, the closest to the "natural" single-bond angle of a carbon, 109.5°; thus cyclohexane is the most stable of the cycloalkanes. Cyclohexane can adopt either a chair or a boat conformation, as shown in Figure 12.6.

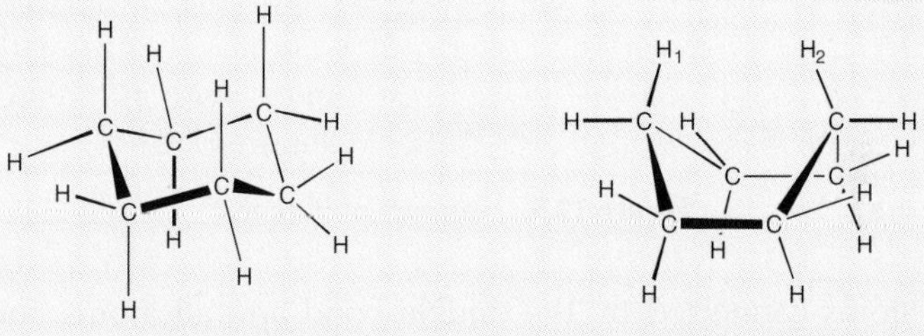

chair form   boat form

Figure 12.6. Cyclohexane.

The chair form is more stable, since in the boat form, two of the hydrogen atoms (H_1 and H_2 in the figure) are very close and repel each other, an interaction that is called **steric hindrance**. The two forms are close enough in energy and the energy barrier between them is low enough, that a single conformation cannot be isolated.

Most alkanes are obtained from petroleum, whose C_6–C_{12} fraction goes into gasoline, while larger molecules are used for kerosene, diesel fuel and oils. Because carbon and hydrogen have similar electronegativities, CH bonds are essentially nonpolar, and consequently alkanes and alkyl groups have very small dipole moments. A result of this is that alkanes are insoluble in water ("Oil and water don't mix.").

Hydrocarbons containing double or triple bonds are called **unsaturated**, indicating that they can react with additional hydrogen. Those with double, but no triple, bonds are called **alkenes**. The three-dimensional structure of the simplest alkene, ethylene, is shown in Figure 12.7. The

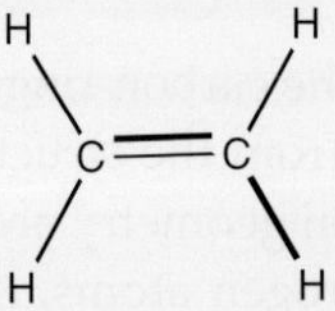

Figure 12.7. Three-dimensional structure of ethylene.

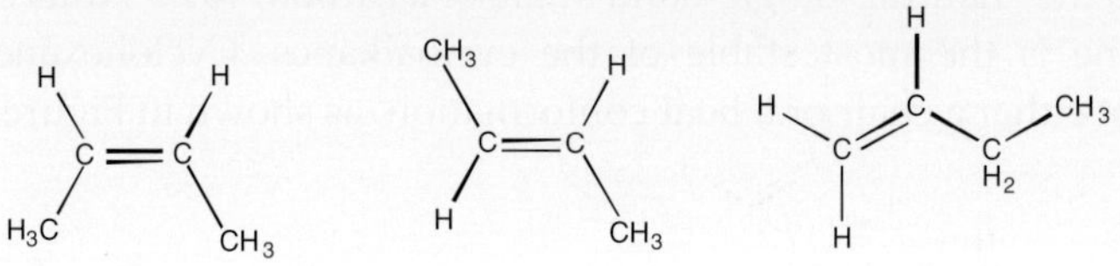

cis-butene-2   trans-butene-2   butene-1

Figure 12.8. Isomers of butene.

carbon atoms connected by the double bond and all atoms bonded to these carbon atoms lie in the same plane. There is no rotation around a CC double bond.

The presence of double bonds in a molecule may result in isomers due to the position of the double bond in the carbon chain. In addition, different isomers can result depending on whether substituents are on the same side (***cis-***) or opposite sides (***trans-***) of the double bond. The three structural isomers of butene are shown in Figure 12.8.

Unsaturated hydrocarbons with triple, but no double, bonds are known as **alkynes**. The simplest alkyne, C_2H_2, has the common name acetylene and is used to produce the very hot flames used in welding. In an alkyne the two carbon atoms and the single substituent attached to each carbon are all on the same line. Thus, acetylene is a linear molecule. The double and triple bonds in hydrocarbons are often responsible for their reactions.

Hydrocarbons may have several double or triple bonds or combinations of double and triple bonds. Particularly noteworthy are molecules with regions containing alternating single and double bonds. Such bond groupings are called **conjugated**, and are responsible for the intense colors of many dye molecules. For example, Figure 12.9 gives the skeletal formula of beta-carotene, a precursor to vitamin A and a pigment that gives many fruits and vegetables their color.

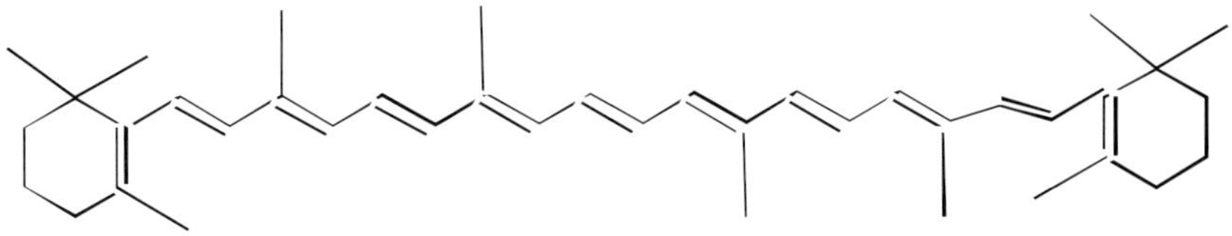

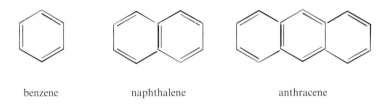

Figure 12.9. Structure of beta-carotene

benzene   naphthalene   anthracene

Figure 12.10. Polycyclic aromatic compounds.

or

Figure 12.11. Resonance structures of benzene.

Benzene, C_6H_6, is the simplest of an important, very stable, group of hydrocarbons that we call *aromatic*. These molecules contain $4n + 2$ carbon atoms, in one or more rings. All these carbon atoms lie in the same plane. The structures of three simple aromatic compounds are shown in Figure 12.10. Aromatic compounds, containing more than one ring of carbon atoms, are known as *polycyclic aromatic compounds*. They are often powerful carcinogens.

Figure 12.10 shows the double bonds in a particular position in each of these aromatic compounds. An alternative structure can be drawn in each case by switching double and single bonds. The two possible arrangements of bonds of benzene are shown connected by a double-ended arrow in Figure 12.11. The actual distribution of the electrons in the molecule is an average of those in these two *resonance forms*. This results in a distribution that is *delocalized* over the carbon ring, as indicated by right-most structure, where the electrons are indicated by a circle. Aromatic molecules are very stable due to electron delocalization.

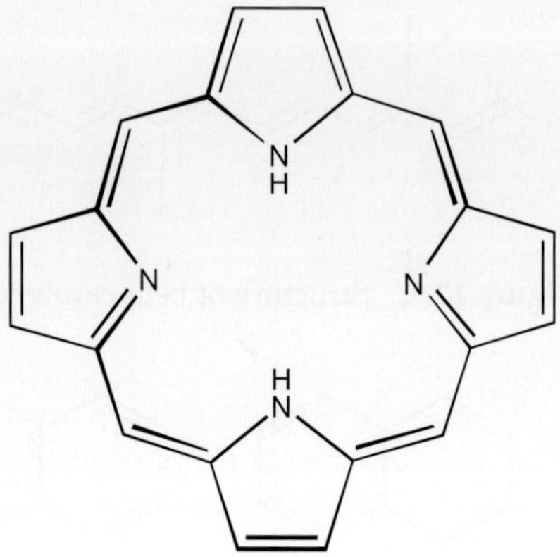

Figure 12.12. Structure of porphyrin.

Aromatic-type properties are also observed in some *heterocyclic* compounds, in which a CH group is replaced by a nitrogen, oxygen or sulfur. One such compound is porphyrin, shown in Figure 12.12.

Besides being an unusually stable molecule, the lone pairs of the four nitrogen atoms in porphyrin facilitates its binding to metal ions, such as iron, which occurs in the heme group of the oxygen-transporting protein, *hemoglobin*.

If a single hydrogen atom is removed from a hydrocarbon, a reactive free radical is obtained. These free radicals are often named with the stem of the hydrocarbon from which they are obtained followed by the suffix "yl." Thus, CH_3 is methyl and C_2H_5 is ethyl. However, there are many exception to this rule; i.e., C_2H_3, the free radical obtained from ethylene, is vinyl, and C_6H_5, the free radical obtained from benzene, is phenyl. The names of free radicals are also used for groups of atoms in molecules. Thus, C_6H_5-C_6H_5 is biphenyl. $C_6H_5CH_3$, could be called methyl benzene; its common name is toluene.

Functional Groups

Any hydrogen on a hydrocarbon can be replaced by a terminal functional group to give molecules with a variety of physical and reactive properties. Some of the most important functional groups are listed in Table 12.1.

Table 12.1. Important functional groups

Group	Name
-X	halide (fluoride, chloride, bromide or iodide)
-OH	alcohol
-SH	thiol (sulfhydryl)
-CHO	aldehyde
-COOH	carboxylic acid
$-NH_2$	amine
$-CONH_2$	amide
$-OPO_3^{2-}-$	phosphoryl
-O-	ether
-COO-	ester

These functional groups are terminal, except for the last three.

"*Halide*" encompasses all the group VIIA anions: fluoride, chloride, bromide and iodide. More than one hydrogen in a hydrocarbon may be substituted by a halogen. Such molecules, e.g., dichloroethane and dichloromethane (methylene chloride) are among the most widely used solvents. A group of substituted methanes and ethanes, in which all the hydrogens are replaced by halogens (called *perhalo* compounds) have a variety of uses. The *chlorofluorocarbons* (CFCs), were formerly used in aerosol cans, air conditioners, foams and cleaning agents. However, the stability of CX bonds endowed these molecules with the ability to persist for decades in the atmosphere after they were used. Eventually, these molecules were transported to the upper atmosphere (the stratosphere) where they were photo-decomposed by the ultraviolet component of sunlight. The resulting chlorine atoms pose a threat to a layer of ozone molecules in the stratosphere that protects life on Earth from the deleterious effects of ultraviolet radiation. The use of these compounds is now banned by an international agreement called the Montreal protocol. Occurrences such as this have led chemists to a new paradigm in which they consider the complete life-cycle of a chemical before introducing it into society.

Polychlorinated biphenyls (PCBs) are another class of environmentally dangerous polyhalide molecules. These compounds originally found considerable use in transformers and other electrical equipment. Their use was discontinued when they were found to cause a multitude of health effects. Due to their stability, however, these molecules have persisted in landfills, soils and river sediments, and continue to present a danger to public health.

Alcohols are derivatives of water in which one hydrogen has been replaced by an organic group. Although drinking ethyl *alcohol* (ethanol), C_2H_5OH, has achieved great popularity, all alcohols are poisonous in high amounts. Methanol is much more poisonous than ethanol, and in developing countries it has occasionally been accidently served to guests at festive occasions, with disastrous results. Even though alcohols contain OH groups, these do not ionize, and thus alcohols are not bases. In fact, the hydrogen of most alcohols is weakly ionizing, giving alcohols slight acidic properties. When OH substitutes for a hydrogen on an aromatic hydrocarbon, the resulting compounds are called *phenols*.

Replacing both hydrogens of water by organic groups gives an *ether*. Diethyl ether is a very common laboratory solvent. In 1846 it was, for the first time, used as an anesthetic to render a patient unconscious during an operation.[80] Anyone doubting the usefulness of chemistry should contemplate what it was like having an operation before the use of ether. It is also possible for an oxygen to be attached to two different positions on the same hydrocarbon, forming a cyclic ether. The sulfur analogs to ethers are called thioethers. You can recognize the name of sulfur-containing molecules from the prefix "thio."

Compounds in which two hydrocarbon groups (or two positions of the same hydrocarbon group) are joined by –OO– or –SS– are called *peroxides* or *disulfides*. The former type of molecule may range from somewhat unstable to explosive, while the latter is often important in stabilizing the three-dimensional structure of biologically important molecules, such as insulin. *Thiols* are molecules containing the SH group. These are exceedingly malodorous molecules, exemplified by the protective scent of the skunk.

Carbonyl compounds contain the group >C=O connected to hydrogen atoms or hydrocarbon radicals. They include *aldehydes, ketones, carboxylic acids* and *esters*. Some simple carbonyl compounds are shown in Figure 12.13.

---

[80] Ether has long been replaced by newer anesthetics that permit better control of consciousness and produce fewer side effects like nausea.

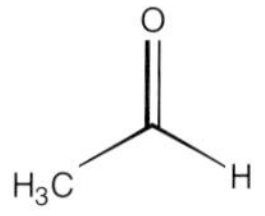

acetaldehyde

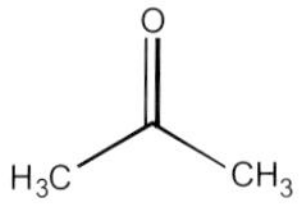

dimethyl ketone (acetone)

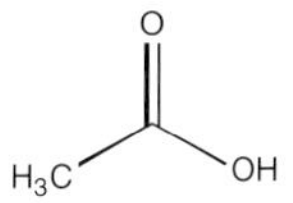

acetic acid

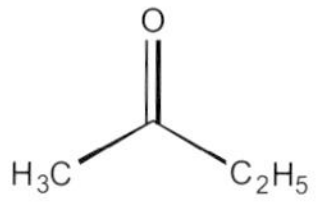

ethylacetate (an ester)

Figure 12.13. Some simple carbonyl compounds.

The simplest carbonyl compound is $H_2C{=}O$, formaldehyde. Formaldehyde is a volatile compound used in furniture and insulation materials. As a result, it is constantly being released into many home environments. Nowadays when we tightly seal the windows and doors of our homes to minimize the escape of heat, indoor formaldehyde concentrations can often build up to the point where they present a potent indoor pollution hazard. Formaldehyde can also occur in the outdoor environment as a product of the incomplete combustion of hydrocarbons. In regions strongly illuminated by sunlight, formaldehyde and other products of incomplete combustion interact with nitrogen oxides in a series of reactions that eventually produce noxious ozone in the air.

Acetaldehyde, CH_3CHO, is an important starting material in many chemical syntheses. Acetaldehyde is produced in the first step of the oxidation of ethyl alcohol in our bodies. In subsequent steps the oxidation continues, to eventually form carbon dioxide and water. When one consumes too much alcohol, the additional oxidation steps cannot keep up with the formation of acetaldehyde, and its concentration in our blood stream builds up. Since acetaldehyde is thirty times as toxic as ethyl alcohol, a severe hangover then ensues.[81]

Molecules with a carbonyl group attached to two hydrocarbon radicals are called ***ketones***. The simplest ketone is acetone (dimethylketone), a very common organic solvent.

[81] *Antabuse,* is a drug used to treat alcoholism. It slows the decomposition of acetaldehyde, making hangovers more immediate and more unpleasant.

Carboxylic acids contain the group

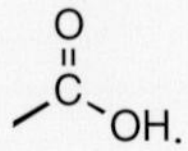

$$\underset{}{\overset{O}{\underset{}{\parallel}}}\!\!\!\!\!\!\!\!\!\!\!\!\!\!\!\!/\!\!\!\!\!-\!\!C\!-\!OH.$$

Carboxylic acids are weak acids, but much more acidic than alcohols,[82] which also have hydrogen attached to oxygen. Acetic acid, the active ingredient in vinegar, is by far the most industrially important carboxylic acid. It is used to synthesize many important compounds, including acetates, which are used as solvents.

When a carboxylic acid group is attached to a long chain hydrocarbon, the resulting compound is known as a *fatty acid*. Fatty acids and similar compounds comprise many of the fats and oils in the foods that we eat. Depending on the nature of the hydrocarbon chain, fatty acids may be saturated (the chain contains all single bonds — illustrated by stearic acid in Figure 12.14) mono-unsaturated (the chain contains one double bond — illustrated by oleic acid) or poly-unsaturated (the chain contains more than one double bond — illustrated by linoleic acid). It is generally believed that saturated fatty acids and unsaturated fatty acids in which the double bonds are in the *trans* configuration contribute to the incidence of coronary heart disease. The latter type of compound is often produced when the polyunsaturated fatty acids found in vegetable oils are partially hydrogenated to produce solid margarine.

In oleic and linoleic acid the stereochemistry around all the double bonds is cis.

$CH_3(CH_2)_{16}CO_2H$ $CH_3(CH_2)_7CH=CH(CH_2)_7CO_2H$

stearic acid   oleic acid

$CH_3(CH_2)_4CH=CHCH_2CH=CH(CH_2)_7CO_2H$

linoleic acid

Figure 12.14. Some fatty acids.

The reaction between a carboxylic acid and an alcohol, eliminating water, is called a **condensation reaction**, and produces an ester, with general formula

---

[82] See question 8.

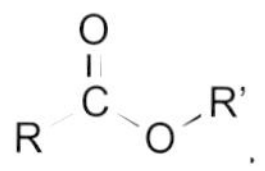

$$R-\underset{\underset{O}{|}}{\overset{\overset{O}{\|}}{C}}-O-R'$$

Esters produce the odors and tastes of many flowers and fruits.

Amines are organic derivatives of ammonia. One, two or all three of the hydrogens of ammonia can be replaced with hydrocarbon groups to form what are called primary, secondary or tertiary amines. Almost all amines are basic, due to the propensity of the lone pair of electrons on nitrogen to combine with a hydrogen ion, leaving a hydroxide ion in aqueous solution.

Amides, of general formula

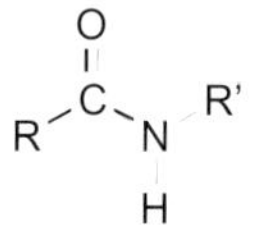

$$R-\overset{\overset{O}{\|}}{C}-\underset{\underset{H}{|}}{N}-R'$$

are similar to amines, but have nitrogen attached to a carbonyl group. Unlike amines, amides are not basic. When the amide group joins two hydrocarbon groups, the bond is called a ***peptide*** bond. Molecules with a few peptide bonds are called polypeptides, and those with many such bonds and biological functionality are known as proteins. Proteins are one example of a class of large molecules called polymers.

Phosphoryl groups ($-PO_2-$) are linked together in biological reactions key to providing energy for processes in the cell. The bonds between these groups are thermodynamically unstable — releasing large amounts of energy when they react with water (hydrolysis). These hydrolysis reactions are very slow, however, so that they are subject to biological control by protein molecules called enzymes.

More than one hydrogen on a hydrocarbon can be replaced by functional groups. For example, in Figure 12.15, formulas are given for a molecule containing two hydroxyl groups, ethylene glycol, and a molecule containing three hydroxyl groups, glycerol.

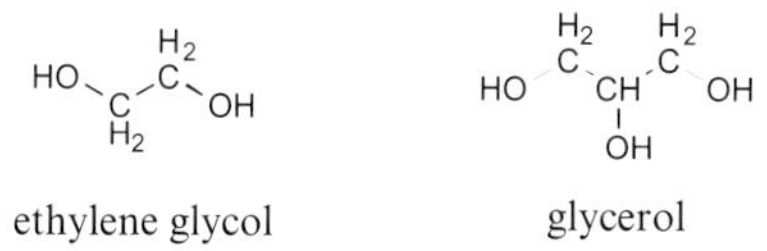

ethylene glycol   glycerol

Figure 12.15. Poly-hydroxyl compounds.

Ethylene glycol is commonly used as an all-weather antifreeze in automobile radiators and compounds of glycerol are of great importance in biological systems. In compounds with more than one functional group, these can be the same or different and can be attached to the same or different carbons of the hydrocarbon. Probably the most important type of compound of this type is the amino acid, where an amine and a carboxylic-acid functionality both replace hydrogen atoms on the same carbon atom. Such molecules are the building blocks of proteins in our bodies. Twenty different amino acids are employed to make up almost all the proteins in living creatures. Amino acids will be discussed in considerable detail in a later chapter.

Polymers

"Plastics!" — probably the most famous one-liner in movies — was the prophecy for the wave of the future given to Dustin Hoffman in the movie *The Graduate*. Derided by young people in the sixties and seventies, this advice was nevertheless prescient about the growing use of these materials in our society. Currently, millions of tons of the materials are used annually. Plastics, which are structural materials, are one type of ***polymer*** or ***macromolecule***. Other uses of polymers are as adhesives, coatings (e.g., paints) and foams. In addition, many biologically important molecules are polymers. Giant polymer molecules are formed by linking together a large number of simple ***monomer*** molecules. For example, polyethylene, the simplest polymer, shown in Figure 12.16, is formed by linking ethylene molecules.

Each polyethylene molecule is comprised of hundreds to thousands of ethylene monomers. Note, that in contrast to ethylene, polyethylene is not an alkene — it has no double bonds; it is made up of repeating ***methylene*** ($-CH_2-$) groups. We can write the formula of polyethylene as R_1-$(CH_2$-$CH_2)_n$-R_2, where R_1 and R_2, the chain-terminating groups, are determined by the particular process in which the ethylene, C_2H_4, is polymerized. In a sample of polyethylene, n, the number of ethylene monomers comprising the polymer molecule, usually has some variability; not all of the molecules are of the same length. A macromolecule, such as that drawn

$$n(C_2H_4) \longrightarrow \cdot CH_2\text{-}(CH_2\text{-}CH_2)_{n-1}\text{-}CH_2\cdot$$

Figure 12.16. Formation of polyethylene.

in Figure 12.16, is called a *linear polymer*. Actually, due to the tetrahedral structure around each carbon atom, the molecular structure can range anywhere from a chain that is fairly extended to one that is mainly "balled up." Pulling on a sample of polyethylene can extend the chains, giving the substance *elastomeric* (rubber-like) properties. Polyethylene is an inexpensive polymer that is produced in huge quantities worldwide. It finds use in plastic bags, bottles, toys and electrical insulation.

Polyethylene is the simplest of the vinyl polymers. Other members of this family, along with their monomers and some of their uses are shown in Figure 12.17.

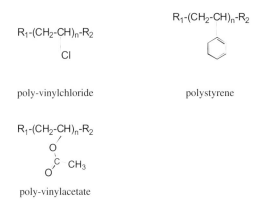

Figure 12.17. Vinyl polymers.

In contrast to linear polymers it is also possible to synthesize *branched polymers*, in which linear chains are occasionally *cross-linked*, such as shown in Figure 12.18.

A sample of such a polymer is one huge macromolecule. Because of the crosslinking, branched polymers are less elastometric than their linear counterparts. They are often used as structural materials.

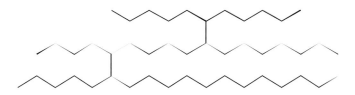

Figure 12.18. Branched polyethylene.

Summary

Because carbon forms four bonds and extended long chains, there are a huge number of carbon-containing compounds, which are the subject of organic chemistry. Organic molecules can be thought of as hydrocarbons, with substitution of hydrogen by a functional group. The functional groups determine the reactivity of the molecule and the hydrocarbon backbone determines the arrangement of the groups in space. Alkanes are hydrocarbons with only single bonds, while alkenes and alkynes have double and triple bonds, respectively.

Among the most important functional groups are –OH (alcohols) and >C=O (carbonyl compounds). The latter include aldehydes, ketones, carboxylic acids and amides. Amide bonds link together amino acids to form the proteins of life.

Simple monomers can be linked together to form large polymer molecules. These can be linear or cross-linked and can incorporate different functional groups.

Questions

1. List the symmetry operations (see Chapter 2) that leave methane unchanged.
2. Fill in the carbon and hydrogen atoms for the two butane isomers in Figure 12.4.
3. Over how many carbon atoms does the conjugation (alternate single and double bonds) extend in the beta-carotene molecule shown in Figure 12.9?
4. Why are there no *cis* and *trans* isomers of butene-1, shown in Figure 12.8?
5. Cycloalkanes can also have cis-trans isomers. Draw three-dimensional formulas for the *cis* and *trans* isomers of dimethylcyclopropane, C_5H_{10}.
6. Draw a six-carbon hydrocarbon of each of the following types: alkane, alkene, alkyne, cycloalkane, cycloalkene, cycloalkyne, and aromatic compound.
7. Polynuclear aromatic compounds have isomeric forms bases on different arrangement of their rings. Draw two polynuclear aromatic

isomers of anthracene with different arrangements of their rings from that shown in Figure 12.10.

8. The observation that carboxylic acids are more acidic than alcohols (have a greater tendency to form a carboxylate anion and a hydrogen ion) is explained by there being resonance forms for the carboxylate anion. Show the resonance forms of the acetate ion, $CH_3CO_2^-$, the carboxylate anion obtained from the ionization of acetic acid. How does the additional stability resulting from resonance in the acetate ion increase the acidity of acetic acid over that of methanol?

9. Successive intermediates in the oxidation of ethane, C_2H_6 are ethyl alcohol, acetaldehyde, acetic acid and carbon dioxide. Show that these compounds have increasing ratio of oxygen to hydrogen.

10. Draw a skeletal formula of oleic and linoleic acid (Figure 12.14), showing the cis structure around the double bonds.

13 Entropy

Entropy is time's arrow.

Sir Arthur Eddington

Previous chapters have dealt with order in the universe; this chapter deals with disorder. Scientists have a name for disorder; they call it ***entropy***. Entropy is the opposite of order. In fact, order is sometimes called ***negentropy***. The word entropy has made its way into our general vocabulary. When our room is messy or our lives disordered, we might say we have "high entropy." Entropy is related to ***information***, the more information we can have about something, the less entropy it has and the more ordered it is. In a library, books are arranged with low entropy and high order, according to the Dewey decimal system. This arrangement gives us sufficient information to easily find a particular volume. As shown in Figure 13.1, if an earthquake shakes the shelves, books fall to the floor, the collection has less order and more entropy. It becomes difficult to find a book.

Regarding the kindergarten class we discussed in Chapter 2: when the children are in their seats, they are highly ordered. We have information about them, knowing where each one is located. When the children are running around the room, they have less order and more entropy.

Figure 13.1. San Jose library, after earthquake of 2007.

The Second Law of Thermodynamics

The most important application of the concept of entropy is to specify which processes occur in nature. We have already discussed conservation rules, which tell us a lot about this subject. Processes can only occur if they conserve energy (actually mass-energy), momentum, angular momentum, charge, etc. By definition, however, a conserved quantity is the same at the beginning and end of a process. A conservation rule can tell us nothing about the direction of a process, since the quantity is conserved in both directions. Newton's laws of motion are also useless for this purpose. They work equally well in the forward and backward direction. If at the end of a frictionless process, we reverse all velocities and let time run backwards, all bodies return to their initial conditions. From the point of view of Newton's laws, whether the earth revolves around the sun in a clockwise or counter-clockwise direction is only dependent on the initial direction of its angular momentum.

The natural direction of processes is determined by the second law of thermodynamics. Application of this law is straightforward for isolated systems. A system is *isolated* if it doesn't interact with its surroundings. For example, I'm sure you'd agree that if the teacher of the kindergarten class has to leave the classroom to receive an important telephone call, when she returned the students will not all be in their seats. The order of the isolated class will have decreased in her absence; the entropy of this isolated system would have increased. This is the basic principle we are looking for:

> **Entropy increases in isolated systems.**

In the above example, the class is the system, which is isolated in the absence of the teacher.

The *second law of thermodynamics* is in agreement with everything observed in the macroscopic world; it's a distillation of all of our experience. Like the first law of thermodynamics, the conservation of energy, it cannot be proved.

If the teacher has trained her class well, she could at least expect that the students remain in the classroom during her absence. There would be limits to their positions. Since limits are one type of order, their entropy would not be as large as if they were *dispersed* throughout the building. Greater dispersal corresponds to more disorder and higher entropy.

It is important that we apply the Second Law only to *isolated systems*, systems not interacting with their surroundings. We know that when the teacher returns and interacts with the class, it is no longer isolated and order will usually be rapidly restored. What relevance is the second law to this process? We can use the Second Law in such cases by considering the entire universe. As far as we know, there are no interactions between our universe and any other universe, if indeed such alternative universes exist.[83] Since the universe is isolated, the Second Law tells us that it is moving in the direction of increasing entropy. The entropy of the universe increases as time increases—or entropy is time's arrow, as in the quote by Sir Arthur Eddington at the beginning of this chapter. The second law of thermodynamics is unique in pointing out a

---

[83] The idea of multiple universes has recently received some popularity among cosmologists.

direction for increasing time; time increases in the direction in which things become more disordered. We probably have an intuitive feeling for the Second Law. If I showed you a cartoon movie in which Humpty Dumpty was put back together again, you'd immediately suspect I was playing the reel backwards.

In applying the Second Law, we can divide the universe into two parts: the system that we focus our attention on, and everything else, which we call the surroundings. System plus surroundings equals the universe. Since the entropy of the universe always increases, we can also say,

> **In any process there is net increase of entropy.**

A pitfall of this statement of the Second Law is the little word, "net," which means that in the process considered, entropy (disorder) and negentropy (order) produced anywhere in the universe, both in the system and the surroundings, must be algebraically combined. Only the net entropy change has to be positive. A diagram of this procedure is shown in Figure 13.2.

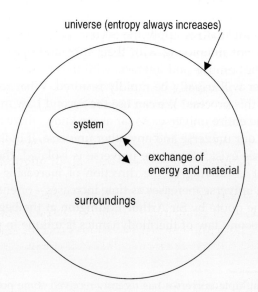

Figure 13.2. Entropy changes in system and surroundings.

Entropy is a concept we use when talking about large numbers of things or events. The Second Law can be violated if we're looking at just a few things or events. For example, consider the process of throwing six-sided dice. Specification of a particular side of a die facing up is a high-information, low-entropy situation. In multiple rolls of a die, we would not be overly surprised if we obtained *one* in three consecutive throws.[84] However, if only *one* was rolled on a hundred consecutive throws, we would presume the die was loaded. We don't talk about the entropy of a single electron or atom. Macroscopic objects, however, are composed of huge numbers of particles. The concept of entropy is very useful in dealing with the macroscopic world.

Ludwig Boltzmann (1844–1906) explained the Second Law with a theory based on *probability*. According to Boltzmann, low-entropy (high-information) states, such as all *ones* on a large number of dice, do not occur because they are very improbable. High-entropy (low-information) states, such as one sixth of the dice showing *one* — **without specifying which dice are so arranged**, are much more probable, because there are many more of such states. In the macroscopic world, in which we are dealing with systems that contain the order of Avogadro's number of particles (6×10^{23}), appreciable deviations from the most probable distributions are never observed.

A very important example of entropy as probability is the tendency of concentrations to be equal in different parts of a system. If we open a valve between a vessel containing gas A and one containing gas B, the gases will mix until their concentrations are equal in the two vessels. This is the highest probability situation because it maximizes the number of ways (called configurations) by which it can occur. Just how much more probable is the equal concentration state than the segregated state, is shown in the supplemental reading below.

Supplemental Reading — Counting Configurations

What is the relative probability that the A molecules are all in the left-hand container compared to their being spread through both containers? Let's assume that the containers each have a volume of

---

[84] Since the throw of each die is an independent event, we would expect such an occurrence once in $6^3 = 216$ times.

1.0 liter. At ambient conditions (25°C and 1.0 atmosphere pressure) there are about 0.05 moles or 3×10^{22} molecules of gas in a liter. It's twice as probable that a single A molecule is somewhere in the combined 2.0 liter system than it is in the 1.0 liter left-hand container. Probabilities for molecules are independent and therefore multiplicative. This means it is $2^N = 2^{3 \times 10^{22}}$ times as probable that the A molecules are spread out over 2.0 liters than their being confined to 1.0 liter. This is an inconceivably large number.

Of course, an arrangement where the A molecules can be anywhere in the 2.0 liter system includes some configurations in which there are noticeably more A molecules in one container than the other. These are very rare, however, and with little error one can consider the system with the valve open as one in which A and B molecules are equally spread over the two containers.

Boltzmann realized that, since entropy and probability both increase with time, these quantities should be related. However, if we want entropy to be an extensive property like energy, something proportional to the amount of material in a system, it can't just be equal to or proportional to probability. For example, if we have twice as many A molecules in our system, the probability ratio calculated above becomes 2^{2N}. The size of the system appears in an exponent, rather than as a multiplicative factor. The function that extracts an exponent from a number is called the logarithm.[85] Thus, Boltzmann proposed that entropy was proportional to the logarithm of the probability of a state of the system.[86]

In small macroscopic systems, with much less than Avogadro's number of particles, deviations from the most probable distribution are sometimes observed, and are called *fluctuations*.

An example of fluctuations is **Brownian motion**, which can be observed by using a microscope to look at a small particle suspended

---

[85] Log 10 = 1, log 100 = log 10^2 = 2, log 1000 = log 10^3 = 3, etc…

[86] Boltzmann's relation S = k log W — where S is entropy and W is the number of configurations — is carved on his gravestone in the central cemetery of Vienna.

in water. Although the most probable condition for a particle is quiescence, in which it is bombarded equally on all sides by water molecules, in fact it will be observed to jump around. This is due to only a limited number of water molecules hitting the particle in a given time interval, allowing for noticeable fluctuations in the forces on its opposite sides.

Since energy depends on position and velocity of particles, it has inherent entropy. When we speak of a macroscopic body having large kinetic energy, we mean that all its constituent atoms are moving in the same direction with the same speed. We have a lot of information about these atoms and their motion is very ordered. Likewise, a boulder sitting at the top of a mountain has potential energy due to every atom of the boulder being located in that place. We have a lot of information about the position of these atoms; their position is very ordered. When we refer to the kinetic or potential energy of a macroscopic body, we are referring to energy that has low entropy.

Thermal energy is the "jiggling" of the atoms of matter. A single parameter, called the temperature, gives the average energy of jiggling. However, there is a ***distribution*** of energies, with atoms having more or less energy than the average. Atoms are moving in all directions and continually exchanging energy. There's a lot we don't know about their individual motion. As a result, thermal energy has low information and high entropy. Just like energy — by which we mean the quantity of energy — is a state function, entropy — the quality of energy — is also a state function. Energy is conserved, while entropy always increases.

A given amount of thermal energy has lower entropy (higher order) at high temperature than at low temperature. This is because an atom (and a chemical bond) can hold more energy at high temperature. At low temperature the energy has to be dispersed among many more atoms or chemical bonds. We define ***heat*** as thermal energy transferred from the system to the surroundings. According to Clausius (1822–1888), heat transferred to the surroundings must be divided by its temperature in Kelvin degrees to find the increase in entropy of the surroundings.

> **When heat flows to the surroundings at temperature, T, the entropy of the surroundings increases by the heat divided by T.**

Order and Disorder

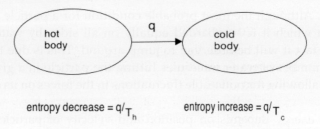

Figure 13.3. Entropy change on heat transfer.

The entropy change in heat flow is shown in Figure 13.3.

Entropy Changes in Simple Processes

Using the entropy attributed to kinetic, potential and thermal energy, we can show how entropy increases in simple processes in agreement with the second law.

If a bullet is stopped, its kinetic energy is converted to higher entropy thermal energy. Initially, the thermal energy is shared between the bullet and the surroundings, but if we wait for the bullet to return to ambient temperature, almost all of its initial kinetic energy is converted to thermal energy of the surroundings. A very similar scenario occurs when a boulder falls off a cliff and converts its potential energy into thermal energy of the surroundings. In both these cases almost all the entropy change ends up in the surroundings.[87]

The Second Law also explains the natural direction of heat flow from higher temperature to lower temperature. At lower temperature, the same thermal energy must be dispersed among more atoms, giving it higher entropy.

An example where the entropy of a system decreases is the freezing of liquid water to ice. Ice is more ordered than liquid water and has lower entropy. However, in order for water to freeze, an amount of energy equal to its heat of melting must flow out of the system.[88] This heat flow

---

[87] There might be a small entropy increase in the system due to distorting the bullet or breaking the boulder.

[88] Since the heat of melting must flow into the system to change ice into water, the same heat must flow out of the system to change water into ice.

produces an entropy increase in the surroundings equal to the heat divided by the temperature. Only if the surroundings are colder than 273 K (0°C), is the entropy increase of the surrounding larger than the entropy decrease of the system — resulting in the freezing of water.

Atoms involved in chemical reactions may increase or decrease their entropy. For example, the combining of hydrogen atoms to form H_2 molecules decreases entropy. In the molecular form, each atom is not as free to move around; it must remain attached to another hydrogen atom. However, under most conditions, hydrogen atoms do combine. This is because the combination releases a large amount of energy (the energy of the newly formed chemical bond), which flows to the surroundings. If the surroundings are not too hot, the entropy increase of the surrounding is sufficient to overcome the entropy decrease of the atoms that are combining.

A very common type of reaction is oxidation. It is carried out at 37°C (98.6°F) in our bodies and at much higher temperatures in flames, where it is called combustion. Oxidation invariably releases energy, since it converts high-energy fuel and oxygen molecules to low energy (much more stable) CO_2 and H_2O (in the case where the fuel is a hydrocarbon). We say that oxidation reactions are *exothermic* reactions as differentiated from *endothermic* reactions, which require energy. The energy released increases the entropy of the surroundings. The large increase in entropy due to oxidation reactions in our bodies can be used to drive the process needed for life.

We can now explain how the kindergarten teacher restores order (negentropy) to her class. In raising her voice or giving a stern look to the students, she uses energy obtained from oxidation of food. This entropy increase of metabolism is sufficient to balance the entropy decrease resulting from increased order of the students.

Free Energy

If your head is spinning from thinking about both the system and the surroundings, you will be relieved to find out that there is a criterion for spontaneity that deals only with the system. This involves the *free energy*, a quantity that can decrease due to an energy decrease in the system (which increases the entropy of the surroundings) or an entropy increase of the system. Mathematically, $F = E - TS$, where F, S and T are the system's free energy, entropy and temperature, respectively. In terms of this quantity:

> **The free energy of a system will always decrease when it undergoes a spontaneous process.**[89]

The entropy part of free energy becomes more important as temperature increases. This is why a substance changes from its most ordered phase, a solid, to the more disordered liquid and eventually the gaseous phase as temperature is increased.

Entropy and the Quality of Energy

If energy (or at least mass-energy) is conserved, why is the media buzzing with topics like: How can we protect our energy supplies? or, Will there be enough energy for the future? Since we don't use up energy, why can't we just reuse it, like we do with copper or scrap iron? The answer to this question has to do with the quality of energy. By quality of energy, we mean how useful it is.

Energy is unusual in that some forms of it are more useful than others. Gravitational potential energy of the water at the top of Hoover dam turns into kinetic energy of the falling water, which can turn a generator and produce electricity. These conversions can be almost completely efficient — they do not waste any energy. Electricity can do almost anything for which it provides sufficient energy. For these reasons we identify kinetic, potential and electrical energy as very high quality. The thermal energy of the water at the bottom of the dam is almost useless and is therefore low quality.

In general, potential and kinetic energy are more useful than thermal energy, especially when the thermal energy is not much hotter than the surroundings. Another very useful form of energy is the chemical energy of high-energy compounds. Of intermediate usefulness is thermal energy

---

[89] To be precise, this only applies to a system at constant temperature and pressure, as would hold for a system exposed to the atmosphere.

at high temperature; a fraction of this is capable of being transformed for useful purposes. The most useful forms of energy are those with low entropy. Since in any process we conserve energy and increase entropy, a corollary of the Second Law is

> In any spontaneous process there is net transformation of energy or arrangements from more useful to less useful forms.

The Second Law shows that the current energy crisis is more accurately described as an entropy crisis. The amount of energy remains constant, but we are rapidly using up our sources of useful energy — those with low entropy.

The Second Law had its genesis in the early days of the industrial revolution, when Sadi Carnot (1796–1832) analyzed steam engines to find the ultimate efficiency these machines could achieve. The purpose of a steam engine is to transform thermal energy at high temperature to mechanical energy. Carnot realized that even if the engine could eliminate all friction and other energy-degrading processes, it could not be perfectly efficient. Now we say that this is because mechanical energy is lower in entropy than thermal energy. In order to be in accord with the second law, simultaneous with producing mechanical energy, the engine has to waste some energy by depositing it in the surroundings at lower temperature.

Cells in our bodies use concentration differences of hydrogen, sodium, potassium or calcium-ions across membranes[90] to perform many of life's processes. Such differences are low entropy, and must be maintained by chemical pumps driven by oxidation reactions that continually add entropy to the surroundings.

Entropy of the Universe and Earth

We have previously noted that the entropy of the universe is continually increasing because it is an isolated system. Since the usefulness of energy

---

[90] Membranes are oily layers in which ions do not dissolve. The ions can only pass through pores in the membrane whose conductance is controlled by chemical gates.

decreases as its entropy increases, the energy of the universe is becoming less useful over time. The eventual fate of the universe is sometimes described as one of *"heat death,"* where its energy has been degraded to the point that nothing useful can occur. In this state, all bodies will be at the same, very low, temperature. They will have lost their kinetic energy and reside in states of lowest potential energy. Substances will have been converted to the lowest energy forms. The Second Law, however, does not deal with the rate of processes and so gives no idea of how long it would take to approach such a state. Some physical processes are very slow. Undoubtedly, humankind has many more imminent crises to deal with before facing the eventual heat death of the universe.

An entropy analysis of the earth is very revealing. Life is continually increasing the order — decreasing the entropy — of Earth. A tree concentrates carbon from the atmosphere. Humans construct cities. Where is the entropy increase required to compensate for the entropy decrease of system Earth over time? The answer to this question lies in the energy balance of the Earth. To a good approximation the energy of Earth remains constant; the energy it receives from the sun equals the energy it radiates out to space. The energy from the sun, however, flows from a very hot body and thus consists of much lower entropy than the energy radiated out to very cold outer space. The net entropy increase of the two reservoirs in the surroundings (the sun and outer space) more than compensates for the entropy decrease resulting from increasing order on Earth.

Information and Entropy

We are said to live in the information society, where having information is more important than having things. (Maybe we only say this because we have lots of things.) We store information in books in libraries, and on computer media, where it can be more easily retrieved or manipulated. Just as architectural and engineering blue prints store the information to produce our buildings and machines, living things need blue prints to provide their instructions for growth, metabolism and reproduction. This information is stored in the repeating sequence of DNA molecules present in every cell of our body. Our brains need information for conscious and unconscious thought. Information in the brain is stored in the synaptic connections between neurons.

The concept of information is related to the concepts of order and entropy. To see this, consider a CD on which computer information is stored. We measure information in computers in terms of *bits*, one bit being a choice between two possibilities (called a binary choice.) A 1.0 Gbit (10^9 bits) contains an ordered array of 1×10^9 domains, each of which can be in a state that scatters laser light (that we call "0") or a state that reflects it (that we call "1"). The CD stores information in an alphabet with only two letters, but because it contains so many letters, it can store a huge amount of information.

The number of possible configurations of a CD is huge. Since the characteristic of each bit of the disk is independent of the state of all the other bits, there are 2^{10^9} possible states of the 1.0 Gbit disk. A song is represented by a particular configuration of the bits on a CD. It has high information content and low entropy. We can add a certain amount of randomness (noise) to these bits and still recognize the song. But if the randomness, or entropy, gets too large, we have lost our information and can no longer recognize the song. Thus, entropy and information are inversely related, when entropy is large, information is small; the configuration approaches that of a random distribution. The second law of thermodynamics states that interactions with the surroundings (weathering, dust deposition, etc…) tend to lose information and increase entropy, making our song less recognizable over time.

Summary
ଔ ଓ ଔ ଓ ଔ ଓ

Entropy is a measure of disorder. The Second Law of Thermodynamics tells us entropy continually increases in isolated systems. When systems interact with their surrounding, it is the net change of entropy — the algebraic sum of the entropy change in the system plus the surroundings — that must increase with time.

Systems have low entropy (high order), if they are represented by just a few configurations and high entropy if they are more random. Mechanical, electrical and chemical energy have low entropy, while thermal energy has high entropy, especially if it is at low temperature. The Second Law also requires that there is net energy degradation in any process.

Questions

1. The entropy increase of the surroundings is the thermal energy added to the surroundings divided by its temperature. What does this tell you about the SI units of entropy?
2. At a casino, you notice that a particular roulette wheel has landed on black five times in a row. What is the probability of this? Would you wager your lunch money on the next spin? Would you bet on red or black?
3. Why does a steam engine have to "waste" energy by depositing heat in the surroundings at low temperature?
4. An indigenous person constructs a hut, using bricks produced by mud dried in the sun. Discuss the entropy changes in the system (the mud) and the surroundings in this process.
5. Energy has quality as well as quantity. Can you think of anything else that has quality as well as quantity?
6. Use the idea of entropy to explain why water will spontaneously boil when the ambient temperature is greater than 373 K (100°C).
7. Use the idea of entropy to explain why particles in a gaseous cloud will agglomerate to form stars and planets.
8. Consider a process where a coin sitting on your desk spontaneously jumps into the air, while the desk and the coin simultaneously cool. Does this violate the first law of thermodynamics? Does it violate the second law of thermodynamics?
9. Creationists sometimes argue that the continual increase of order on Earth violates the Second Law of Thermodynamics and thus indicates the presence of a deity. What is wrong with this argument? Does it being wrong rule out the possibility of a deity?

14 Equilibrium and Steady State

What type of systems can exhibit the highly ordered characteristics of life? In this chapter two types of systems, which we call equilibrium and steady-state, will be considered for this purpose.

Equilibrium

Consider an insulated container (one which does not permit any heat flow through its walls) into which 50 mL[91] of cold water (at 10°C) and 50 mL of warm water (at 30°C) are added. Initially this is a very complicated system, with regions of colder and warmer water and thermal currents mixing the two. After a time, these currents disappear and the system reaches a much simpler state that we call *equilibrium*. In equilibrium the system no longer changes and it can be described by just a few parameters. The temperature (very close to 20°C), pressure and the amount of water in the container (100 mL) completely specifies the system. Even though the system still has thermal energy, the energy has been completely degraded and no longer produces thermal currents.

In an isolated system, change can occur if and only if entropy can increase. Since change can no longer occur at equilibrium,

---

[91] mL is a milliliter or a cubic centimeter.

Entropy is a maximum in isolated systems at equilibrium.

At equilibrium, we have the maximum disorder compatible with the parameters describing the system, such as its energy or temperature. Nothing useful can occur in a system at equilibrium, because the system's entropy is already a maximum and cannot increase further.

From our microscopic definition of entropy in the previous chapter, the equilibrium state corresponds to the state with the maximum probability — the state that can be described by the maximum number of configurations. There are many more ways of nearly equally distributing the system's energy among its molecules than there are if we require that certain molecules have more energy than the others.

Although macroscopic change ceases at equilibrium, things are still happening on the microscopic level. For example, in the equilibrium between liquid and gaseous water at its equilibrium *vapor pressure*, evaporation of molecules from the liquid is balanced by condensation of molecules from the gas phase onto the liquid. This can be verified by injecting some heavy water, HDO, into the liquid phase. In a very short time the heavy water will be distributed between both the liquid and the gas phase.

Equilibrium corresponds to the state of maximum entropy and maximum disorder. However, order still exists in a system at equilibrium, because its state must be consistent with the conditions imposed upon the system. For example, water molecules at equilibrium below 0°C form the highly ordered structure of ice. The water molecules have a low average energy and achieve their disorder by "jiggling" around their ordered positions in the solid ice. Because water is a single *component* system — a single chemical species, its equilibrium state is quite simple. Structures such as snowflakes are a result of how the solid water was formed. They disappear over time as the system approaches equilibrium.

More complex equilibrium states can be obtained with multi-component systems. For example in the two-component system, oil and water, the oil, being less dense, forms a layer on top of the water. Oil and water do not mix, since oil molecules are nonpolar hydrocarbons and water molecules are polar. It is not energetically favorable for water molecules to separate and lose their attractive interactions by associating with oil molecules. When an oil-water mixture is shaken, an emulsion forms, as is seen with salad dressing. However, this is not the equilibrium state, and allowed to stand, the components of this *suspension* separates. Some suspensions, although unstable, can persist for a very long time. For example, suspensions of gold particles in water prepared by Michael Faraday[92] can still be seen in the British Museum. The long life of these suspensions results from the electrical charge on the gold particles, which opposes their *coagulation*. Long-lived suspensions are called *colloids*. Milk, a suspension of butterfat globules in an aqueous solution, is another example of a colloid.

Interesting structures can be formed at equilibrium with particular combinations of compounds, for example, with surfactants and water. *Surfactants* are long molecules that are hydrocarbon-like on one end and either polar or charged on the other end. The formula of sodium dodecylsulfate, a surfactant that is commonly used in detergents is given in Figure 14.1.

The polar or charged ends of these molecules are called ***hydrophilic***, because of their attraction to water molecules, while the hydrocarbon ends are called ***hydrophobic***, because of the lack of such attraction. At low concentrations, surfactant molecules can form single layers (*monolayers*) on top of water and keep the water from evaporating — a technique useful for water conservation. At higher concentrations, surfactants form *micelles* in water. These are tiny droplets, with their hydrocarbon tails loosely held in their interior, as shown in Figure 14.2.

Most soaps and detergents contain surfactants, which form micelles that dissolve nonpolar grease and dirt particles in their hydrocarbon interior — a process called *emulsification*. Bile salts, produced by the liver and stored in the gallbladder, are our body's detergents, which emulsify fats, so they can be better broken down by digestive enzymes.

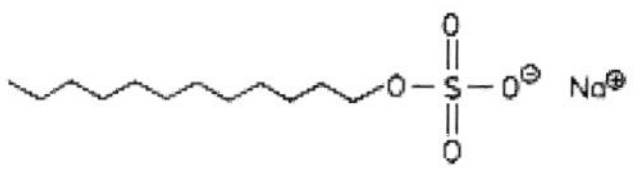

Figure 14.1. Sodium dodecylsulfate.

---

[92] We discussed Faraday (1791–1867) in relation to the laws of electromagnetism.

210   *Order and Disorder*

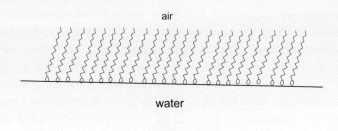

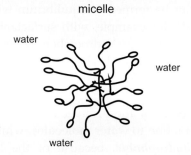

Figure 14.2.   Structures of a surface monolayer and a micelle.

The ability of a micelle to suspend a grease particle in water is shown in Figure 14.3.

Phospholipids are surfactants with two hydrocarbon tails and a charged phosphoryl group. They often form **double layers**, with their hydrocarbon "tails" intertwined and their charged ends exposed to water or aqueous solutions. These double layers can curve around to form completely enclosed **vesicles**, with aqueous solutions both inside and outside the enclosure, as shown in Figure 14.4. When stabilized by non-polar molecules such as cholesterol and penetrated by proteins, such structures make up the **plasma membranes** that enclose all cells.

Chemical Reactions at Equilibrium

Chemical reactions are driven to an equilibrium state with definite proportions of reactants and products. For example, if we dissolve a little of the weak acid, acetic acid (the active ingredient in vinegar), in water, some of

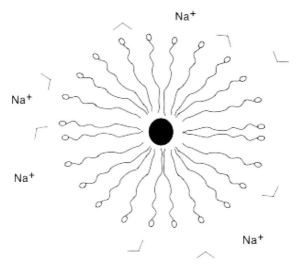

Figure 14.3. Sodium dodecylsulfate micelle suspending a grease particle in water.

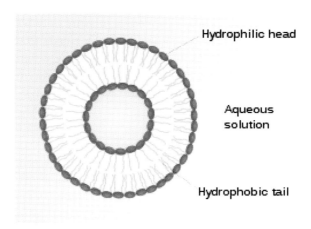

Figure 14.4. A vesicle formed with phospholipid molecules.

the acid molecules ionize, giving H^+ and acetate ($CH_3CO_2^-$) ions. However, as soon as the concentrations of these ions build up in solution, some of them recombine to reform the acid. We indicate an equilibrium with a double-headed arrow, so for this reaction we write

$$H_3CCO_2H \leftrightarrow H^+ + H_3CCO_2^-.$$

In many cases the equilibrium state for a chemical reaction lies so far toward products that we say that the reaction goes to ***completion***. One example of such a reaction is ***neutralization***, the combining of H^+ and OH^- in an aqueous solution to form water. If we add equal amounts of a strong acid (HCl) and a strong base (NaOH) to water, the H^+ and OH^- ions immediately combine and the concentrations of these ions are reduced to a very low level, 10^{-7} moles/liter. This corresponds to a pH of 7.0, which is characteristic of a ***neutral*** (neither acidic nor basic) solution.

As a chemical reaction proceeds, reactants are consumed and products are formed, and the net rate of approaching equilibrium slows. We say that the reaction ***asymptotically*** approaches equilibrium. We have seen this behavior when we discussed radioactive decay in Chapter 3. Radioactive decay goes to completion in a ***first-order process***, i.e., the rate of decay is proportional to the amount of the original isotope remaining. Many chemical reactions require the interaction of more than one molecule to occur, and will approach equilibrium more slowly than a first-order process.

There are great variations in the speed with which systems approach equilibrium. For example, a spark converts a mixture of two parts hydrogen gas and one part oxygen gas almost instantaneously to water vapor. However, the rate with which diamond converts to graphite is so slow that this process can be completely ignored, except under extreme conditions like those existing deep beneath the surface of the Earth.

Steady State

Not every system that has reached an unchanging state is at equilibrium. For example, a metal bar heated at one end by a flame and cooled at the other by running water will reach a temperature distribution along the bar that doesn't change with time. Although the bar is not changing with time, it is not isolated. Heat is added to and removed from the bar by the surroundings. It is transferred along the bar by ***conduction***, where vibrating molecules transmit their vibration energy to adjacent molecules. The entropy of the bar is constant, but the entropy of the surroundings is continuously increasing, since heat is being deposited in the surroundings at a lower temperature than it is being removed from the surroundings. Systems whose unchanging state requires interactions with the surroundings are said to be at ***steady state***.

Heat transfer can produce more interesting structure in liquids, where it usually occurs by thermal *convection*, rather than by conduction. This is because a heated parcel of liquid expands, becomes lighter than surrounding parcels, and then rises.[93] With a small temperature difference between the top and the bottom of the liquid, the convection currents in the liquid are gentle and disorganized. As the temperature difference is increased, organized structures that transfer heat more efficiently develop. Near hexagonal cells are formed, with circulation as shown in Figure 14.5. This type of circulation is called Benard convection.

With more intense heating of the bottom of the water, other interesting phenomena occur. As anyone who has ever impatiently watched water being heated on the stove has observed, bubbles of gas first form at various places on the bottom of the pan. As the temperature rises further, bubbles form spontaneously within the liquid. This latter state, called *boiling*, is very complicated. It is impossible to predict just where the next bubble will form. Finally, if a cover is placed on the pan, it will bounce up and down to release the buildup of steam pressure in the pan. The

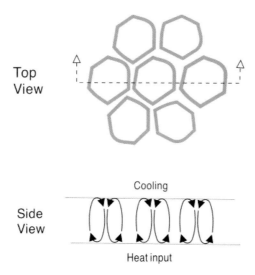

Figure 14.5. Benard convection.

---

[93] When the liquid is in a gravitational field.

bouncing of the cover, under some conditions may even be regular in time. We can say:

> **In far-from-equilibrium systems spatial and temporal order may develop, as well as disorder.**

Another example of steady state is a block of dry ice (solid carbon dioxide) placed in the outside air. Carbon dioxide evaporates from the block into the air. Near the block there is a high concentration of carbon dioxide and far away a very low concentration in the ambient air. The distribution of carbon dioxide in air is constant as carbon dioxide *diffuses* away from the block. Diffusion is movement of molecules from where their concentration is high to where their concentration is low by random movement. Its direction is opposite to the concentration *gradient*.[94]

A steady state is more difficult to describe than an equilibrium state. Usually temperature, pressure and concentrations cannot be assigned at steady state, because these variables vary throughout the system.

Probably the simplest example of spontaneous generation of patterns in a steady-state system is so-called phantom traffic jams, slowdowns of dense traffic on unobstructed highways in the absence of accidents. Here positive feedback results from cars braking when the brake light of the car in front of them goes on. Passing through the jam, cars speed up, only to have the situation recur further down the road. The system is steady-state, since in the absence of cars entering the highway with fuel and leaving the highway having used up some of their fuel, all traffic would come to a standstill — an equilibrium state. Behavior is very dependent on the parameters of the system. Below a certain traffic density, no jams occur. At a particular density, the system changes its nature as jams begin to occur. Such a qualitative change in behavior is called a *bifurcation*. With further increases in density, the frequency and separation of jams continues to vary.

---

[94] This example is actually more complicated, since the low temperature of the block will set up air currents that transfer carbon dioxide by the process of convection, which is bulk motion, in contrast to the molecular motion of diffusion.

Heated water and traffic jams are single component steady-state systems. Multi-component steady-state systems can produce even more complex patterns in space and time than single-component systems. Especially interesting are systems undergoing chemical reactions. Of course, if such systems are not to approach equilibrium, reagents must continually be added and products removed from them. In Figure 14.6, two spatial wave patterns that develop in reacting mixtures are shown.

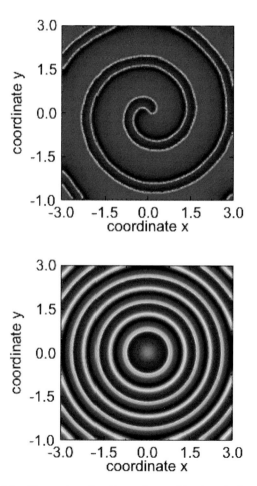

Figure 14.6. Examples of patterns formed in chemical reactions.

Spontaneously developing spatial patterns in chemical reactions are called Turing patterns, after Alan Turing, who proposed them as a mechanism for initiating the development of biological structures.[95]

Many of the reacting systems that show spatial variations similar to those in Figure 14.2, can also show temporal variations. These usually occur when the system is mixed sufficiently rapidly that spatial variations are suppressed. The periodic behavior is particularly dramatic when the concentration variations produce color changes. Such *oscillating* chemical reactions are very popular demonstrations in chemistry courses. If reagents are continually added and products removed from the reacting mixture, the oscillations continue indefinitely.

Living Systems

Over a short period of time, most living things don't change much and can be considered steady-state systems, continuously taking in nutrients and discarding waste products. In addition, if a living system is not at the same temperature as its surroundings, it continuously exchanges energy with its surroundings. We will see that the degree of structure of even the smallest living thing, say a single-cell organism, is incredibly more complex than that of a chemical reactor operating at steady-state, which is also continuously exchanging chemicals and energy with its surroundings. What is it that gives living systems the ability to show such a tremendous degree of organization?

The primary difference between living and non-living systems is their degree of organized complexity. Not only are a huge number of physical and chemical processes occurring in all living systems, but these processes are coordinated to a very high degree. Chemicals are synthesized just when they are needed; energy is stored until it is needed. This

---

[95] Although he never won a Nobel prize, Alan Turing was chosen by Time magazine to be one of the most important people of the twentieth century. Besides his work on morphogenesis (the spontaneous formation of structure) discussed above, he is often credited with laying the groundwork for the computer. During World War II he developed methods for breaking German codes that were instrumental in Allied victories. Turing was prosecuted in Britain for being a homosexual, which essentially ended his career. Soon after he committed suicide.

is because the processes in a living system are under a very high degree of control, with their speeds regulated by numerous catalysts. It is as if the living system is being directed by computer, using a massive program stored in its memory. Of course, specifying a computer program to control the structure of organisms requires the storage of an enormous amount of information. We will discuss how living systems achieve this massive information storage in Chapter 18.

Earth as a Steady-State System

At the lowest approximation, our planet is not changing. Viewed from space, it looks pretty much the same from one day to the next.[96] However, we know that Earth is interacting with its surroundings; it is continually absorbing solar radiation and, less obviously, radiating energy to space. Moreover, we know that if these interactions change, there would be profound changes on the planet: e.g., it would get much colder or hotter — uncomfortable conditions for life as we know it. Thus Earth as a whole can be approximated as a steady-state system.

Since at steady state the temperature of the Earth doesn't change, the first law of thermodynamics tells us that the energy influx to the Earth from the sun must equal the outflow by radiation to space; these are the only two appreciable interactions of the Earth with its surroundings. However, as we learned in the last chapter, in order to find entropy flows, we must divide the energy flow by the (absolute) temperature. Since the surface temperature of the sun (ca. 6000 K) is much higher than the average surface temperature of the Earth (ca. 288 K) entropy flows are not in balance, much more entropy is flowing out of the Earth than is flowing into it. Where is this entropy coming from? It is a necessary part of all the processes on Earth that create order (e.g., growing plants and building cities). We know that each of these processes must also create disorder — create entropy, so it can occur without violating the second law of thermodynamics.

---

[96] Probably the only observable difference would be changes in the position of clouds.

Summary

Equilibrium in is a state of maximum entropy, where nothing is changing in an isolated system. Although nothing useful can be accomplished by an equilibrium system, it can show structure, especially if it consists of more than a single component. For example, in aqueous solutions, surfactants, long molecules with hydrophilic and hydrophobic ends, can form monolayers, micelles and vesicles.

Steady-state systems also do not change with time, but they are continually exchanging energy and matter and producing a net entropy increase in the surroundings. With even small deviations from equilibrium, interesting structure can develop in single-component systems subject to heat transfer or multicomponent systems with diffusion or chemical reactions. With greater deviation from equilibrium, chemical reactions, flows or diffusion can produce spatial patterns. Organisms are approximately steady-state systems incorporating a huge number of chemical reactions. They are highly regulated and under the control of stored instructions. The Earth is a steady state system, with continuous energy input from the sun and output to space.

Questions

1. Is each of the following systems at equilibrium, at steady-state or neither? If it's steady state, what are its energy or matter transfers to the surroundings?

    (a) a lake
    (b) a river
    (c) a city
    (d) a fish in the ocean
    (e) a star

2. Describe in detail what happens when an ice cube is placed in a glass of water. Is the system approaching equilibrium or steady state?

3. Would each of the following be more likely to be associated with an equilibrium or a steady-state system?

    (a) temperature
    (b) convection

(c) vapor pressure
(d) chemical reaction

4. Draw each of the following:
   (a) a surfactant molecule
   (b) a micelle
   (c) a double layer
   (d) a vesicle

5. Why is it necessary that the suspended particles in an aqueous colloid be charged in order for the colloid to be stable (exist for a long time without coagulating)?
6. Are there any chemical systems that you have observed to form interesting spatial patterns?
7. What do you think produces the periodic behavior in systems such as the flashing of fireflies or the beating of your heart?

15 Introduction to Life

Life is a whim of several billion cells to be you for a while.

Groucho Marx

Not all scientists agree on the requirements for life. However, we will take as our working definition:

> **Life is a complex, homeostatic system that has the potential for action, growth and change.**

Life is, or has produced, much of the order that we see around us.

Complexity

An *organism*, something that is alive, is incredibly complex — much more complicated and sophisticated than an enterprise of human society, such as a modern chemical factory. Even a *cell*, the simplest form of life that is capable of independent existence, carries out the synthesis of a multitude of different chemicals and interacts with its environment in many ways. We have discussed examples, such as periodic patterns, of how order can arise in simple mechanical and chemical systems. Life in all its variety, however, is much more complicated than these systems, indicating that

> **The degree of complexity of life requires that it be under the direction of an elaborate set of instructions.**

To provide these instructions, organisms require the storage, use, transmittal and reproduction of huge amounts of information.

There are several hundred different types of cells in the human body. The cell that we will use for comparison with a chemical factory is the **beta cell** in a tissue called the islet of Langerhans, in the pancreas, an organ of our body. A photograph of a beta cell, taken with an electron microscope, is shown in Figure 15.1.

It has been demonstrated that beta cells can live outside the pancreas, both in culture media and when transplanted into the liver. The beta cell

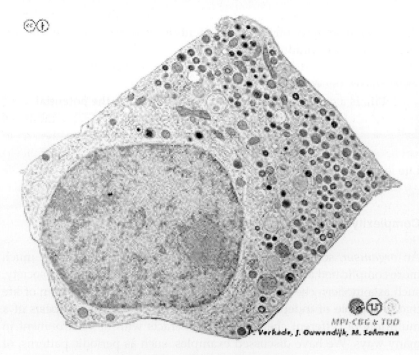

Figure 15.1. Beta cell of the pancreas.

is a chemical factory, whose output is a protein molecule, *insulin*, which is a *hormone* that participates in regulating the level of *glucose* in blood. We will discuss proteins and their components, amino acids, in more detail in the next two chapters.

Glucose is the body's preferred fuel; it's like gasoline for a lawn mower. Insulin is essential for transporting glucose from your blood into the cells of your body. Without insulin, your cells make use of other fuels, such as proteins or fats. But just like a lawn mower doesn't run well on diesel fuel or kerosene, your cells won't run well without glucose, their preferred fuel. Insulin stimulates numerous types of cells to increase their rate of use of glucose (*metabolism*) and also activates cells in the liver and muscles to store glucose as the polymer *glycogen*, and other cells to store glucose as fat. The islet of Langerhans also contains alpha cells that produce *glucagon*, another hormone. Glucagon has the opposite effect to that of insulin; it stimulates the production of glucose from stored glycogen and fats.

A cell is surrounded by a *plasma membrane*, which forms its protective barrier and allows for controlled transport of material into and out of the material of the cell — its *cytoplasm*. Cells are either *prokaryotes* (mainly bacteria) or *eukaryotes* (everything else), the difference being that eukaryotes have a large separate compartment, called the *nucleus*, where genetic material — the cell's instructions encoded in the molecule DNA is stored. A diagram showing a number of features of a typical eukaryote cell is shown in Figure 15.2.

Both insulin (with 51 amino acids) and glucagon (with 29 amino acids) are much more complicated molecules than those synthesized in industrial factories by conventional chemical techniques.[97] In order to select the proper amino acids from among the 20 different varieties in the cell, and join them together in a precise sequence, the beta cell employs catalysts — protein *enzymes* — which perform the synthesis with amazing speed and accuracy. Enzymes constitute most of the tools and machines of the cell. Unlike a factory, that buys most of its tools and machines externally, a cell manufactures these in-house. A cell is also very good at recycling the components of its tools. Once an enzyme is no longer needed, it is disassembled, and most of its amino-acid components are reused to synthesize other proteins.

At the shipping dock of the beta cell, oxygen, minerals, salts, glucose, amino acids and other substances are delivered by the blood, rather than

---

[97] Biotechnology companies synthesize insulin by using bacteria as factories.

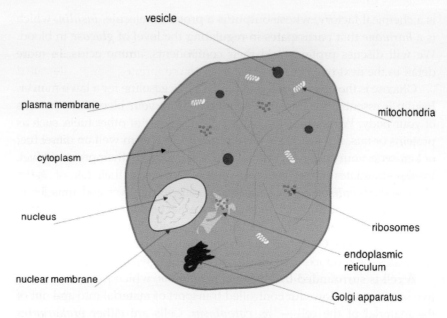

Figure 15.2. Some components of a typical eukaryote cell.

by Fedex, as they are in the industrial factory. Glucagon and insulin are packaged in vesicles, which are much like polyethylene wrapping, for shipment through the cell membrane into the blood. Just as in a modern-day chemical factory where the production and shipping of chemicals is directed by market analysis, the beta cell does its own market analysis with glucokinase, an enzyme whose action has the beta cells release more insulin at higher glucose concentrations. Further regulation of insulin production results from signals generated in the liver, gut and brain. After a meal, when blood glucose concentration is high, the brain sends out signals that cause the pancreas to produce more insulin and less glucagon. It also signals other organs to slow down their functioning while food is being digested. This is why we often feel like taking a nap after a large meal.

In a factory, different operations (tool room, assembly, painting, packaging) are carried out in different areas. Similarly, a cell has different substructures, called ***organelles***, which carry out different operations. Among the organelles shown in Figure 15.2 are: the nucleus, where the organism's instructions are stored; the endoplasmic reticulum and the ***ribosomes***, where the molecular tools — the enzymes (and other proteins) — are synthesized; and the ***Golgi complex***, where cell products are packaged. These different structures are joined by filamentous "conveyor belts" along which materials are transported within the cell.

Unlike the factory, which buys most of its energy from the local electric utility, our cell produces its own energy, in the form of a molecule called adenosine triphosphate (ATP), from internal power plants, called **mitochondria**, which are fueled by glucose. The mitochondria carry out the large number of chemical reactions required for the production of ATP with impressive speed, efficiency and selectivity. For maximum efficiency, as well as historical reasons, the mitochondria store their own instructions.

Our cell factory is enclosed within a wall, but the wall of the cell, the plasma membrane, is much more complex than that of the chemical factory. Rather than having a single loading dock, the cell wall is penetrated by numerous protein structures that allow selected chemicals to pass in and out of the cell, while inhibiting the passage of water and salts, in order to maintain a good "working environment" in the cell.

The *nucleus* of the cell, surrounded by its own plasma membrane, is its central office. Here is where the orders are processed for any chemicals the cell needs for its operation. The nucleus is like that in an antiquated factory, in that it sends its instruction throughout the remainder of the cell (its cytoplasm) by means of chemical messenger boys (called **messenger RNA**), rather than high-speed data lines.[98] However, since the distances within a cell are small, messenger boys work just fine. The nucleus also contains a complete set of instructions for expansion of its activity, either by growing the cell or generating another identical insulin-glucagon factory. Actually the central office of our cell plant is very crowded. Its filing cabinets contain, not only a complete set of instructions for the beta cell, but instructions for several hundred other types of cells in our bodies as well. Somehow its messengers know how to select only instructions concerning the beta cell for transmittal to the cytoplasm.

Of course, our living chemical plant only produces its product, insulin, in microscopic quantities, but insulin is such a powerful chemical, that the amount produced in the islet of Langerhans is sufficient for most individuals. In just about all other respects, as we have seen, the sophistication and complexity of this single living cell exceeds that of the chemical factory.

If a single cell is as, or more, complex than a chemical factory, imagine how much more complicated is a multi-cellular organism, such as a fruit

---

[98] Some bacteria, which are single-celled organisms, have a rudimentary electrical communication network, not involving their nucleus.

fly or ourselves. Such an organism is more akin to the entire defense industry, where different factories produce different products: some explosives, some tanks, some uniforms. An additional capability of an organism is that, not only can it create essentially identical copies of itself, but also, over time, improved copies better adapted to survival in its changing environment. It's as if the organism contained all the Department of Defense research laboratories!

Supplemental Reading: Diabetes

It is very critical that glucose levels in the blood remain in the appropriate range. Too little glucose (hypoglycemia) and a person feels weak; too much glucose (hyperglycemia) and he can "burn" up critical organs, such as the heart, kidneys and retinas of the eyes. This latter condition occurs with untreated *diabetes*. Paradoxically, a diabetic might feel weak at the same time that he is destroying organs, since diabetics do not effectively transport glucose into cells. There are two types of diabetes: In Type 1, or juvenile-onset diabetes, the beta cells produce inadequate amounts of insulin.[99] In Type 2, or adult-onset diabetes, the body does not make efficient use of and regulate the amount of its insulin.

Until the 1920s, diabetes was an almost always-fatal wasting disease. In some cases, the disease could be controlled by use of a near-starvation diet, which maintained a tolerable level of blood glucose. All that changed with the discovery of insulin by Fredrick Banting (Nobel prize in medicine and physiology in 1923). In 1922 the first patient was relieved of the horrible symptoms of the disease by injection with insulin from a cow's pancreas. Nowadays, rather than making use of insulin from cows or pigs, diabetics primarily use human insulin, manufactured by inserting human genes into bacteria (recombinant technology). Insulin is injected, since, as for all proteins, the molecule is broken down into its constituent amino acids in the human digestive tract. Some advances have been made with a form of insulin that is administered as a nasal spray.

---

[99] Usually, because they have been attacked by the immune system.

Homeostasis

Homeostasis is the property that enables a complex system that we call life to

> **maintain a relatively constant set of internal conditions with variable external conditions.**

While nonliving systems may be complex or homeostatic, rarely are they both. For example, the weather is complex, but not homeostatic. Under the annual variation of sunlight, the weather changes considerably (although heat transfer greatly moderates these changes). It even changes on a day-to-day basis with constant solar illumination. An example of a homeostatic system is a nuclear reactor, which is designed to maintain a constant rate of the nuclear fission reaction regardless of external conditions. This is accomplished by means of negative feedback, using graphite rods whose position is varied to absorb excesses of the neutrons that induce the reaction. However, since a nuclear reactor needs to control only a single reaction, its complexity does not come close to rivaling that of a living system.

An organism controls many of its variables, such as pressure, pH and the concentrations of many ions and compounds. Individual cells of multi-cellular organisms are not completely homeostatic and usually can only survive in *culture media* at controlled temperature. The concept of homeostasis in a complex system implies a means for sensing these variables and systems for controlling them. The instructions for these negative feedback loops must be of necessity so involved, that they require the storage and transmission of huge amounts of information. Such information is stored in the cell nucleus.

We have discussed the body's control of blood glucose concentration. We have seen how this "glucostat" operates by signals in the pancreas, brain, liver and gut. The glucostat also provides signals for satiety and hunger and thus establishes a natural set point for weight. Each of us has a somewhat different set point for our glucostats, which can only be overridden by exercise of extreme will power.

Our bodies also regulate the concentration of solutes in our blood (a quantity called **osmolality**). When our brain senses that these concentrations are too high, it releases a hormone called vasopressin that promotes reabsorption of water from urine. Thus, less urine volume is excreted and more water is kept in the blood. In a similar manner, increased osmolality generates a feeling of thirst, the purpose of which is to increase water input and dilute the blood.

A considerable degree of homeostasis is necessary for living organisms, since most of the chemical reactions essential for life are exquisitely sensitive to variables such as temperature, concentrations and pH. As one example, aquatic animals live in an environment that is either fresh water — which contains much lower concentrations of salts than their bodily fluids — or salt water that contains much higher concentrations. Such differences in concentration across the cell membrane, which is **semipermeable** (allows the passage of water, but not of solutes) would usually rupture it or shrink the cell, due to the passage of water to equalize concentrations — a process called *osmosis*. Somehow, aquatic animals can regulate the rate at which water passes through their cell membranes. As another example, many plants that live in desert environments, where they could rapidly dry out, conserve water by opening and closing pores in their leaves, so they only exchange material with the environment during the cooler evening and nighttime hours.

Homeostasis entails a sizeable energy cost. Maintaining constant temperature (***thermostasis***), for example, is particularly costly, as you have probably surmised, if you've looked at your gas or electric bill lately. However, since most of the chemical reactions of life work optimally at a particular temperature, a warm-blooded creature can be much more efficient than a cold-blooded creature, whose temperature is essentially that of the surroundings.[100] Reptiles, for example, are cold blooded; they are limited in their habitats and become sluggish in cold weather. This is why you might see an alligator sunning itself on a rock when its cold. Normally thermostatic organisms, such as ourselves, sometimes allow their temperature to rise modestly. These fevers may help the organism fight bacterial infection by raising body temperature above the range that is optimum for bacterial reproduction.

---

[100] In hot weather, cold-blooded creatures can actually be warmer than warm-blooded creatures.

> **Supplemental Reading: Extremophiles**
>
> As anyone who has tried to keep plants from growing out of cracks in his driveway will attest, life on Earth can flourish in unusual places. *Extremophiles* are creatures (almost exclusively microbes) that can live under extreme conditions. Life (usually bacteria) has been found to flourish in the driest deserts, miles beneath the oceans and solid Earth and in brine and acid pits. Some extremophiles accomplish this by amazing feats of homeostasis. An example of these are *acidophiles* — bacteria that can thrive in exceptionally acidic environments. They accomplish this by means of chemical pumps that remove hydrogen ions from their cytoplasm. This *active transport* requires energy to move the hydrogen ions against their concentration gradient. It can maintain internal hydrogen-ion concentrations in the cell that are, in some cases, as much as a factor of 10^5 lower than the external concentration. A key requirement is that the environment must provide more energy for the organism than is consumed in maintaining homeostasis.
>
> *Thermophiles* are organisms that can live at very high temperatures. They are found near geysers and hot-water vents in the oceans; some can survive (and in fact, prefer) temperatures as high as 110°C. They do this, not by maintaining lower internal temperatures, but by employing very stable proteins and mechanisms for repairing molecules that are not damaged by high temperatures. Because the material streaming out of hot water vents are rich in nutrients, some scientists have proposed that bacteria living in such environments were the earliest living creatures on Earth.

Action

The capability for action, responding to stimuli, is a test that we often apply in determining whether something is living or not. Does that object on the floor move when we tap next do it? If not, it is probably dirt or a dead, rather than a living, bug. Does that plant give out an aroma to attract insects or does it drop its leaves or seeds on the ground? If not, it is probably an artificial plant. At the very minimum, does that object take in and give out chemicals to its environment, i.e., does it have a metabolism? Is it at steady-state, rather than equilibrium? We freely apply these tests, although we know that in some cases, such as for viruses, seeds or spores, we might have to wait a while before the potential for action is attained. We will not consider viruses, seeds and spores as living, even though they have the potential for life.

Growth and Reproduction

Some organisms are composed of a single cell, while others, like ourselves, are composed of trillions of cells.[101] Cells carry out most of the processes of life. They take in nutrients and eliminate waste products. They generate energy, usually consuming oxygen and releasing carbon dioxide in the process. Cells grow, and when they reach a certain size, they often divide. This requires that they duplicate their genetic material, the instructions that control the processes of the cell. When the cells of a single-cell organism divide, two organisms result. In multicellular organisms, cells are highly specialized, with different types of cells growing and dividing at different rates. For example, red blood cells are replaced at a prodigious rate of about 100 million per minute whereas brain cells hardly ever divide. That is why you can donate blood, but you can't donate brain. The cells in the outer layer of our skin never divide; they slough off and are replaced by new layers of skin produced from a layer of *stem cells* at the base of the skin.

All multicellular creatures eventually die. They must therefore reproduce in order to maintain their species.

> **Multicellular organisms reproduce sexually or asexually.**

In *asexual* reproduction, offspring result from the cells of a single individual, whereas in *sexual* reproduction they result from special cells, called *gametes*, of two parents. Some creatures, such as the mule (an infertile

---

[101] Groucho's quote at the beginning of this chapter greatly underestimates the complexity of our bodies, which contain between ten and a hundred *trillion* cells.

hybrid of a female horse and a male donkey) cannot reproduce, although they do grow.

Asexual reproduction is ideally suited to produce a large number of individuals in a single environment, particularly when, as in sponges and corals, individuals remain connected in a colony. In asexual reproduction, each individual has an identical copy of its parent's genetic material — it is a *clone* of its single parent.[102,103] The advantage of asexual reproduction is that it minimizes the expenditure of energy.

In sexual reproduction, each of the two gametes that combine to form a new individual has only half of the genetic material of the parent from which it derives. Sexual reproduction is considerably wasteful of energy. For example, a human male produces about 30 million sperm cells a day. Thankfully, there is only a miniscule chance that one of these will successfully produce a new individual. The advantage of sexual reproduction is that it produces new combinations of genetic material. These new *genotypes* produce individuals with new characteristics, called *phenotypes*, which may provide them an advantage under unfavorable living conditions.

Constancy and Change

Changes occur within the life cycle of single individuals, between generations and within a *population* (a species living in a particular habitat) over time. Living things change in form as they grow. The changes in humans are not as obvious as those in, say, frogs — which exist as tadpoles confined to water in their early life. However, the physical form of an infant human differs from an adult in more ways than just size.[104]

A new child is born, and relatives crowd around, noting similarities between her and one or the other of her parents. Similarity between generations is one of the hallmarks of biology. However, children do not

---

[102] Even in single-cell organisms, exchange of informational material occasionally occurs between different cells.
[103] Scientists have learned how to produce clones of some normally sexually reproducing animals, such as sheep.
[104] This difference was not recognized by many pre-Renaissance painters, who depicted infants as just small adults.

necessarily resemble their parents in every trait. As examples, brown-eyed parents might have a blue-eyed child or normal parents might have a child with a devastating disease, such as sickle-cell anemia. How can we explain such observations?

The experiments of Gregor Mendel, an obscure nineteenth century Austrian monk, first brought some insight into the question of how some characteristics are inherited. Mendel, who was a combination of a biologist and a mathematician, performed his experiments on common garden peas, which had a number of traits (e.g., short or tall plants), each of which appeared in two distinct forms.

Peas reproduce by sexual reproduction, in which pollen from anthers (the male sexual organ) falls onto the stigmas (the female sexual organ). Starting with groups of plants that always bred true for a characteristic (i.e., the offspring of tall plants were always tall and those of short plants were always short), Mendel investigated what happened when he crossbred the plants with opposite characteristics. The accepted theory of the day was that the substance of heredity was blended in the offspring. This predicted, for example, that a major fraction of the offspring of one tall and one short plant would be medium-sized. This was not what Mendel observed; as shown in Figure 15.3, all of the first-generation plants of this crossbreeding were tall.

An even more interesting observation was made concerning the second-generation hybrids, i.e., the offspring of two first-generation hybrids. In the second-generation, Mendel found once again, no medium height plants, but short plants reappearing in 25% of the cases. Obviously, the genetic material specifying "shortness" had not disappeared, but had only been masked in the first generation. Mendel explained his results by hypothesizing that for each of the inheritable traits, there were two forms

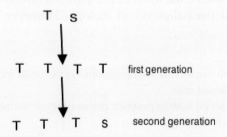

Figure 15.3.   Phenotypes observed by Mendel.

(that we now call *alleles*) of the inheritable factor (that we now call *genes*), and

> **In sexual reproduction, one unchanged gene is passed randomly from each parent to progeny.**

Since Mendel's original plants always bred true, they must have had two copies of the same allele, i.e., they are *homozygous*, and express the *phenotype* (observable characteristic) of that allele. If we call the allele for tall height H and for short height *h*, the true-breeding plants had HH or *hh*, *genotypes* and expressed the tall and short phenotypes, respectively. Mendel's first-generation hybrid plants received one gene from each of its parents, and therefore all had the *heterozygous* H*h* genotype. Since these first-generation plants were all tall, Mendel reasoned that H, the tall allele, must be *dominant*, and *h*, the short allele, must be *recessive*. In other words, as long as a plant had a single allele for tall height, it was tall; there was no blending of this characteristic.

Alleles are passed randomly from first-generation plants to their second-generation progeny. This means that in the second generation a genotype will occur in proportion to the number of ways it can be produced by association of the first-generation alleles. These possibilities can be inferred from the matrix of combination of alleles shown in Figure 15.4.

As can be seen from this figure, there are three genotypes (HH, H*h* and *h*H) that would express the tall dominant phenotype and only one genotype (*hh*) that would express the short recessive phenotype, giving a ratio of one short to three tall plants, or 25% short plants, very close to what was observed by Mendel.

Mendel's beautiful results were fortuitous, due to his decision to study traits of the garden pea, each of which was determined by a single gene in a simple dominant-recessive relationship. Such transmission is called

	Hh	hH
Hh	HH	Hh
hH	hH	hh

Figure 15.4. Matrix of progeny genotypes from H*h* parents.

Mendelian genetics. Results for each of the traits Mendel studied were similar to those found for the height of the plants, indicating that the genes for each trait were transmitted independently of the genes for the other traits. Notwithstanding the elegance of Mendel's work, when he presented it in front of the Historical Society of Brunn, Austria in 1865, it only received polite applause. Science was not ready for his results at that time, and they lay in obscurity until the turn of the century, when they were rediscovered and received the acclaim of scientists worldwide.

Most traits are not transmitted by Mendelian genetics. For example, snapdragons exhibit *incomplete dominance*, where the combination of an allele for a red flower with one for a white flower produces a pink flower. Other traits, such as skin color in humans, are determined by multiple genes. In the latter case, a range of colors can result from different combinations of the alleles.

Individuals that can interbreed and have fertile offspring are a *species*. An isolated group of such individuals are a *population*. The genetic change of a population over time is called *evolution*. While mixing of alleles can account for changes of traits between generations, it doesn't explain changes in traits of a population or the emergence of new species. Such processes require changes in the frequency in which an allele occurs in a population or the generation of new alleles. The frequency of genes in a population can occur by the method of controlled selection, or *breeding*, which has been used in agriculture for millennia, or can occur by similar natural processes, in what is called *natural selection*. Natural selection results from certain genetic variations among individuals of a population having an adaptive advantage in their habitat that provides them with an advantage in passing on their genes to the next generation. Thus under changing conditions of temperature, rainfall or the presence of predators, new phenotypes may become favored. Charles Darwin's observations of variations between populations of species in the Galapagos Islands called attention to the importance of natural selection.

Because of their rapid multiplication, evolution is most easily observed in populations of bacteria, which over our lifetime have evolved to be resistant to many of the antibiotics with which we previously controlled them. On a longer time scale, fossil evidence reveals about a doubling in brain size for hominid (human-like) species over the past two million years. If various populations of a species evolve differently over time, they may no longer be able to interbreed, and thus become distinct species.

Life is such a dominant factor on our planet that it does more than just react to changes in the physical world; life can also bring *about* changes in the physical world. Life and the physical world have evolved together. For example, atmospheric oxygen, so essential for our survival, would not exist were it not for evolutionary processes that produced **photosynthetic** plants. We will see that the strong similarity between all life forms at the molecular level suggests that all life evolved from a single cell. By the processes of growth, asexual and sexual reproduction and gene modification, the descendents of this cell now comprise the wonderful variation in life forms that we see around us.

Energy for Life

All the activities of life: maintaining order and complexity, homeostasis, growth, reproduction and action, require energy. All living things must devise mechanisms for extracting energy from their environments. This energy must be stored and be made available on vastly different time scales, just an instant for a surprised grasshopper to months for a hibernating bear. Almost all life forms use the same molecules to store energy: ATP, for energy on the time scale of seconds; simple sugars and amino acids, on the time scale of minutes to hours; and complex carbohydrates, fats and glycogen for longer-term storage. Our cells can only make use of energy in the form of ATP. Since ATP is produced from fuels such as glucose by reaction with oxygen, if we stop breathing, or if our blood stops circulating, cells will begin to die in a matter of seconds.

The ultimate source of energy for all of life is solar energy or the gravitational, chemical and nuclear (radioactivity) energy present at the formation of the Earth. Organisms that can directly use such energy are called *autotrophs*, or primary producers. Their source of carbon is carbon dioxide. They are either *photoautotrophs*, such as plants or cyanobacteria (sometimes called blue-green algae), which generate carbon compounds by photosynthesis, or *chemoautotrophs*, bacteria that synthesize carbon compounds from carbon dioxide using the energy stored in certain inorganic compounds. All other organisms are **heterotrophs**, which obtain their high-energy carbon and nitrogen-containing nutrients by consuming the primary producers.

Organisms are classified by their *trophic level*, which indicates the number of types of organisms that stand between them and the primary producers. For example, in my rose garden, the roses would be the first trophic level, the aphids (*herbivores*) that eat rose leaves, would be the second level, while the ladybugs (*carnivores*) that I add to the garden to eat the aphids would be the third level. Sparrows that eat the ladybugs would be the fourth level and a hawk that eats the sparrow would be the fifth level.

Organisms can generate energy either ***aerobically***, by reacting high-energy compounds with oxygen, or ***anaerobically***, by employing other oxidizers or reducing reactions. Oxygen is generally poisonous to species of the latter type. *Clostridium botulinum* is a poisonous bacterium that prefers an anaerobic environment. It is sometimes produced in food that has been canned without adequate sterilization.[105]

When Earth was formed, it contained a lot of free iron, which rapidly oxidized (rusted). As a result, the early atmosphere of Earth could not have had appreciable oxygen until almost all this free iron was removed. Because of this, scientists believe that the first life on Earth was anaerobic, consuming the small amounts of carbon-containing organic compounds produced by lightning and other stimuli. Without oxygen in the Earth's atmosphere, there could also not have been any ozone. Since ozone protects life from the harmful ultraviolet radiation from the sun, early life had to be confined to the oceans, or even more likely to stagnant pools of water.[106]

Classifying Life

Understanding the variety of life forms would be completely overwhelming, if they were not classified in some orderly manner. The science of describing, naming and classifying life forms is called ***taxonomy***. Classification is particularly important because it reveals close relationships between different life forms, indicating that one may have evolved from the other or that they share a common ancestor.

Taxonomists, not without some controversy, divide all of life into five kingdoms, called Animal, Plant, Fungus, Protist and Monera. We are, of course, all familiar with animals, plants, and fungi (including yeast, molds and mushrooms). The Protist kingdom contains a variety of,

---

[105] One of the toxins of this bacterium is used dermatalogically as the drug *Botox* to remove wrinkles.

[106] Oceanic life would have had to deal with currents, which periodically transported it near the surface, where ultraviolet radiation was intense.

mainly microscopic, creatures, such as amoeba and algae. In these first four kingdoms, members of which are called eukaryote**s**, genetic material of the cells is isolated in a separate compartment called a nucleus. The Monera, consists of single-celled bacteria, called prokaryotes, in which the genetic material is only loosely held together. Prokaryote cells are much smaller than typical eukaryote cells and have a rigid cell wall.

The classification of organisms continues with sequential subdivision of the kingdoms into phyla, subphyla, classes, orders, families, genera and finally species. For example, humans are of the animal kingdom, the chordate (with spinal chords) phylum, the vertebrate (with spine) subphylum, the mammalian (with mammary glands) class, the primate (with flexible hands and feet) order, the hominidae family (humans and great apes) and genus homo, of which only the species, *homo sapiens* — ourselves — currently survives. There are probably of the order of ten million different species currently existing on our planet, less than half of which have been described to date.

The complexity allowed by having many cells, provides organisms with the potential for new functions and new ways of prospering in their environments. This complexity comes with a price, however, since multiple-celled organisms generally reach maturity and reproduce at a much slower rate than single-celled organisms. In addition, organisms with many cells have stringent requirements for functional order — the various cells must coordinate and work together. While some multicellular organisms (e.g., slime molds) can consist simply of agglomerations of identical cells, in most multicellular organisms, cells are arranged in tissues, which in turn form organs and organ systems. These systems comprise the individual. Individuals are not islands, however, and different types of organisms often reside together in an ecosystem. For example, friendly bacteria occupy our gut and skin. Other organisms, such as ants, can only live in communities. In fact, in many ways, an anthill, or beehive can be considered as a single super organism. Humans, of course, can survive in communities of various sizes, ranging from isolated families to communities of nations.

Summary
 octoso octoso octoso

A living organism is more complex than most of the enterprises of society. It is largely homeostatic and capable of action, growth and change. A cell, consisting of a plasma membrane surrounding cytoplasm and organelles such as mitochondria, is the simplest form of life. Except for bacteria, cells store their instructions in an organelle called the nucleus. Cells can duplicate their instructions and divide, individuals can reproduce either asexually or sexually. Asexual reproduction produces clones identical to their parents, while sexual reproduction produces individuals who may have qualities that uniquely prepare them to survive in their habitat. Life requires the continual use of energy. Almost all organisms use the same molecules to store energy on different time scales. The primary source of energy for life is the sun, but it must be processed through different levels of life to be useful for different organisms. Taxonomists divide life into five families and further divide it until reaching the species, a population of interbreeding organisms.

Questions

1. The Mars explorer satellite has just returned from its mission and has collected a green powder. It is obviously of interest to determine whether this is a living material. Devise some experiments that would help you to answer this question.
2. For glucostasis, blood glucose concentration must be maintained at stable equilibrium.
    (a) Discuss the mechanisms that provide negative feedback to maintain blood glucose at its proper value.
    (b) Is this a global equilibrium or a local equilibrium, in which deviations from the proper value that are too great could drive the system to a different state?
3. What are the organs that are involved in regulating blood glucose levels in humans?
4. In this chapter we compared a cell to a chemical factory. Take some operations carried out by a chemical factory, such as: security, equipment repair or modernization, and show how a cell, an organism or a species might carry out such operations.
5. What are some advantages of storing genetic material (DNA) in a separate compartment (the nucleus), as done in eukaryotes?
6. What are some advantages of having genetic material (DNA) in the cytoplasm of a cell, as in prokaryotes?

7. Propose some mechanisms by which our body maintains thermostasis, keeping it at constant temperature.
8. Many warm-blooded animals with dense hair do not sweat. How then do they maintain constant body temperature when the surroundings are warmer than their body?
9. What would be the ratio of short to tall peas in a third generation of Mendell's experiment?
10. What is the trophic level of a human who eats meat? Of one who is a vegetarian?

16 The Components of Life

What are little boys made? Frogs, and snails and puppy dog tails, And that's what little boys are made of.

Nursery rhyme

Life is a hierarchy. An organism, like ourselves, is composed of *organ systems*, such as the circulatory system, which contains the heart and other *organs*. The heart is assembled from different types of *tissues*, one of which is the myocardium, which contains muscle cells. A cell contains many different structures, including its cytoplasm, plasma membrane, nucleus and other organelles. Each of these is made up of small and large molecules. The large molecules, such as proteins, are formed by linking together smaller molecules. These small molecules, the basic components of life, are discussed in this chapter.

Most buildings are built from just a few components, such as bricks, boards and panes of glass, arranged in different ways. In a similar manner,

> **All forms of life on Earth are constructed from very similar molecules.**

Although a variety of small molecules are important in organisms, most of these can be classified as being of a few different types. These are: amino acids, which comprise proteins; lipids, which make up cell walls and fats; nucleic acids, which constitute the information-storing apparatus of the organism; and sugars, which comprise carbohydrates and are components of structural and information-storing molecules and some proteins.

Water

One molecule, water, comprises over 50% of the mass of most living creatures on Earth, including ourselves. The properties of water that make it a uniquely suitable medium for life are: it is a small, polar molecule that dissolves, but does not react with many molecules. Other molecules do not dissolve in water, allowing the segregation of cell membranes and organelles. The rapid migration of hydrogen ions through water is also important for life processes. Water readily forms hydrogen bonds and is attracted to many surfaces, permitting aqueous solutions to rise up in tubular channels, such as exist in plants.

Liquid water has such favorable characteristics for supporting life that scientists take its presence on other bodies in our solar system as an indication they may harbor or have harbored life.[107] It is possible, however, that there are life forms that can exist in the absence of water.

Amino Acids

Amino acids are hydrocarbons that contain an amino ($-NH_2$) group and a carboxylic-acid ($-CO_2H$) group. The amino acids important in biological systems are α-amino acids, in which both the amino group and the carboxylic acid group are attached to the same carbon atom, called the alpha-carbon. The formula of a general alpha-amino acid is shown in Figure 16.1.

---

[107] Scientists are quite sure that at some time a considerable amount of liquid water existed on Mars.

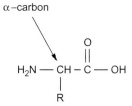

Figure 16.1. An alpha-amino acid.

"R" is the *side group* that makes each amino acid unique.

The nature of the side group, R, is key to understanding the properties of these amino acids and the proteins that are synthesized from them.

> **The side groups of amino acids are of three types: non-polar, polar and charged.**

Just about all proteins in living organisms are synthesized from only 20 α-amino acids.[108] The structures of these amino acids, identified by their names, with their three-letter and one-letter abbreviations, and classified according to type, are shown in Figures 16.2–16.4.[109]

The *non-polar amino acids*, shown in Figure 16.2, have non-polar side groups that are not attracted to water. In globular proteins, these amino acids often congregate on the inside of the structure.

Cysteine is just like serine, except with a thiol group, SH, in place of the hydroxyl group, OH. When incorporated into proteins, two cysteine molecules often lose their hydrogens and form weak disulfide bonds (-S-S-).

The *polar amino acids*, shown in Figure 16.3, have side groups that interact with water help solubilize the proteins that they constitute.

---

[108] Amino acids may be modified after they are incorporated into proteins.
[109] There are two additional amino acids that are occasionally used by organisms.

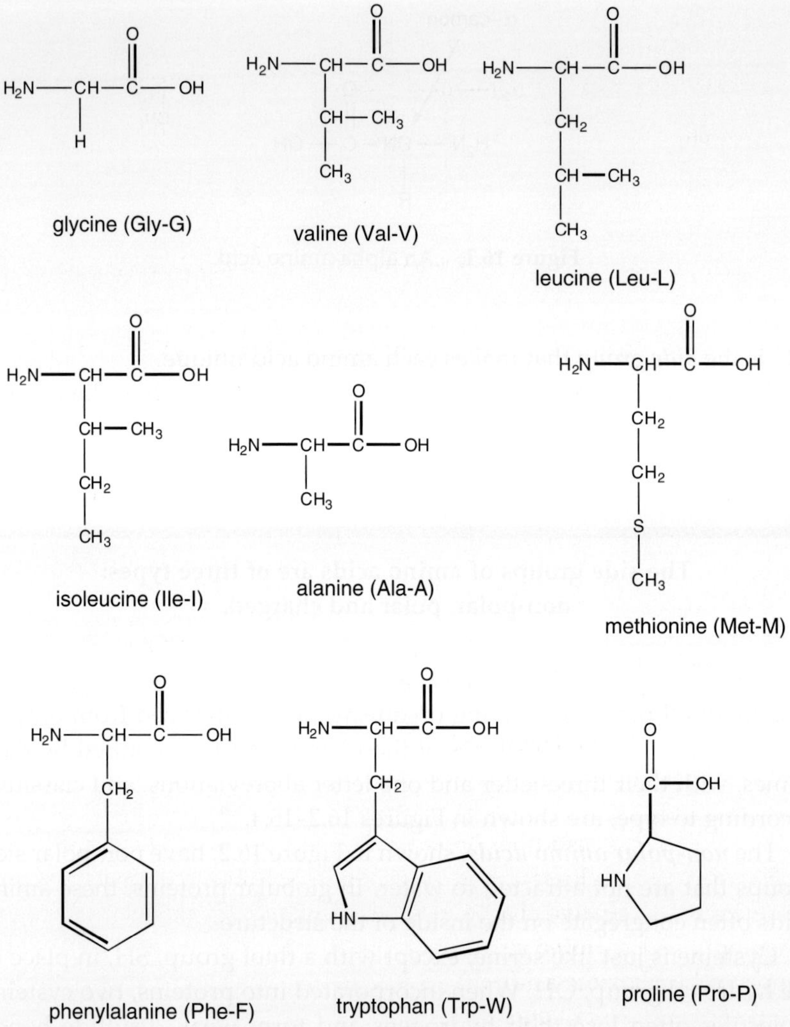

Figure 16.2. Non-polar amino acids.

The *charged amino acids*, shown in Figure 16.4 have side groups that are ionized at pH values generally found in human physiological solutions (7.0–7.5). These groups are also ionized when the amino acids are incorporated into protein molecules.

The positively charged NH_3^+ group in lysine results from a basic NH_2 group that picks up a hydrogen ion from water. A similar process occurs

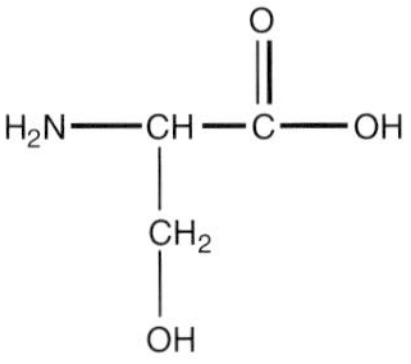

serine (Ser-S)

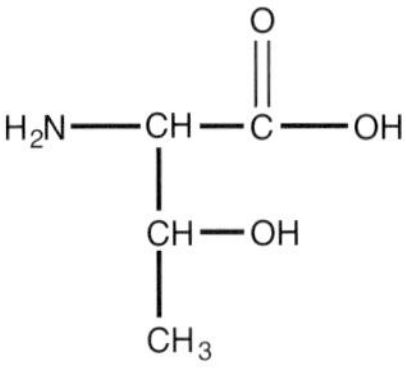

threonine (Thr-T)

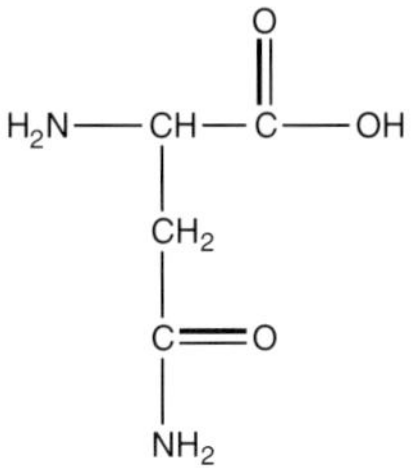

asparagine (Asn-N)

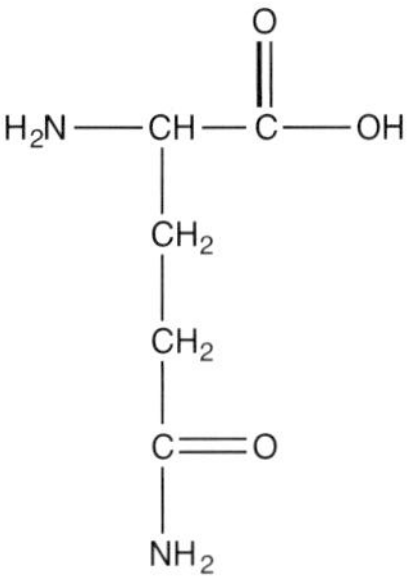

glutamine (Gln-Q)

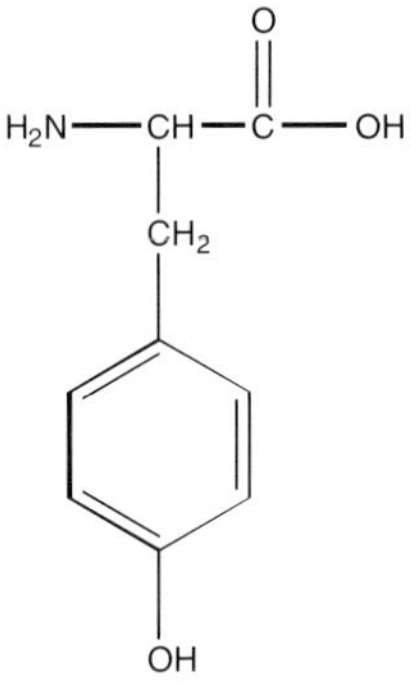

tyrosine (Tyr-Y)

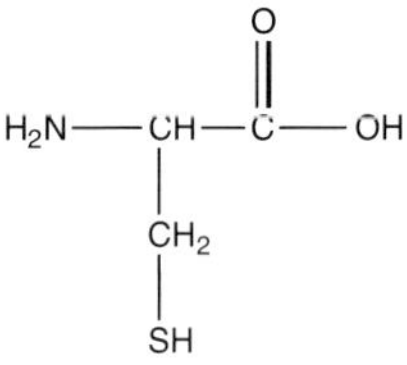

cysteine (Cys-C)

Figure 16.3. Polar amino acids.

with arginine and histidine. Aspartic and glutamic acids have carboxylic acid groups that release a hydrogen ion in solution, leaving these side groups with a negative charge. Besides these side groups being charged, when unlinked α-amino acids are in physiological aqueous solutions, the amine and carboxylic acids that reside on the α-carbon are both in their charged forms ($-NH_3^+$ and $-CO_2^-$).

Humans employ the twenty amino acids listed above, but only eleven can be synthesized in our bodies. The other nine, called the ***essential amino acids***, are: leucine, isoleucine, methionine, valine, phenylalanine, tryptophan, threonine, lysine and histidine. These must be included in our diets. Because lysine and tryptophan are found in very limited amounts in plant proteins, vegetarians must be very careful that their diets include adequate amounts of these two amino acids.

The simplest amino acid, glycine, has R equal to H, and therefore two hydrogen atoms on the α-carbon. All other amino acids have only a single

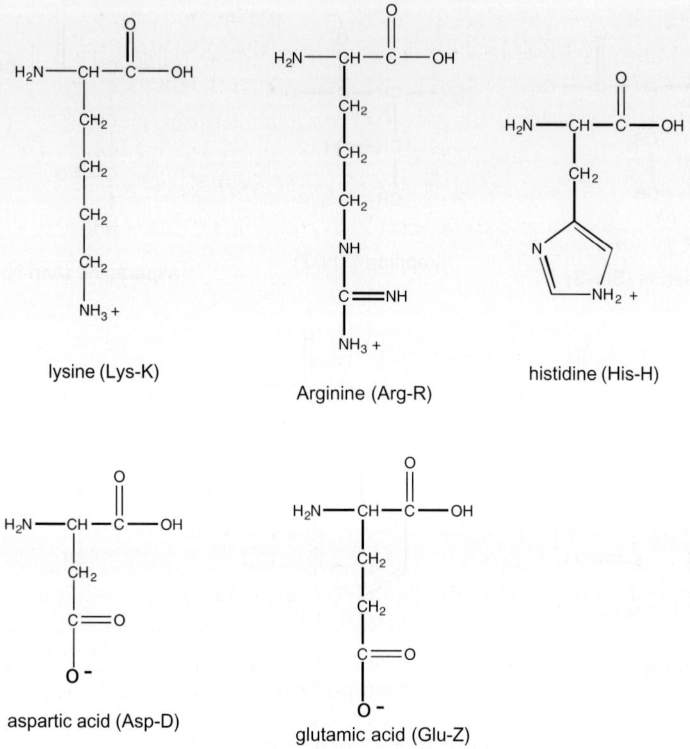

[Structures of lysine (Lys-K), Arginine (Arg-R), histidine (His-H), aspartic acid (Asp-D), and glutamic acid (Glu-Z) shown]

Figure 16.4. Charged amino acids.

hydrogen on this carbon, which is therefore substituted by four different groups. The three-dimensional tetrahedral structure of such an amino acid is shown in Figure 16.5, with two different arrangements of its substituents.

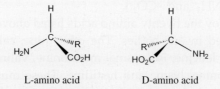

L-amino acid   D-amino acid

Figure 16.5. Three-dimensional structures of a general amino acid.

Are these two structures identical or different molecules? Note that they are mirror images of each other. Their relationship to each other is the same as your left and right hands. Just as for your hands, there is no way to rotate these molecules so as to superimpose one on the other. We say they are *chiral* molecules. They are two different species, which we indicate as the D- and the L-amino acids. The difference between these is that when you view the central carbon atom from hydrogen, in the clockwise direction, the order of the groups is: CO_2H, NH_2, R for the L isomer and CO_2H, R, NH_2 for the D isomer. Molecules that come in mirror image forms are called *optical isomers* or *enantiomers*. Such molecules usually have the ability to rotate the plane of polarization of light (the plane that contains its electric field). Optical isomers occur whenever a carbon atom is attached to four different chemical groups.

If you put your two hands on a flat surface, they both interact in the same way with the surface. The same thing happens if you put your two hands around a sphere or a cube, both non-optically active objects. In fact enantiomers will experience identical interactions with any non-optically active entity. When an optically active substance is synthesized from non-optically active reactants, the product is a *racemic mixture*, which contains equal amounts of the two enantiomers. However, if such a mixture is allowed to form crystals, each of the enantiomers will more strongly interact with its own type, and the molecules will segregate into separate crystals of the two mirror image forms, as shown in Figure 16.6. In some

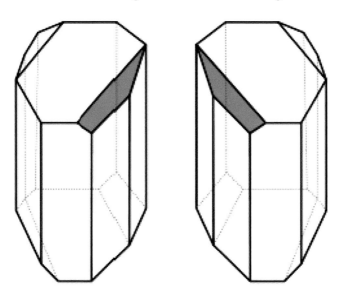

Figure 16.6. Crystals of tartaric acid enantiomers.

cases, these crystals are large enough so that they can be separated by hand. This is how Louis Pasteur (1822–1895) first discovered optical isomers; he found that solutions of the separated crystals of a salt of tartaric acid (HO$_2$CCHOHCHOHCO$_2$H) rotated the plane of polarization of light in opposite directions.[110]

Optical isomers only interact differently with chiral objects. For example, with your right hand you can shake other right hands, but not left hands. It is interesting that living things on Earth only contain and make use of L-amino acids. Most of our enzymes can only interact with this isomer. If we traveled to another planet, where life was based on D-amino acids, our bodies couldn't digest foodstuffs, and we would probably starve to death. It is easy to see that once a particular optical isomer becomes favored in biological systems, its preponderance can expand and propagate. The mechanism whereby this preference would occur in the first place is currently an active field of research. Perhaps it's just the result of a fluctuation.

Sugars

In addition to being the primary energy-providing fuel of most life, sugars play an important role in structural and information-storage molecules and some proteins. Sugars, the simplest carbohydrates, can be drawn as cyclic ethers in which several of the carbon atoms contain hydroxyl (alcohol) groups.[111] We will be most interested in sugars containing five or six carbon atoms, which are called *pentoses* and *hexoses*, respectively. The formulas of D-glucose, the most important sugar in our bodies, and D-fructose, a major component of corn syrup, are shown in Figure 16.7. D-glucose is a hexose and D-fructose is a pentose.

In these molecules, almost all of the carbon atoms are attached to four different groups, and thus there are many different optical isomers of these molecules. Each of these variations has distinct chemical properties.

---

[110] In some cases, 1:1 crystals of the two isotopes are formed. This is interesting, since when a nearly racemic mixture crystallizes, the predominance of the isomer present in slight excess is greatly enhanced in the small amount of solution in equilibrium with the crystals.

[111] The structures of sugars can also be written as linear aldehydes or ketones.

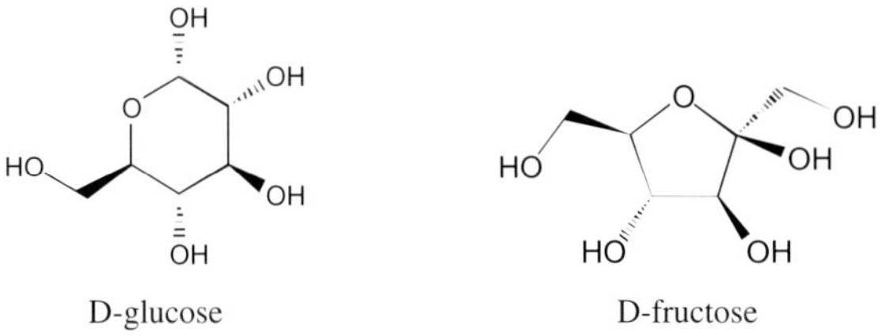

The Components of Life 249

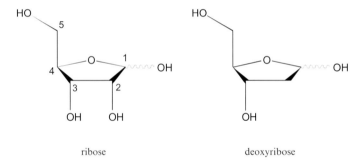

Figure 16.7. Formulas of several sugars.

Sugars in which one of the OH groups has been replaced by a hydrogen atom are known as ***deoxy sugars***. The relationship between ribose (part of RNA) and deoxyribose (part of DNA) is shown in Figure 16.8, where the numbers are used to designate the different carbon atoms of these molecules.

Figure 16.8. Formulas of ribose and deoxyribose.

Sugars in which an OH group is replaced by an NH_2 group are known as amino sugars, and are important in biological systems. Glucosamine, an amino sugar often taken as a supplement to improve joint health, is shown in Figure 16.9.

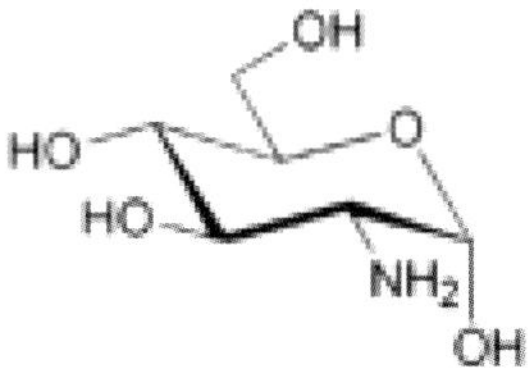

Figure 16.9. Glucosamine.

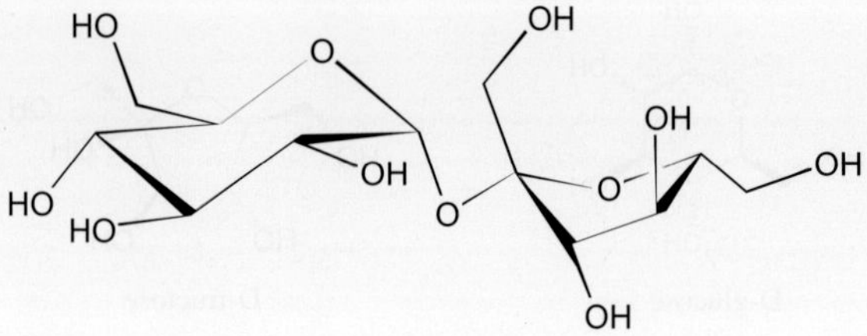

Figure 16.10. Formula of sucrose.

Many biologically important molecules are formed by linking simple sugars (monosaccharides) through an ether bond. For example, common table sugar, sucrose, shown in Figure 16.10, is a ***disaccharide***, formed by combining one glucose and one fructose molecule in this manner.

Polysaccharides, polymers formed from sugar molecules, will be discussed in the next chapter.

Lipids

The term *lipid* encompasses many biological molecules that have low water solubility. In general, such molecules have extensive regions of unsubstituted hydrocarbon. ***Fats***, one class of lipids, are esters of glycerol (Figure 12.14 of Chapter 12) with three fatty acids. The fatty acids can be completely saturated, containing as much hydrogen as they can, or they can contain varying degrees of unsaturation (multiple bonds). Animal fats, containing a large fraction of saturated hydrocarbon chains, are usually solids, while vegetable oils, contain larger amounts of unsaturated hydrocarbon chains and are usually liquids. Fats made from cis fatty acids, such as oleic and linoleic acids, discussed in Chapter 12, are much more healthy for our circulatory systems than those made from trans fatty acids.

Phospholipids, with two fatty acids and a phosphoryl group attached to glycerol, were discussed as surfactants in Chapter 14. The structures of a saturated fat and a phospholipid, both formed from stearic acid are shown in Figure 16.11.

Steroids are molecules based on a particular four-ring hydrocarbon. Some common steroids are shown in Figure 16.12.

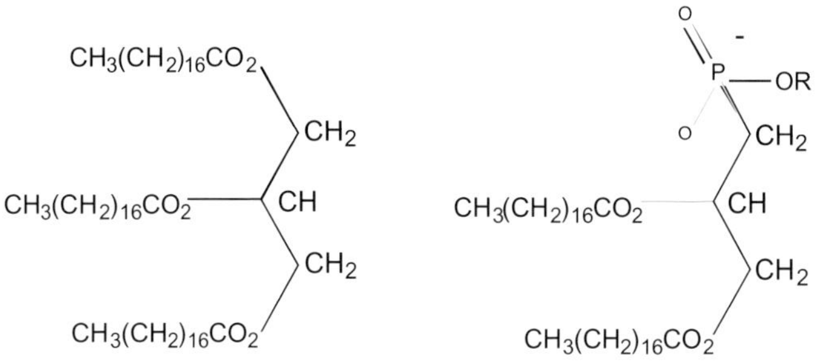

Figure 16.11. A saturated fat and a phospholipid.

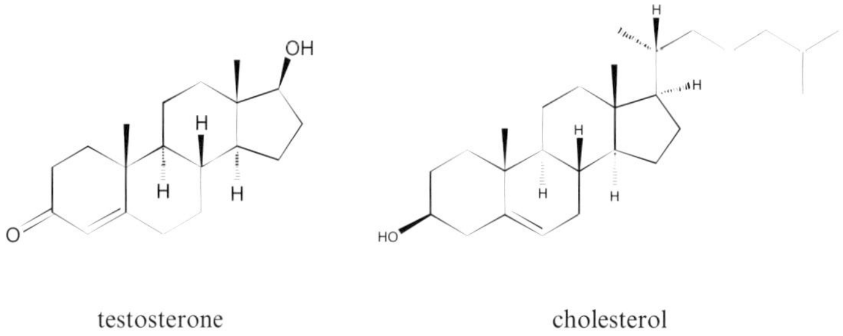

Figure 16.12. Some common steroids.

Cholesterol is used to provide strength and rigidity to cell membranes. Unfortunately, it can also deposit in blood vessels, clogging them and impeding blood flow. Other steroids, such as testosterone, are hormones, which control many metabolic pathways in biochemical systems.

Information-Coding Bases

There are five molecules with basic characteristics that are components of *DNA* and *RNA*, the molecules that store and transfer the genetic information that directs the processes of life. The formulas of these *information-coding bases* are given in Figure 16.13.

Adenine and guanine, with double rings, are called *purines*, while the molecules with single rings are known as *pyrimidines*. These molecules are called bases, because in aqueous solution, hydrogen ions attach to the lone pair electrons on nitrogen, much as they do with ammonia. The

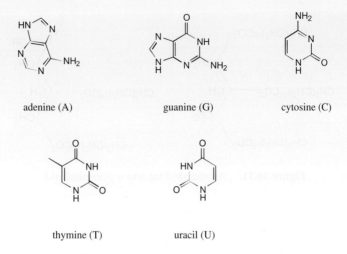

Figure 16.13. The information-coding bases.

hydrogen-bonding possibilities of these molecules are key to their biological functions.

High-Energy Molecules

Many of the activities of life involve endothermic reactions, those that require energy.[112] These include: providing motion and synthesizing large molecules. In almost all cases, these processes proceed by coupling them to one of just a few energy-releasing chemical reactions. The most important of these is the hydrolysis of *ATP* (adenosine triphosphate) to form ADP (adenosine diphosphate) or AMP (adenosine monophosphate), as shown in Figure 16.14.

ATP is formed from the base, adenine, attached to the sugar, ribose (This combination is called adenosine.), which is attached to three phosphoryl ($-PO_3^{2-}$) groups. The bonds between the phosphoryl groups in ATP are high-energy bonds — due to the electrostatic repulsion of these charged moieties. However, hydrolysis of these bonds requires surmounting

---

[112] More exactly, these reactions are endergonic; they require *free energy*.

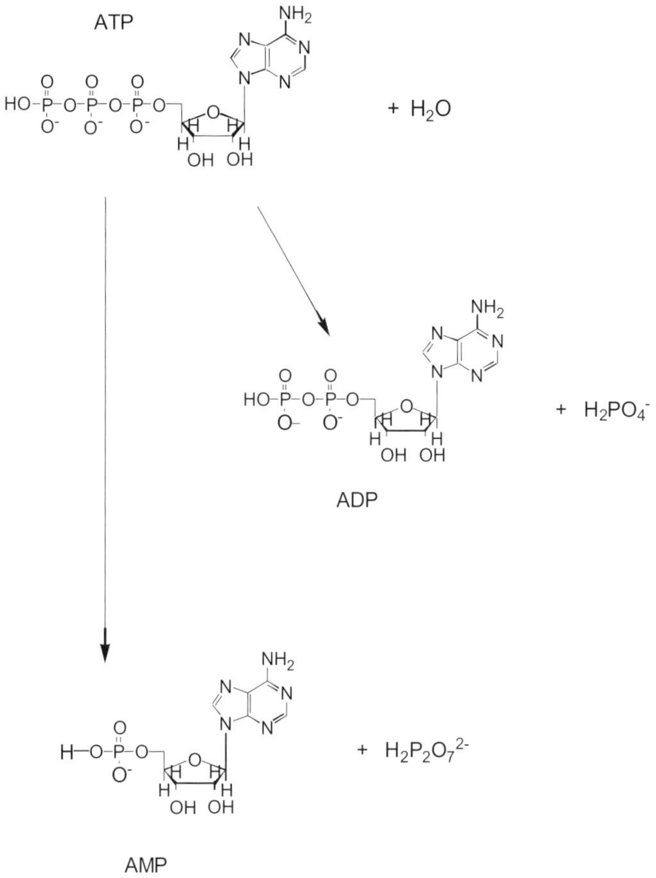

Figure 16.14. Hydrolysis of ATP.

a considerable energy barrier, so that in the absence of a suitable enzyme catalyst, the ATP molecule is kinetically stable. Coupling with an endothermic reaction is usually achieved by the phosphate produced in ATP decomposition being transferred to a enzyme that catalyzes the endothermic reaction. Incorporating this highly charged group usually results in a considerable conformational change in the enzyme.

The molecule cyclic-AMP (cAMP), widely used in the body for intracellular signaling, is shown in Figure 16.15. cAMP is synthesized in the body from ATP.

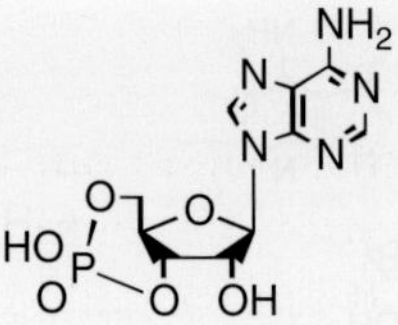

Figure 16.15. cyclic-AMP.

The multiple uses of the adenine moiety in organisms greatly simplifies their requirements for synthesizing organic compounds.

Neurotransmitters

There are a large number of small molecules that carry signal within neurons (the information-carrying cells of the nervous system) or between neurons and their target cells (other neurons or muscle or glandular cells). A few *neurotransmitters* are shown in Figure 16.16.

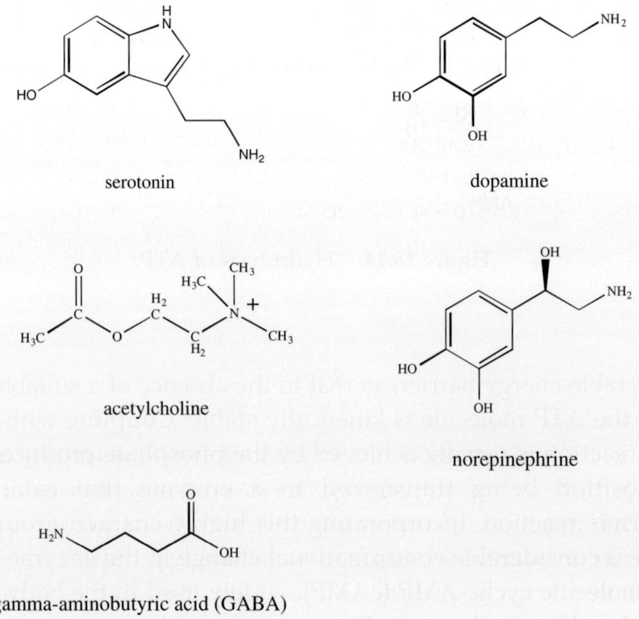

serotonin

dopamine

acetylcholine

norepinephrine

gamma-aminobutyric acid (GABA)

Figure 16.16. Neurotransmitter molecules.

Origin of Building Blocks

Just as you would be foolish to try to put together a building without bricks, planks and electrical wire, life could not have begun its evolution on Earth until a sufficient repertoire of building-block molecules had been formed. When Earth condensed from the rocky material and gases of the solar nebula, it was too hot for such molecules to be present. Upon cooling, water condensed into oceans and smaller bodies of liquids, and inorganic materials dissolved in these. Compounds such as N_2 and CO_2 with low water solubility remained mainly in the vapor phase along with water vapor. In addition, H_2, NH_3 and CH_4 may have been present. There probably was no appreciable concentration of O_2, since free oxygen would rapidly react with the abundant iron on the Earth's newly formed surface. Could such materials have been converted to components of life, such as amino acids, lipids and nucleotide bases?

In 1953, Stanley Miller performed the first experiment addressing this question. Miller circulated a mixture of H_2, NH_3, CH_4 and water vapor and subjected it to repeated electric sparks, to mimic the effect of lightning on the Earth's primordial atmosphere. After just a week, he analyzed his mixture and found that it contained considerable concentrations of a suite of amino acids. In 1961, Juan Oro performed a similar experiment, and showed that when hydrogen cyanide (HCN) was added to the mixture, the nucleotide base, adenine, was produced in large concentration. Similar experiments have been performed using other types of energy, such as UV radiation. Undoubtedly, if building-block molecules can easily be formed in such short-term experiments, there is little difficulty explaining their creation over the hundreds of millions of years of the pre-biotic Earth.

Summary

Besides water, organisms are mainly composed of a few types of small molecules that may be linked together in polymeric forms. Proteins are synthesized from 20 L-enantiomers of α-amino acids, which may be nonpolar, polar, or charged, depending on their side group. Nine of these are essential; they must be included in our diets. Simple sugars, mainly hexoses and pentoses, are combined to form carbohydrates and often linked to proteins and cell membranes. Lipids, molecules with low water solubility, include energy-storing fats, as well as phospholipids, the major components of cell membranes. Five basic molecules with unique hydrogen-bonding properties are the key to information storage in life. Adenine, one of these bases, is part of ATP, which is used to supply the energy for many reactions in cells and cAMP, which is used for intracellular signaling. A number of small neurotransmitter molecules transfer signals in the nervous system. Experiments have shown that the components of life can be formed by energetically exciting small gaseous molecules probably present in the atmosphere of early Earth.

Questions

1. Show that tartaric acid ($HO_2CCHOHCHOHCO_2H$) contains carbon atoms attached to four different groups.
2. Draw a three-dimensional structure of L-leucine.
3. Draw a three-dimensional structure of glycine and show that it is identical to its mirror image and thus does not have optical isomers.
4. Count the carbon atoms in the formula of D-glucose shown in Figure 16.8 and show that it is a hexose.
5. Write down molecular formulas for D-glucose and D-fructose from the structural formulas shown in Figure 16.8.
6. Draw two optical isomers of glucose besides the D-glucose isomer in Figure 16.8.
7. Which of the neurotransmitters shown in Figure 16.16 is an amino acid? Is it an α-amino acid?
8. Our bodies often simplify synthesis by using the same moiety in a number of different molecules. Find a component that is a large part of a number of biologically important molecules.

17 Molecules of Life

> ...*we have failed to discover any aspect of life, ... whose explanation appears today to lie beyond the ultimate capabilities of physical science ...*
>
> Dean Wooldridge

The statement above summarizes current scientific understanding of life: that it requires no principles other than those commonly used to describe inanimate matter. Chemistry and physics are considered sufficient to explain life, which is comprised of ordinary atomic and molecular building blocks, interacting in usual ways. It differs from non-living materials only in its complexity, order and information content.

How can the relatively small number of components of life that we have discussed in the last chapter give rise to all the variety of life that we observe on this planet — from bacteria and mushrooms to elephants and redwood trees? Moreover, how can so few chemical compounds accomplish what is necessary for the functions of all organisms, namely homeostasis, action, growth and change, as well as the activities specific to particular species? The answer to these questions is that, in addition to employing the components of life as individual small molecules, living systems also use them as building blocks, linked together to form a staggering number of different polymeric molecules. Some of these polymers are used by many different species, while others are unique to a particular species, and may vary slightly among individuals of the species. In this manner results the endless variety of the world around us.

Just as in a house, a certain organization of simple rectangular bricks produces a fireplace, capable of confining a fire and directing its combustion products outside, new properties emerge when the components of life are polymerized. Probably the reason all life uses the same building blocks is because the means of synthesizing these molecules was developed very early in evolution. Thereafter, it was much easier to achieve change by linking the components in different arrangements, rather than by developing the means to synthesize and employ new components. Compared to the organization of a building, the organization of life is both less and more complicated. Less complicated, because in most biological molecules the arrangement is one-dimensional, i.e., that of a linear polymer,[113] while in buildings, bricks, tiles, etc. are arranged in two- or three-dimensional patterns. More complicated because, as we will see, the patterns of life are much more complex than those used in buildings. In this chapter the structure of proteins, polysaccharides; and polynucleotides — the most important polymeric materials of life — will be described.

Proteins

Primary structure

A protein is a linear polymer of amino acids joined by peptide bonds. A peptide bond, R-NH-CO-R', is formed when the $-NH_2$ group of one amino acid reacts with the $-COOH$ group of another amino acid, eliminating water (called a condensation reaction). A chain of amino acids joined in this way is called a ***polypeptide***. The approximately 100,000 different proteins in our body make up about half its dry weight. A typical protein is made up of about 300 amino acids. Even though almost all living things make use of only 20 different amino acids, there are still 20^{300} different proteins of this size — an inconceivably large number! The order of amino acids in a protein is known as its ***sequence*** or its ***primary structure***. For example, the primary structure of insulin, using the single-letter designation of amino acids, is shown in Figure 17.1.

---

[113] Biological molecules often involve linear chains linked together to produce a non-linear structure.

Molecules of Life 259

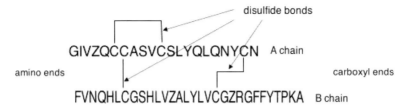

Figure 17.1. The amino-acid sequence of human insulin.

Insulin is comprised of two linear chains, the 30-amino acid B chain and the 21-acid A chain, joined in two places by disulfide bonds at cysteines. The sequence was determined by Frederick Sanger (Nobel prizes in chemistry in 1958 and 1980).

Some proteins serve as structural materials in skin, fingernails and other tissues and provide motion by their incorporation in muscles, ligaments and tendons. Others are enzymes that play a central role in catalyzing and regulating the speed of practically every reaction in all organisms. Structural proteins are generally *fibrous* (extended) and contain repeating sequences, which lowers their information content compared to enzymatic proteins, which are *globular* (roughly spheroidal). Repeating sequences are much less common in enzymatic proteins, and therefore they have much higher information content than structural proteins.

Similar proteins perform the same task in different species. In many cases, a small change in the sequence of a protein within or between species does not inhibit its ability to perform its function. For example, human diabetics have been treated with insulin obtained from cows and pigs. Bovine insulin differs from the human species in three amino acids, and porcine insulin in only one. Such changes usually involve replacing an amino acid with one of similar electrical properties, for example, non-polar alanine for non-polar valine, and are called *conservative replacements*. In some cases, however, replacement of even a single amino acid in a protein can substantially influence its ability to carry out its function. An example of this is sickle-cell anemia, a disease prevalent in groups of African ancestry. In those with this disease, hemoglobin, the oxygen-transporting protein in blood, differs from normal hemoglobin by replacement of a single amino acid; a polar glutamine is replaced by a non-polar valine. While the substituted protein is still adept at transporting oxygen, its modified structure results in the hemoglobin molecules forming fibers, which deforms the shape of red blood cells (erythrocytes). The deformed cells tend to clog smaller capillaries at various positions in the body, causing great pain and death of tissue. The phenotype of sickle cell anemia is only expressed in individuals who are homozygous for the modified allele. The gene for sickle-cell anemia has probably persisted in the human population because even in the heterozygous form it provides resistance to malaria, a mosquito-born disease that is endemic in Africa.

Although the peptide bond is marginally unstable — at equilibrium only about 1% of a peptide will not be hydrolyzed — hydrolysis of peptides is very slow. This allows formation and destruction of proteins in our bodies to be under enzymatic control and proteins to be synthesized and destroyed as needed.

Secondary structure

The three-dimensional native structure of a protein is usually discussed at successive levels of organization: its secondary, tertiary and quaternary structure. The *secondary structure* of a protein is the short-range structure of its peptide backbone, determined by its sequence. Proteins with no recognizable secondary structure are called *amorphous*. The two most common motifs in short-range protein structure are helices and pleated sheets. A typical helix (called the *alpha helix*), shown in Figure 17.2, has a pitch of 0.54 nm, with 3.6 amino acids (called residues) per turn.

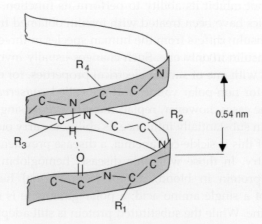

Figure 17.2. Typical protein alpha helix.

This structure is an optimum geometry for the formation of a strong hydrogen bond between an N-H group of one amino acid and a C=O group of a second amino acid located four amino acids away on the same polypeptide chain. The fibrous protein keratin, the structural material of human hair and nails, contains a large proportion of the two smallest amino acids, glycine and alanine, which give keratin a hierarchy of structural elements. First, pairs of keratin molecules wind around each other to form protofibrils. The protofibrils then gather together in groups of nine protofibrils around a central two protofibrils, which forms a microfibril. The microfibrils in turn are packed together to form macrofibrils, such as a human hair. In human hair there is extensive cross-linking between different keratin molecules through disulfide bonds at cysteine residues. When hair is treated to produce a "permanent wave," these bonds are split, the hair is curled and then the bonds are reformed.

Collagen is a fibrous protein of great strength that forms the basis of connective tissues such as bone, cartilage and teeth. It also provides much of the strength of skin and blood vessels. Although there are varieties of collagen among different species, they are all characterized by polypeptides having a very high percentage of two amino acids, glycine and proline. After the polypeptide is assembled, many of the prolines are converted to hydroxyproline, an example of *post-translational modification* of amino acids. Such transformation is essential for stabilizing collagen's helical structure, which is somewhat different from the alpha helix. The strength of collagen fibers results from three of these helical molecules wrapping around each other to form a tropocollagen fiber held together by hydrogen bonds, as shown in Figure 17.3, and also by cross-linking of these molecules at cysteine residues. There are number of diseases that are characterized by abnormal bones and tissues that result from an inability of collagen to form such cross-linked trimers.

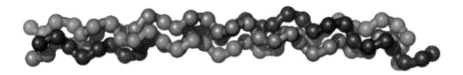

Figure 17.3. A tropocollagen fiber.

In bones the toughness of collagen fibers is augmented by the deposition of crystals of calcium phosphate. In a living bone, these crystals are continually being dissolved and reformed. Osteoporosis, a disease characterized by very weak bones, results when the rate of dissolution of the crystals exceeds their rate of deposition for a considerable period of time.

In teeth, calcium hydroxyapatite, a salt of calcium, hydroxide and phosphate ions is deposited in the collagen matrix. If fluoride salts are included in one's diet during tooth formation, tooth enamel also contains calcium fluoroapatite, which strengthens teeth, since it is less soluble than calcium hydroxyapatite. Fluoride for hydroxide substitution can also be obtained by topical application of soluble fluorides by dentists. Overexposure to fluoride during tooth formation years, however, produces dental fluorosis, a condition that mottles and weakens teeth as well as having similar effects on bones.

An alternative motif for the secondary structure of proteins is the *beta-pleated sheet*. As shown in Figure 17.4, in a beta sheet, hydrogen bonds form between neighboring parallel polypeptide chains, rather than within a single chain, as in the helical structures.

Globular proteins often contain segments of helices or beta sheets, with the helical segments usually only being three or four turns long. Since the beta sheets in globular proteins form from anti-parallel chains within the same molecule, these protein structures must also incorporate *loops*, so the chains can reverse direction, as shown in Figure 17.5.

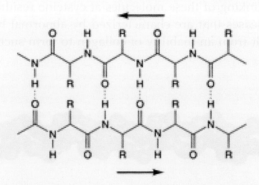

Figure 17.4. Structure of a beta-pleated sheet.

Figure 17.5. Direction of chains in beta-pleated sheet.

Tertiary structure

The *tertiary structure* of a protein is the three-dimensional arrangement of its coils, sheets and amorphous regions, as well as the positioning of its side groups. Whether the protein is involved in enzymatic catalysis of a chemical reaction or as a receptor or transporter of a substance, its function is intimately related to its tertiary structure. Globular proteins are usually arranged with nonpolar amino acids in their interior and charged amino acids on their surface. Polar amino acids can be either in the interior or on the surface, but when they are in the interior, they are usually involved in hydrogen bonding that gives the protein its shape. Disulfide bonds, -S-S-, between cysteine amino acids in the interior parts of a protein also serve to stabilize its three-dimensional structure. The tertiary structure of insulin is shown in Figure 17.6.

Quaternary structure

Globular protein molecules usually have a negative charge, which causes them to repel each other and avoid sticking together, allowing the monomer

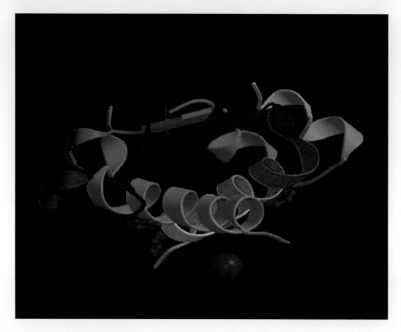

Figure 17.6. Structure of insulin molecule. A number of helical regions can be seen in this protein molecule.

to carry out its biological function. In some cases, however, an active protein molecule is comprised of a number of protein subunits that enhance each other's effectiveness in their biological function. This is the case with the oxygen-transporting protein, hemoglobin, shown in Figure 17.7.

Hemoglobin is comprised of four subunits, two sets of two identical molecules (red and blue), each bonded to an oxygen-containing heme molecule (green). The hemoglobin subunits have an *allosteric effect* on each other, i.e., binding of an oxygen molecule to the iron atom of one of the four subunits produces a subtle change in shape of the other subunits that enhances their ability to bind oxygen. Such change allows hemoglobin to more smoothly take up oxygen in the lungs and deposit it in oxygen-depleted tissues.

Insulin molecules tend to produce dimers in solution. In the presence of Zn^{2+} and Ca^{2+} in beta cells, insulin tends to form hexamers. These forms of insulin are shown in Figure 17.8.

The large hexamer molecules diffuse slowly out of the beta cells, enhancing their ability to store insulin over longer times.

Molecules of Life 265

Figure 17.7. Structure of human, adult hemoglobin.

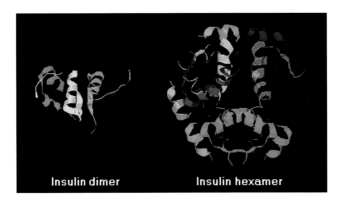

Figure 17.8. Dimeric and hexameric forms of insulin.

Proteins as Enzymes

The tertiary structures of globular proteins are usually flexible, facilitating their action as enzymes or receptors. An enzyme is a catalyst that accelerates a chemical reaction in a living organisms many-fold. Some enzymes increase the speed of selected reactions by as much as sixteen orders-of-magnitude (a factor of 10^{16}) over their rates in pure water. The reactant(s) of the catalyzed reaction are called the *substrate(s)* of the reaction and they bind to the *active site* of the enzyme. The active site often shows a complementary structure (as a key is to a lock) to that of the substrate, facilitating its binding by means of hydrogen bonds. After binding, an enzyme usually undergoes a change of conformation that stresses the substrate(s) and reduces the activation energy of the reaction. Likewise, a protein that enables transport of a polar or a charged species through the cell membrane, firsts binds the species and then undergoes a conformational change that allows its passage through the membrane.

A very common type of enzyme is a kinase, which transfers a phosphate group from a high-energy phosphate compound, such as ATP, to the enzyme. This often results in a change of its shape and permits it to carry out a function, such as opening of an ion channel in a cell membrane. Hundreds of different kinases are known, and they are often are an attractive candidate for drug design.

Many protein enzyme molecules will not catalyze their reaction unless they are bound to a *cofactor*. Cofactors may be small organic molecules called *coenzymes*, or they may be metal ions. For example, hemoglobin, the oxygen-transporting protein in blood, requires an iron atom. Many *vitamins* are coenzymes that we need, but cannot be synthesized in our body.

The secondary and tertiary structure of a globular protein in its native state is a particular low-free-energy structure, usually the structure of lowest free energy in its biological aqueous solution.[114] As the conformation of the protein is varied, its free energy has, in addition to this global minimum, a large number of other local minima. The energy landscape of a protein, which depends on a large number of coordinates, is diagramed as a simple one-dimensional curve in Figure 17.9.

A large protein molecule has so many conformations, that it is unlikely that its structure samples all these conformation in the process of finding the native state. More likely, sections of the protein first attain their individual lowest-energy conformation and then these interact to produce a low-energy conformation of the entire molecule. Thermal

---

[114] That is, the solution at body temperature with the pH, and salt concentration of the organism.

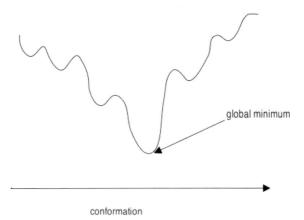

Figure 17.9. Free energy of a protein as a function of its conformation.

energy provides sufficient fluctuations of conformation, so that the molecule does not get "trapped" in a local minimum of free energy.

A number of deadly diseases have been discovered that are due to protein misfolding. These include scrappie in sheep, mad-cow disease in cows and kuru and Creutzfeldt-Jakob disease in humans. These diseases are collectively called spongiform encephalopathologies, since they are characterized by a sponge-like appearance of brain tissues. They all involve the deposition of a structural variant of a natural protein in the brain. The variant has the same primary sequence as the natural protein, but folds with a large amount of beta-sheet structure, which renders it insoluble and resistant to decomposition. Such proteins were called *prions* by Stanley Prusiner (Nobel prize in medicine in 1997), and diseases that are caused by such proteins are called prion diseases. According to current theories, prion molecules act as templates for the folding of normal protein molecules in the brain. The normal protein is directed to the conformation of the template and then finds itself also insoluble and resistant to decomposition; it thus builds up in the brain.

Prions may form spontaneously, or may be transmitted by ingesting material containing the misfolded protein; no bacteria or viruses are involved. For example, mad-cow disease has been attributed to ground up parts of scrappie-infected sheep being included in cow feed. In the human diseases: Kuru transmission, has been ascribed by Carlton Gajdusek (Nobel prize in medicine, 1976) to ritual cannibalism in the New-Guinean tribes in which it has been found, while a variant of

Creutzfeldt-Jakob disease has been attributed to eating meat from mad-cow or scrappie-infected animals. There are also genetic differences between individuals, which result in some people producing variants of proteins that are more susceptible for producing prions.

Alzheimer's disease involves the build-up of insoluble protein particles, both within (neurofibrillary tangles) and outside of (amyloid plaques) brain cells. While the disease may be inherited, it also occurs sporadically as people age, and is the leading cause of dementia in aging human populations. It has tentatively been attributed to the spontaneous conversion of a normal brain protein to an insoluble form that cannot be eliminated from the body.

The structure of a protein permits it to perform its functions in the body. Since the folded structure of a protein is highly ordered, it has low entropy. Protein unfolding, or **denaturation**, produces a random, fluctuating structure, which increases entropy and is favored by heating.[115] Most proteins denature at temperatures considerably below 100°C. For example, when we boil an egg, we denature its protein molecules, and these entangle in their unfolded state. Thermophilic bacteria that live in hot springs or in hot geothermal vents on the ocean floor have been found to survive temperatures as high as 120°C. Although their proteins are constructed from the same 20 amino acids as used by other species, they involve sequences that provide high-temperature stability.

Polysaccharides

Polysaccharides, or glycans, are linear or branched polymers of sugars that are used to store chemical energy for biological systems. They surround and fill the **extracellular matrix**, which provides support for the body's cells. Polysaccharides also regulate the activity and direct the movement of proteins. The properties of polysaccharides depend, not only on the particular sugars of which they are composed, but also on the precise way in which these sugars are linked. As an example of this, **cellulose** and **starch**, two polysaccharides formed from glucose, are shown in Figure 17.10.

---

[115] This effect is countered somewhat by the order of the water solvent increasing when the protein unfolds — producing a decrease in entropy. This is called the **hydrophobic effect**.

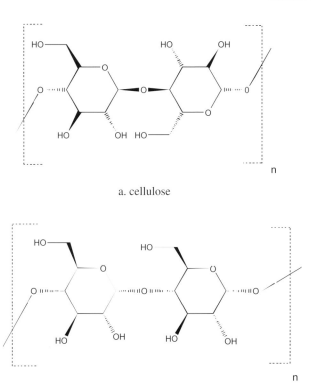

Figure 17.10. Structures of cellulose and starch.

Cellulose, the structural material of plant cell walls, probably comprises half of the carbon in the biosphere. In this polymer, the glucose monomers are linked as shown in Figure 17.5a. Although the intestinal tracts of humans and other vertebrates do not contain enzymes capable of decomposing cellulose into glucose, bacteria that live in the digestive tracts of herbivores, such as cows, can decompose this molecule. This process is quite slow, however, and requires that the herbivores either consume very large amounts of plant material to compensate for its inefficiency, or have a separate stomach where bacterial digestion can occur over a considerable period of time.

Starch is another polysaccharide formed from glucose monomers. In starch the monomers are linked as shown in Figure 17.10b. These linkages are attacked by alpha-amylase, an enzyme in human saliva and stomach and intestinal juices. Starch provides a major component of the caloric input of our species. Starch contains branched, as well as linear, polymer. Glycogen, which is similar to starch, but contains much more branched polymer, is the means by which animals store energy. In humans, in the presence of insulin and glucagon, glycogen stores are built up in the liver and muscles when our blood glucose level is high and are depleted when our blood glucose level is low.

Glycoproteins are combinations of carbohydrates with proteins. Often, part of the protein portion of these molecules is composed of nonpolar amino acids, which facilitates its dissolving in the lipid of cell membranes. Short chains of carbohydrates are attached to the remainder of the protein, stabilizing this part of the molecule in the aqueous extracellular fluid. This arrangement allows the glycoprotein to act as a receptor, collecting particular molecules in the aqueous solution and facilitating their transport into the cell.

The carbohydrates of glycoproteins are composed of a wide range of carbohydrates, arranged in many different configurations. This provides a unique "fingerprint," whereby the cells of the immune system can distinguish between native and foreign cells in the body. For example, carbohydrates attached to human red blood cells are the basis for the classification of blood into A, B, O and AB groups. Blood cells have either A-type or B-type attached carbohydrates, or both (AB) or neither (O) of these. Individuals with A group blood have *antibodies* (protein molecules) that bind cells of the B group, those with B group blood have antibodies that bind cells of the A group and those with O blood have antibodies that bind cells of both A and B group blood. Since antibodies can bind more than a single cell, they will clump together red bloods that they recognize. When giving blood transfusions, it is most important to avoid transfusion reactions — the clumping together of the transfused red blood cells by the recipient's antibodies. While type O blood has no antigens, and individuals with this blood type have been called "universal donors," such blood does contain both A and B antibodies and can damage the recipients red blood cells. Thus precise matching of blood groups for transfusions is preferred.[116]

---

[116] It is also essential to match blood for the presence (+) or absence (−) of another antigen, called the Rh factor. Other blood group antigens can occasionally cause transfusion reactions in cases in which there is no ABO or Rh incompatibility.

Polynucleotides

There are two molecules, DNA and RNA that are involved in the storage and use of genetic information. While the detailed structure and function of these molecules will be discussed in the next chapter, at this point their general chemical structure will be described. DNA and RNA are linear polymers of phosphorylated sugars, substituted with the information-coding bases adenine (A), thymine (T), guanine (G), cytosine (C) and uracil (U). Since the sugars are bridged by phosphate groups, which are the anions of phosphoric acid, these molecules are called nucleic acids. DNA (deoxyribonucleic acid), shown in Figure 17.11, employs the sugar deoxyribose, and incorporates only the first four bases, A, T, G and C.

Chemical analysis has shown that in the DNA polymer, which can contain up to billions of subunits, the concentration of adenine is indistinguishable from that of thymine and the concentration of guanine is indistinguishable from that of cytosine. This is a result of the molecule consisting of two strands of linear polymer, which were first shown to form a

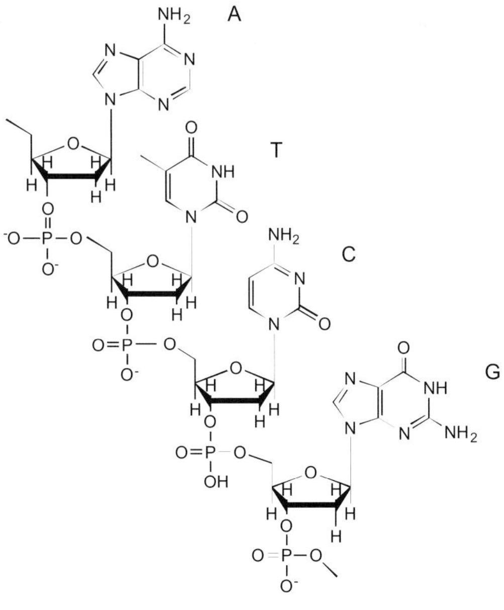

Figure 17.11. Structure of one possible section of a strand of DNA.

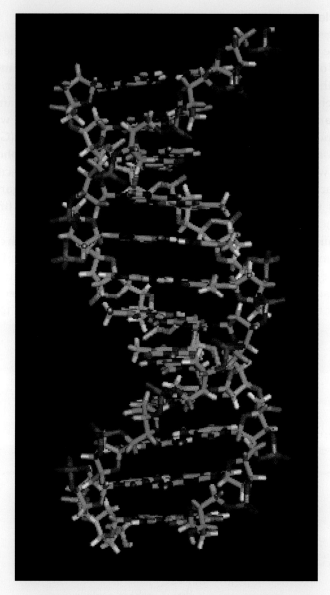

Figure 17.12. Double helix structure of DNA.

double helix shown in Figure 17.12 by James Watson and Francis Crick (1962 Nobel Prize in Medicine). The two strands are held together by hydrogen bonds, which are only strong between A and T or between C and G bases. The two strands are said to be complementary to each other.

There are several different types of RNA (ribonucleic acid), all of which differ from DNA in several features. First of all, RNA molecules are much shorter than those of DNA and usually consist of a single, rather than a double, strand of phosphate-linked, base-substituted sugar. The sugar in RNA is ribose, which differs from deoxyribose, employed in DNA, by having an OH functional group on the sugar. Finally, in RNA, the bases employed are adenine (A), uracil (U), guanine (G) and cytosine (C); RNA uses uracil instead of thymine.

At physiological pH (for humans ca. 7.3), the phosphate groups of nucleic acids are negatively charged. In order to minimize the repulsions between negatively charged phosphate groups, these molecules tend to assume an extended configuration with the phosphate groups weakly associated with positive ions in solution.

Molecular Machines

The multi-step processes carried out in cells would often be much too slow if they relied on sequential encounters of substrates with different protein molecules. As a result, groups of proteins are frequently bound together, possibly on a cellular filament or nucleic acid molecule, to form a *molecular machine*. For example, ribosomes, the cellular particles that construct proteins from the genetic instruction on mRNA, are comprised of 55 protein molecules assembled on ribosomal RNA (rRNA). Conformational changes in proteins induced by energy-rich molecules, such as ATP, insure that protein synthesis in the ribosome proceeds only in the forward direction.

Summary
03 80 03 80 03 80

The great variety of tasks required for life are accomplished by linking together molecular building blocks in a huge number of different linear arrays. The primary structure of a protein is the sequence of α-amino acids that are joined by peptide bonds in the protein molecule. Branching is obtained by disulfide bonds. Intramolecular interactions organize the protein chain into motifs, such as helices and sheets (the secondary structure), which in turn fold it into its stable tertiary structure in solution. Association of similar protein monomers produce a quaternary structure. The function of a protein is directly related to its folded structure.

Simple sugars combine to form polysaccharides, which are used for storage of energy and molecular recognition. DNA and RNA, the molecules used for storage and transmission of cellular information, are formed by attaching bases in a precise order to long sugar-phosphate chains. Strong hydrogen bonding between the bases is very specific and results in the double-helix structure of two complementary strands in DNA.

For their function in a cell, many different molecules may be organized into a molecular machine, whose operation is driven by the consumption of high-energy molecules.

Questions

1. List two globular proteins and two fibrous proteins. What are their functions in our bodies?
2. How many different tri-peptides can be synthesized from the amino acids present in our bodies?
3. What are some amino acids that you might expect can substitute for valine with little loss of protein function?
4. At what amino acid does cross-linking of protein chains usually occur?
5. As discussed in Chapter 14, equilibrium structures are those that minimize free energy, which is energy minus temperature times entropy ($F = E - TS$). Use this idea to explain why a protein that has a highly ordered structure at physiological temperature (37°C) will become disordered (denatured) as temperature is increased.
6. Considering that tooth enamel is strengthened by fluoride in drinking water, but excess fluorine can also produce adverse effects on teeth and bones, what do you think of communities adding fluoride to their water supplies?
7. If a reaction takes 100 years to occur in pure water, how long would it take in the presence of an enzyme that increased its rate by sixteen orders-of-magnitude?
8. What type of antigens and antibodies does each of the following blood groups contain: A+, AB−, O+?
9. What is the structure of the strand of DNA that is complementary to the one drawn in Figure 17.11?

18 Information Storage and Transmission in Life

People are DNA's way of making more DNA.

E. O. Wilson

How can we account for the occurrence of the extremely complicated systems we call life? It can't just be that the structure and operation of an organism is detailed in a set of blue prints or instructions, since that would require another organism capable of reading the prints and following the instructions. In the past two chapters we have discussed some of the many components and molecules of which life is composed. These species interact through an extensive set of reactions. In Chapter 14, we saw that even simple sets of chemical reactions can produce patterns in time and space. Using computer simulation, scientists have demonstrated that larger sets of chemical reactions, in which the product of one reaction catalyzes another, can exhibit even more complicated structures. We will take this as our model for the complexity of life. In Chapter 11 we discussed enzymes, proteins that exert remarkable control over the speed of biochemical reactions. It is through the speed of reactions that life is controlled.

How does a living organism know just what proteins — what sequences of amino acids — to synthesize, and how does it control these syntheses in order to have just the right amount of these proteins available where and when they're needed? The production and availability of these proteins are ultimately controlled by a group of molecules we call the genetic information of the organism. In eukaryotes, organisms that include plants and animals, genetic information is stored in a separate organelle, called the cell nucleus. This is known, since in cloning experiments on eukaryotes, a nucleus transplanted from a cell of individual A into a enucleated (with the nucleus removed) reproductive cell of individual B gives rise to an individual phenotypically identical (with the same physical properties) as individual A. Thus, in eukaryotes, the nucleus must provide the information necessary for producing and maintaining A, while the cytoplasm of B provides the energy and raw materials necessary for following those instructions. Moreover, since this experiment can be performed using a variety of cells from organism A, we can say that

> **The nucleus of every cell of multicellular eukaryotes contains all the genetic information of the individual.**

An exception to this is the gametes, the sperm and egg that fuse in the first step of producing a new organism. Gametes contain only half of the genetic information of an individual. In sexual reproduction a full complement of genetic information is restored by combination of sperm and egg. The nucleus, surrounded by a lipid-bilayer membrane, provides a protective environment that minimizes changes in the genetic information.

Prokaryotes, such as bacteria,[117] reproduce prolifically, and rapid accessibility of their information is more important than its stability. These organisms don't have a nucleus and distribute their instructions throughout the cell. Since eukaryotes generally reproduce much less rapidly, the extra security obtained by having DNA stored in the nucleus makes sense, even though transfer of information to the rest of the cell (the cytoplasm) takes time.

Originally, scientists debated the chemical nature of life's information-storage medium, since it was found that the nucleus contained both protein and DNA, as well as the freely circulating *nucleotides* (an information-coding base, connected to a sugar, which is connected to a

---

[117] *Archaea*, previously thought to be types of bacteria, are also prokaryotes. They were classified as a separate domain of life in 1990, based on the similarity of some of their chemical pathways to those found in eukaryotes.

phosphate group) from which DNA is constructed. Experiments in which genetic material treated with either DNA-destroying or protein-destroying enzymes was added to enucleated cells, indicated that it was the DNA that was essential to transmit genetic instructions. How is a DNA molecule, based on the simple plan discussed in the previous chapter, i.e., a linear molecule with variation among only four different bases, capable of such a monumental task?

Since all life on Earth has adopted DNA for storage of information,[118] it is worthwhile to ask: what is it about the structure of this molecule that makes it so well adapted for this task? Similar to the requirements for information-storage in computers, the requirements in a cell are that the information is secure, easily read, reliably copied, capable of modification and, of course, that it has sufficient information-storage capability. We will explore each of these factors.

Security of Information

As shown in Chapter 17, DNA is a very durable molecule. The backbone of DNA is composed of a chain of very stable sugar molecules, deoxyribose,[119] connected by strong covalent bonds through phosphate groups. One of four nitrogen-containing bases: adenine (A), thymine (T), guanine (G) and cytosine (C), protrude from each sugar. The nitrogens and oxygens of the bases are reactive, but these are protected inside the double-helix structure of the molecule. The structure of the DNA molecule is like a twisted ladder, with the sugar-phosphate chains the sides of the ladder, and the bases (one from each chain) forming its rungs. Each base forms hydrogen bonds only to its *complementary* base (adenine to thymine and guanine to cytosine). With the reactive bases securely tucked away in the center of the double helix, they are not easily attacked by components of the surrounding solution.

Even though the hydrogen bonds between two bases are much weaker than covalent bonds, the large number of such bonds, along with interactions of the π electrons of adjacent planar bases, stabilizes the double helix structure of DNA.

---

[118] Some viruses store their genetic information as RNA.
[119] Unlike ribose in RNA, deoxyribose when incorporated into DNA has no protruding -OH groups to function as points of attack.

278 Order and Disorder

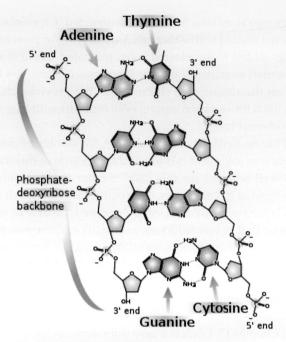

Figure 18.1. Hydrogen bonds in DNA.

Information Storage Capacity of DNA

Since information in DNA is stored in a particular sequence of bases (A, T, G and C) in the DNA molecule, the genetic language has only four letters. It takes a sequence of three successive bases on DNA to code for one amino acid in a protein. All genetic words are therefore three letters long, and there are $4^3 = 64$ such words. DNA words are different from language words in that triplets are read in sequence; they are not separated by markers, such as spaces or commas.[120] Protein size varies from less than 100 to several thousand amino acids. The base sequence required to code for the amino acids in a protein is the sentence of the genetic language and is called a *gene*. A gene also includes additional segments which are not translated into proteins, called *introns*, as well as other segments called promoters and enhancers that regulate the production of the protein.

---

[120] This is why omission or insertion of even a single base in DNA is such a deleterious mutation.

Genes are organized in linear segments of DNA that, with associated proteins, are called *chromosomes*. Each species has a particular number of chromosomes. For example, most cells in humans are *diploid* and have 46 chromosomes, 23 inherited from each parent. In 22 of these pairs, called *autosomes*, the members of the *homologous pair* (one inherited from the father and one from the mother) are very similar. The *sex chromosomes*, which make up the last pair, are of two very different types. One, called the *X-chromosome*, is similar in length to the autosomes, while the other called the *Y-chromosome* is unusually short. Individuals may have two X-chromosomes, in which case they are females, or one X and one Y-chromosome, in which case they are males. The human reproductive cells (gametes) are *haploid* and contain only 23 chromosomes. Throughout most of the life of a cell, individual chromosomes are not distinguishable, since their material is in a stretched-out state called *chromatin*, which extends throughout the nucleus and intertwines with the chromatin of other chromosomes.

Humans (*homo sapiens*) have 20,000–25,000 genes that code for different proteins. An average protein contains about 1000 amino acids, each selected from the twenty amino acids of which proteins are composed. Thus a sequence of up to 25 million amino acids specifies a particular human. Human DNA contains about three billion bases. Since it takes three bases to code for an amino acid, a DNA molecule can determine a sequence of a billion amino acid — more than enough to determine the proteins of an individual.

The DNA in a human cell, if stretched out, would be about a meter long (about 5 cm for each chromosome). All this DNA is packed into a nucleus of dimension of a few microns by tight coiling of the molecule around proteins called *histones*. Thus DNA achieves volumetric storage of information in a form that can be read in a linear manner.

Reading DNA (Transcription and Translation)

The first step in using the information stored in a eukaryote's DNA is to export it out of the nucleus into the cytoplasm, where the machinery, raw materials and energy for protein synthesis are available. The DNA double helix is too large to penetrate the nuclear membrane, so movement is accomplished by first transferring the information to a *messenger-RNA* (mRNA) molecule, in a process called *transcription*, diagramed in Figure 18.2.

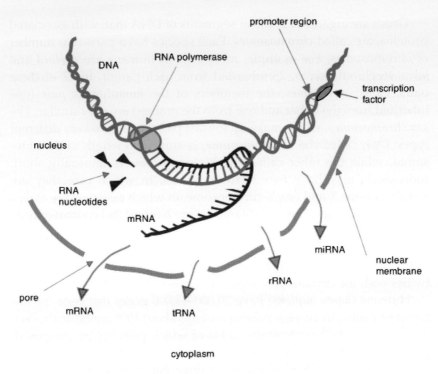

Figure 18.2. Transcription.

Details of Transcription

In *transcription*, a section of the DNA double helix, coding for a particular protein (a gene), is activated by proteins, called *transcription factors*, which attach to *promoter* sequences on the gene. An enzyme, RNA polymerase, then attaches to the proteins and moves along the gene, separating the coils of the double helix and synthesizing another nucleic acid called messenger-RNA. The sequence of mRNA is complementary to that of the DNA in the gene, except that it uses the base uracil (U) in place of thymine (T). Like T, U is complementary to adenine. In addition, mRNA employs a somewhat less stable sugar, ribose, in place of the deoxyribose of DNA. Even though an mRNA molecule may contain up to a thousand bases, it is much shorter than DNA. The mRNA initially formed is called *pre-mRNA*. Is its processed in the nucleus to remove information corresponding to introns and to place a "cap" on the 3' end of the molecule. Finally, in a process called polyadenylation, a section of RNA containing only the base adenine is added to the 5′ end of mRNA. This "tail" allows the molecule to pass through a pore in the nuclear membrane and enter the cytoplasm. Transcription is also controlled by signaling proteins that bind to enhancer sequences, which may be located anywhere on DNA.

Other regions of DNA code for RNAs that do not produce proteins. Called tRNA, rRNA and miRNA, the purpose of these molecules will be discussed below.

In the cytoplasm, ribosomes, consisting of ribosomal RNA (rRNA) and several dozen proteins, *translate* the base sequence of messenger RNA into the appropriate amino acids in the protein that is being synthesized. It has been found that

> **All life on Earth employs a very similar genetic code.**

The *genetic code*, a relationship between the base sequences in RNA and the amino acids of proteins, is shown in Figure 18.3.

In the diagram, the 20 basic amino acids used by life are given in the outer ring.[121] RNA bases are read from the inner ring outward to give *codon* for a particular amino acid.

Example. What protein would be coded for by the nucleic acid sequence.

AUGCAAGCUACGGUGUGCUGA?

Solution: Start – Glu – Ala – Thr – Val – Cys – Stop

As can be seen from the diagram, there is redundancy in the code; more than one sequence codes for almost all amino acids. Often this redundancy is due to a change in the third letter of the code. There are also codons that specify the beginning and the end of protein molecules. Protein synthesis begins at the amino end of the protein.

---

[121] At least two additional amino acids are coded for by the DNA in bacteria.

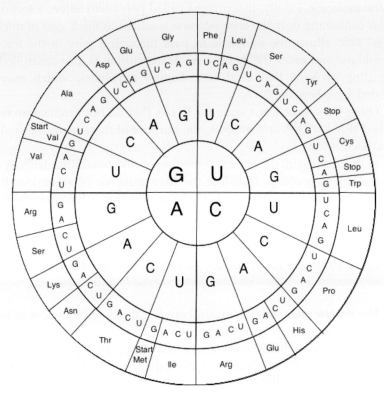

Figure 18.3. The universal genetic code.[122,123]

Details of Translation

The mechanism of translating the information encoded on RNA into a protein molecule is outlined in Figure 18.4.

Translation takes place at a molecular machine called a ribosome, a complex of rRNA and a number of proteins. The translators of the genetic code are small molecules called ***transfer-RNA*** (tRNA). There are 61 different tRNA molecules that bind to a particular amino acid on one

---

[122] There are slight variations in the genetic code in protists and organelles, such as chloroplasts and mitochondria.

[123] The first AUG or GUG codon signifies the start of protein synthesis, but precludes initiation at subsequent AUG or GUG codons before a preceding stop.

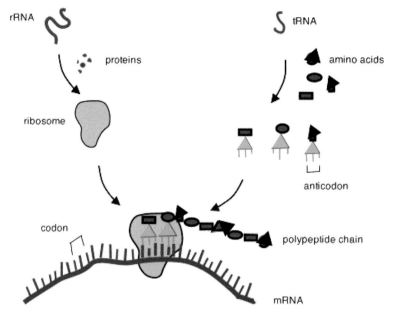

Figure 18.4. Translation.

end and contain the bases complementary to its codon (its **anticodon**) on the other end.[124] As the ribosome moves along mRNA, the anticodon of the proper tRNA molecule binds to the codon on mRNA and links its amino acid to the protein chain being produced. In eukaryotes, translation proceeds at speeds of 6–7 amino acid residues per second.

Supplemental Reading: Gene Regulation

It is not only necessary that proteins are synthesized, but also that they are synthesized when and where they are needed. Just about every cell in our body has a complete set of genes. That means, for example, that the taste buds on our tongues contain genes for

---

[124] Some cells have fewer than 61 tRNAs, because the third base in the anticodon (called the wobble base) is often capable of bonding to several different bases in the codon.

generating enzymes that can produce hair and nails. Tongue cells do not produce hair and nails because the relevant genes are not active for cells in the tongue. Functional diversity from genetic uniformity is achieved in a number of ways. One of these is by *epigenetic* modifications of DNA. These can be stable modifications of one of the bases of DNA due to interactions with chemicals in the environment, such as those that are produced in particular organs of the body. A very common epigenetic modification of a gene is the substitution of a methyl group for one of the hydrogens of a cytosine base on the gene. Such substitution effectively "shuts off" the gene; it will no longer produce its protein. Although necessary for the proper development of an organism, epigenetic modification of cells accumulates over time, and is one of the causes of cancer.

Stem cells, in particular those of human embryos before implantation in the uterus, are *pluripotent*, i.e., they can differentiate into any type of cell. Medically, such cells are of interest for possible regeneration of damaged tissues, such as in Parkinson's disease and spinal cord injuries.

When a gene is actively producing mRNA for its particular protein, we say that the gene is *expressed*. The degree to which a gene is expressed is usually modified at the level of transcription. Each gene has, in addition to a segment that codes for a particular protein, other segments, displaced from it by different distances on the same strand of DNA, where regulation of the amount of mRNA produced by the gene is controlled. These segments are called promoters and inhibitors and are activated by binding transcription factor proteins. In contrast to epigenetic modification, promotion or inhibition is a temporary change of gene expression. As a possible example, when sufficient amount of a particular protein has been synthesized, some of it might find its way back into the nucleus, where it can "turn off" its gene. Since mRNA has a limited lifetime in the cytoplasm, synthesis of the protein soon stops. Expression is not an on-or-off process; a gene may be expressed to produce its mRNA at different rates by various control processes.

Expression can also be controlled at the level of translation, by proteins interacting with mRNA or by short segments of RNA, called *microRNA* (miRNA), which bind to mRNA and stop its translation.

Copying DNA (Replication)

The double-helix structure of DNA is uniquely suited for copying (replicating) this molecule. Under the influence of a group of enzymes, the hydrogen bonds holding the two strands of a small part of the DNA molecule are broken, the strands are separated, and each strand serves as a template for the assembly of complementary nucleotides. Freely circulating nucleotides in the nucleus then attach to their complementary bases and an enzyme, DNA polymerase, links these up into a new chain. The reproduction proceeds along the chain and the result is two new daughter DNA molecules, each containing all the genetic information of the original. The DNA replication apparatus is very fast and very accurate: about 1000 nucleotides are replicated in a second, with only about one error for every billion nucleotides inserted. Even this very low error rate, however, is unacceptable, and the cell is provided with a set of enzymes that proofread and repair these errors. Proofreading is facilitated by the double-helix structure of DNA. The repair enzyme can move along the DNA molecule checking for the proper pairing of bases and correcting these when they do not match. The product of replication consists of two double-helix molecules, called *sister chromatids*, which are held together by a number of proteins.

Reproduction

Reproduction in life is carried out at three different levels: that of the genetic material (replication of DNA), that of the cell, and that of the individual. Cellular reproduction (and asexual reproduction of organisms) occurs by a process called *mitosis*, diagramed in Figure 18.5.

In mitosis, a eukaryote cell that has already replicated its DNA is signaled to divide by a protein called mitosis-promoting factor (MPF). This protein is common to all eukaryotes. Upon receiving this signal, the cell's nuclear membrane disintegrates and the chromatin segregates into tightly wound-up chromosomes. Each chromosome consists of the two sister chromatids, now connected only in a central region called their *centromere*. The chromosomes line up at the midplane of the cell and are connected to newly formed filaments, called the *spindle apparatus*, which pull the chromatids of each sister pair to opposite ends of the cell. At the same time, the other organelles in the cell are divided up more-or-less equally. The process

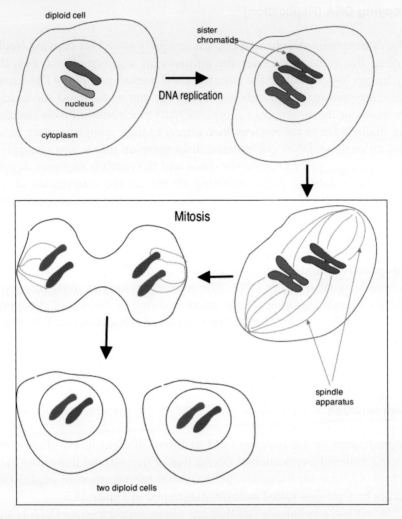

Figure 18.5. DNA replication and mitosis.

of cell division is completed by forming a new cell membrane for each half of the original cell and encasing its chromatin in a new nuclear membrane. Epigenetic changes of DNA are maintained throughout mitosis, so that, for example, liver cells give rise only to liver cells and not to lung cells.

The production of reproductive cells, called *meiosis*, is more complicated. Replication of DNA in this case is followed by two successive divisions of the cell, so that each reproductive cell (a *gamete*) ends up with only half of the original number of chromosomes. In the first meiotic division, homologous pairs of chromosomes (one from the father and one from the mother) line up. Prior to separation the pair of chromosomes is held in close proximity, and some interchanging of the genes between them occurs, a process called *crossing-over*. This interchange increases the genetic diversity of offspring.[125] In the second meiotic division, the sister chromatids in each chromosome divide, resulting in four gametes from the original cell. Since in females only X-chromosomes are present, all the gametes will have X-chromosomes. However, in males, half the gametes will have X-chromosomes and half will have Y-chromosomes.

Inheritance of Traits

In sexual reproduction a new individual is formed from the *zygote* that results from the fusion of two gametes, one from the mother and one from the father. Almost always, the individual will have one sex chromosome from its mother (always X) and one from its father (ca. equal probability of X and Y). Thus there is about an equal probability of female (XX) and male (XY) progeny resulting.[126] Klinefelter's syndrome, is a condition in which males have an extra X chromosome. Although there are many phenotypical variations of individuals with XXY sex chromosomes, almost all show small testicles and reduced fertility. On the other hand, it has not been established whether males with XYY sex chromosomes have any physical or mental problems.

The transmission of characteristics by DNA provides a chemical mechanism to explain different types of genetic transmission of phenotypes. For example, if a characteristic, such as a tendency for tall plants, is determined by the presence of a single enzyme, it will occur if the gene for that enzyme is functional on either of the pair of homologous chromosomes. Only if the gene is absent on both of these chromosomes will the characteristic not appear. Such a phenotype is transmitted by a dominant-recessive Mendelian mechanism.

---

[125] There is no interchange of material between the X and Y sex chromosomes.
[126] Equal numbers of males and females may be modified by differences in survivability and competitiveness of the X and Y gametes and by differences in survival of the male and female embryos. For example, in humans there are approximately 1.05 live male births for every female birth.

Some characteristics result from the amount of an enzyme available, not just its presence. Among these are the color of snapdragons, which can be white, red or pink; the pink color resulting from only one copy of the red-producing gene. This is called *incomplete dominance*. Most traits result from the action of more than a single protein enzyme and thus are *polygenic*. Among these are skin color and height in humans.

Some traits are *sex-linked*, meaning that the gene that is responsible for them is carried on either the X or Y chromosome. Hemophilia, a disease caused by the deficiency of a blood-clotting protein is of this type. The gene for this protein is carried on the X chromosome. The presence of hemophilia in women requires the absence of the gene on both of the X chromosomes — a very rare occurrence. Much more frequently, women are carriers of hemophilia and transmit the disease to half of their male offspring. There are very few traits determined by genes carried on the Y chromosome.

Mutations

Unrepaired replication errors are one way of producing genetic changes, called *mutations*. Such changes can also be brought about by interaction of DNA with *mutagens*, such as ultra-violet radiation, high-energy particles or mutagenic chemicals. Mutations occur in all cells of an organism, but it is only those that occur in the reproductive cells that can be passed on to the next generation of the organism. Mutations may be of the following types:

Neutral mutations: These involve changes, usually of single DNA base (a *point-mutation*), which, due to the redundancy of the genetic code, do not involve a change of protein coding. They may also result in the substitution of an amino acid with very similar properties for the original amino acid (such as non-polar valine for non-polar leucine). Sometimes mutations may also result in the *duplication* of an entire gene, or part of a gene, at a different point on a chromosome. Since the original gene will continue to produce its coded protein, such mutations may be relatively benign.

Deleterious mutations: Point mutations that change the nature of the coded amino acid, such as the replacement of glutamine (GAG) with valine (GUG), in sickle-cell anemia, can be deleterious. Mutations that result in the addition or deletion of a base from DNA will be particularly serious. They will alter the specification of all subsequent amino acids in a protein, since there are no spaces between amino acids in the genetic code.

Zygotes with harmful mutations will often not develop into viable mutations. In other cases, the mutation may be recessive (indicating that sufficient amounts of the required protein may be produced by the homologous chromosome), and propagate through future generations.

Advantageous mutations: Very infrequently a mutation will produce a protein that is superior in its function to the original protein or a new protein that is capable of new metabolic tasks. The latter can occur, without loss of the original protein function, if it occurs in a section of DNA that has resulted from duplication, with the original gene maintaining its genetic function.

Evolution

Evolution operates at the level of the genes. A gene (or genes) will maintain or expand its presence in a population, if the phenotype determined by the gene (or genes) provides a reproductive advantage for individuals. This is more than the standard selfish "survival of the fittest," since the gene must provide an advantage over many generations. Thus, altruistic behaviors, such as a propensity to nurture offspring, or peacefully resolve group differences, can provide phenotypical advantages.

Genetic Modification

Humans have been experimenting with genetic material for centuries. Such experiments include "unnatural selection," where elm trees resistant to Dutch elm disease have been produced by repeatedly generating offspring from individuals showing resistance to this disease. Additionally, cuttings from plants that normally reproduce sexually are regularly used to maintain favorable characteristics. Animal cells will not generally reproduce asexually. However, by placing DNA into fertilized de-nucleated egg cells, a number of species have been **cloned**, i.e., produced with genes identical to those of a single parent. The first mammal to be produced in this manner was Dolly, a sheep produced in Scotland in 1996. Dolly died in 2003, and her relatively short life has been tentatively attributed to the cloning process.

Several methods have also been developed to insert foreign genes into host DNA. For example, the gene for producing human insulin has been inserted into bacteria, which then act as insulin factories. Such human insulin is far superior and cheaper than material isolated from dogs or cows.

> **Supplemental Material: RNA World**
>
> Since much of the machinery for transcription and replication of DNA involves the use of proteins, which are coded for by DNA, the question naturally arises of how such a "chicken and egg" process could have gotten started. Simultaneous spontaneous generation of both DNA and the necessary proteins would seem very unlikely. The suggestion has been made that this process could be short-circuited, if both information storage and catalytic functions could be combined in the same molecule, and that RNA is a likely choice for such a molecule.
>
> RNA, involving specific base pairing and double helix formation could be used to transmit genetic information. However, since in RNA, the formation of double helices often involves short intramolecular segments, the molecule can fold into interesting shapes required for catalysis. Catalytic activity of RNA was verified in 1980 by biochemists Sidney Altman and Thomas Cech (Nobel prize in chemistry, 1989), who discovered *ribozymes*, RNA molecules that could function as enzymes.

In combining information storage and catalytic functions in the same molecule, RNA probably did not accomplish either of these tasks with optimum efficiency. Further evolution could therefore have produced the more stable DNA molecule and proteins. DNA, with its propensity for forming complementary base pairs between separate strands, would more efficiently accomplish the information-transmittal task, and proteins would be more effective catalysts than RNA.

Summary

The genetic instructions of life are stored on DNA, a very stable molecule, which in eukaryotes is located in the nucleus of the cell. The sugar-phosphate backbone of DNA is held in a double helix structure by hydrogen bonds between the complementary bases adenine (A) - thymine (T) and guanine (G) - cytosine (C). For protein synthesis, DNA is transcribed to messenger RNA (mRNA), a smaller, but similar molecule, which can leave the nucleus. At the ribosomes, codons of three bases of mRNA are translated to particular amino acids in proteins. DNA is replicated by enzymatically unwinding the double helix and connecting complementary nucleotides to each of the strands. In cell mitosis and asexual reproduction the two sister chromatids are drawn to opposite sides of the cell and a new dividing cell membrane is formed. In sexual reproduction, genetic material from parents is mixed and homologous chromosomes are separated before sister chromatids are separated. This produces gametes with half the number of chromosomes of the original cells. Zygotes are new individuals, produced by combination of two gametes, one from each parent.

Questions

1. The Morse code contains two letters (dot and dash). What would be the length of the Morse code word required to specify 38 elements (26 letters, 10 numerals, period and comma) of the English language?
2. In the genetic code, three nucleic acid bases code for 20 amino acids. What is the maximum number of amino acids that can be coded for by four nucleic acid bases?
3. Assume that you can perform manipulations that remove the nucleus from a eukaryotic cell and transplant the nucleus from one such cell to another. Devise experiments to test each of the following hypotheses for eukaryotes:

    (a) The genetic instructions for an entire organism are contained in each and every cell of that organism (except the sex cells).
    (b) The genetic instructions for an organism are stored in its nucleus.

4. List some of the advantages of a double helix structure for DNA.
5. Give one reason for RNA being composed of a more reactive sugar than DNA.
6. It is advantageous to life that it is resistant to mutations (substitution of one nucleotide for another). Show how the universal genetic code is resistant to some mutations.
7. Which amino acid is coded for by only a single codon?

8. Is cloning the same as asexual reproduction?
9. What are advantages and disadvantages for humans reproducing by cloning, rather than sexual reproduction?
10. Discuss how human insulin is manufactured by bacteria. Why would this be superior to insulin obtained from the pancreas of a pig?

19 Beyond the Cell

The foot bone is connected to the leg bone, The leg bone is connected to the knee bone…

from "Dry Bones"

As the child's song above illustrates, we have an inherent curiosity about how our bodies are put together. We want to know what each part does, how they work together and how they change over time. It's also an unfortunate fact of life that almost all of us will at some time be involved with health problems, either our own or of loved ones. Some understanding of the higher levels of organization of our bodies will be useful at those times.

The members of the Animal, Plant and Fungus kingdoms are made up of many cells. Compared to single-cell organisms, the complexity achieved by having many cells, provides organisms with new capabilities to prosper in their environments. This complexity comes with a price, however, since multiple-cell organisms generally reach maturity and reproduce much slower than single-cell organisms. In addition, organisms with many cells have stringent requirements for functional order — the various cells must coordinate and work together. While some multiple-celled organisms (e.g., slime molds) consist simply of agglomerations of identical cells, in most multiple-cell organisms, the cells are of different types.

In our own bodies there are over two hundred different types of cells. There are insulin-secreting beta cells in the pancreas, neurons in the brain, erythrocytes (red-blood cells) and at least twenty types of white-blood cells in our blood. These cells are arranged in tissues, which in turn form organs and organ systems.

Tissues and Organs

A *tissue* is a group of cells arranged together to accomplish a function. The tissue may be composed of just a single type of cell, such as the cells in potatoes that store starch, or of many different type of cells, organized together to achieve one or more functions. Animal tissues are classified as being of four different types: *epithelial tissue* (the stuff of the inner and outer linings of the vessels, cavities and structures of our body), *connective tissue* (the stuff of bone, cartilage, tendons and ligaments), muscle tissue and nerve tissue.

An *organ* is a number of different tissues joined in a structural unit. For example, our skin is composed of epithelial, muscle and nerve tissues, with a large number of specialized cells, such as glandular, pigment and hair cells. Some of the functions of our skin are protection from injury and sunlight and control of water evaporation.

Organ Systems

Some organs are linked together in an *organ system* to achieve a particular task. A single organ may be involved in many tasks and be part of several organ systems. For example, because the liver breaks down toxic substances in the blood, it is part of the circulatory system. It also produces bile, which emulsifies fats in our digestive system. A brief outline of some of the major organ systems of the human body is given in this chapter. The nervous system will be considered separately in Chapter 20.

The digestive system

Humans are heterotrophs, meaning that they cannot synthesize the carbon compounds needed for life from carbon dioxide, and therefore must consume organic compounds produced by other organisms. The ultimate source of these compounds are the autotrophs, plants and bacteria that can carry on photosynthesis. Humans are *omnivores* that can consume both plant and animal matter, as distinguished from those animals that eat only plants (*herbivores*) or only other animals (*carnivores*).

> **The digestive system supplies blood with nutrients from food.**

To do this, food must be ingested, digested and absorbed, and unabsorbed material must be excreted. Absorption occurs in the *small intestine* (*ileum*), through pores in the epithelial lining that connect to small blood vessels. Only small molecules, such as sugars, amino acids and fatty acids can be absorbed. Chunks of food must be broken up, and large molecules, such as starch, proteins and fats, must be decomposed into their components.

The components of our digestive system are shown in Figure 19.1.

Digestion starts in the mouth, where food is chewed and mixed with saliva, to produce a *bolus* (a ball of food and saliva). *Saliva* contains the enzyme amylase that breaks starch into sugars. When the bolus is swallowed, it passes into the esophagus, which by *peristalsis* (a wave of muscle contraction) propels it through the *esophageal sphincter* into the stomach. (A *sphincter* is a structure that normally keeps a passage closed.)

Digestive juices in the stomach are highly acidic (pH = 1–2) and also contain the digestive enzyme pepsin, which degrades proteins into peptides. The walls of the stomach are made of cells that protect it from this corrosive liquid by excreting a thick layer of mucous. When this protective mechanism fails, stomach ulcers occur. The esophagus can be damaged if stomach acid leaks through the sphincter — a process called *heartburn*. Food remains in the stomach for 1–4 hours, during which time muscles in the stomach wall churn its contents and aid in digestion.

The lower part of the stomach is sealed by the *pyloric sphincter*, which opens when the stomach contents are digested, allowing these to be pushed into the upper part of the small intestine, called the duodenum.

At the duodenum, other enzymes, such as trypsin and chymotrypsin, both produced in the pancreas, complete digestion. Enzymes target specific

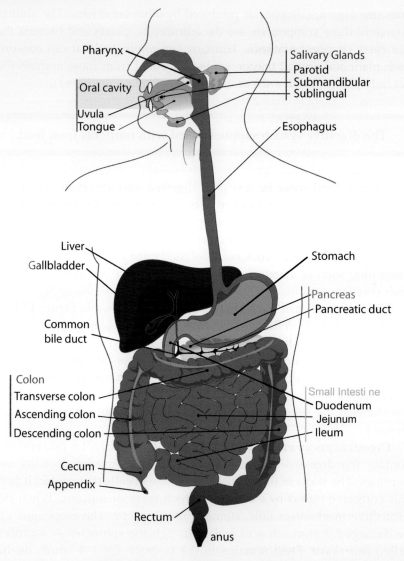

Figure 19.1. Human digestive system.

bonds holding carbohydrates and proteins together. The specificity of these enzymes is indicated by their ability to digest starch, one polymer of glucose, but not cellulose, a somewhat different polymer in which the monomer units are joined by the same bonds in a different orientation

(Chapter 17). Bicarbonate is added in the duodenum to neutralize the digested material. Fats are solubilized in the small intestine by surfactants (Chapter 14) called *bile salts*, which are produced in the *liver* and stored in the *gall bladder* until needed.

During a period of 4–7 hours, food is pushed along the small intestine, a tube of 2 to 3 cm diameter and ca. 8 m in length, by peristalsis. The surface area of the small intestine is increased by a factor of 500 to about 250 m² by extensive folding. The lining of the small intestine delineates the separation between the outside world and our bodies. The inside of the intestinal tract belongs to the outside world, much like the hole in a donut is not part of the donut.

The small intestine is joined to the large intestine at the *cecum*, which contains the appendix, a *vestigial* (no longer of use) organ.[127] The *colon* comprises most of the large intestine. Its major purpose is to allow readsorption of water from digestive wastes before they are expelled as fecal matter through the anus, after 8–24 hours. Roughly half of fecal matter is bacteria, the remainder is undigested food.

The digestive system, like most of the systems in our body, is under control of the brain. When the brain receives the proper signals from the intestinal tract, it initiates a set of actions that result in the production of proper amounts of digestive compounds. Satiety, a feeling of having consumed enough food, results from complex psychological factors, rather than just intestinal signals.

The respiratory system

The nutrients absorbed into our body are made use of by processes involving reactions of oxygen, which produce energy and generate carbon dioxide.

> **The respiratory system supplies oxygen to and removes carbon dioxide from blood.**

---

[127] There is some evidence that the appendix may harbor bacteria that are useful for digestive health.

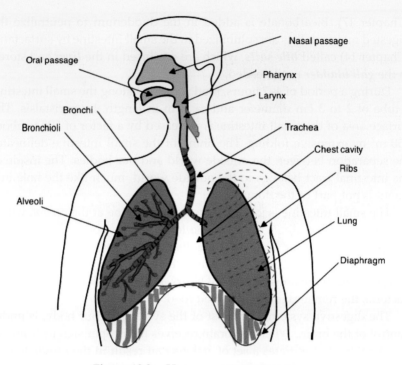

Figure 19.2. Human respiratory system.

A diagram of the human respiratory system is shown in Figure 19.2.

Air is brought into the respiratory system, when the ***diaphragm***, a large muscle at the base of the chest cavity moves down. The reduced pressure in the cavity expands the lungs and draws air into them through the oral and nasal passages, past the ***pharynx*** that contains the vocal cords. A major function of the respiratory system is to cleanse and prepare air before it reaches sensitive tissues in the lungs. The upper respiratory tract moistens the air and adjusts it to body temperature. It also cleans the air by filtration by nasal hairs and by having pollutants stick to and dissolve in mucus membranes. The ***trachea*** contain *cilia*, tail-like projections, which by a waving motion, drive mucous towards the oral passage, where it can be swallowed or expelled by coughing.

The primary ***bronchi*** are two tubes that branch off from the trachea in the lower respiratory tract. Further branching into secondary and tertiary bronchi and ***bronchioles*** lead to the *alveoli*, where gaseous exchange takes place. The branching in the lower respiratory tract is so abrupt that

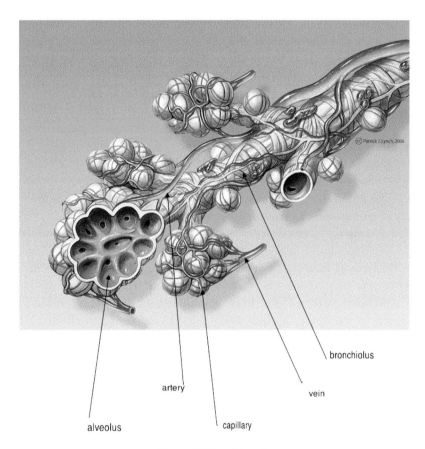

Figure 19.3. Alveoli.

most particulates hit the walls and stick to mucous, this avoids deposition of the particles in the alveoli, from which they are very difficult for the body to remove. Chemical impurities diffuse to the walls in the slower moving gases in this region.

The interface between these respiratory and circulatory system occurs in the alveoli in the lungs, shown in Figure 19.3.

The alveoli are microscopic air pockets surrounded by a thin membrane, through which gases diffuse into and out of tiny capillaries. Oxygen diffuses from its high concentrations in the lungs into the blood and carbon dioxide diffuses from its high concentration in the blood into the lungs. The lungs contain several million alveoli, with a surface area for gas exchange of 60 to 70 square meters. In the incurable disease *emphysema*, the surface area of the alveoli is greatly reduced — usually by long-term irritation due to chemicals and particulates, such as found in cigarette smoke. Victims of this disease are continually gasping for oxygen. While some relief can be obtained by breathing pure oxygen, this is only temporary, since pure oxygen itself produces further damage to the lungs.

The circulatory and excretory systems

> The circulatory system supplies nutrients, oxygen and other substances to and removes waste products from cells.

This transfer of material is achieved by blood circulating in *capillaries*, microscopic tubular vessels, some of which are in the vicinity of every cell. The heart pumps blood to the capillaries through *arteries*. Blood is returned to the heart through *veins*. Blood receives nutrients from the digestive system, oxygen from the respiratory system and is cleansed by the liver and the kidneys.

Blood is an aqueous medium in which the solubility of oxygen is not sufficient to supply tissue cells. Oxygen is transported in blood by red blood cells containing hemoglobin, a molecule made up of a protein incorporating four heme groups, each of which binds a single oxygen molecule. In tissues, the liquid component of blood (blood plasma minus its clotting factors) leaves the arterial end of the capillaries and forms the interstitial fluid, the solution that surrounds and nourishes cells. The fluid picks up carbon dioxide and other waste products before most of it is returned to the venous end of the capillary.[128] Carbon dioxide is transported from the tissues to the lungs primarily in the form of bicarbonate (HCO_3^-) ions. Some of the interstitial fluid finds it way back to the circulatory system via the lymphatic system, a separate group of vessels, which are part of the body's immune system.

The circulatory system also pumps blood through the *liver*, where nitrogenous wastes, are converted to *urea*, $H_2N\text{-}C(O)\text{-}NH_2$, a relatively

---

[128] In the lungs, the fluid deposits carbon dioxide.

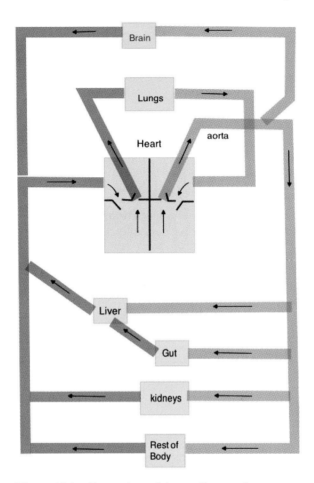

Figure 19.4. Front view of the cardiovascular system. (Oxygenated blood is shown in blue, deoxygenated blood in red.)

non-toxic compound. Urea and excess salts and water are removed from the blood by the *kidneys*, forming urine, which is stored in the *bladder* until excreted through the *urethra*. In cases of kidney failure, this simple separation function of the kidneys can be accomplished by *dialysis*, in which blood is passed through narrow-walled tubing residing in appropriate liquids. The kidneys, however, also perform many regulatory functions (i.e., of blood pressure) in the body, which are not accomplished by dialysis.

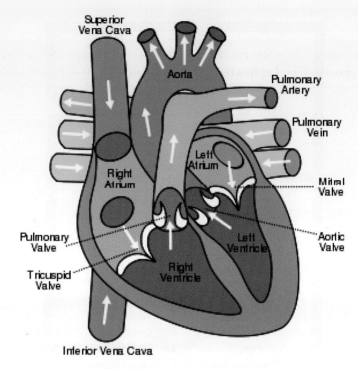

Figure 19.5. The human heart.

The Heart

The heart is the four-chambered pump shown in Figure 19.5.

Pumping a viscous liquid, such as blood, through very small diameter capillaries requires a considerable pressure, which is provided by the heart, which upon contraction raises blood pressure from its resting value, the diastolic pressure, to a higher value, the systolic pressure. Normal values for these quantities are 90–140 mm Hg for the *systolic* pressure and 60–90 mm Hg for the *diastolic* pressure. High blood pressure, *hypertension*, involves pressures continually higher than these values, and can be the cause of many maladies, among which are kidney, heart and blood vessel damage. Abnormally low blood pressure, *hypotension*, may starve critical organs like the brain and heart of oxygen.

The human heart is comprised of four chambers: right and left *atria* and *ventricles*. When the heart relaxes, blood enters the two *atria* (the upper chambers of the heart) and when it contracts, blood is forced out of the two *ventricles* (the lower chambers of the heart). A valve on each side of the heart prevents backflow from the ventricle to the atrium. Deoxygenated blood enters the right atrium (to the left in Figure 19.4) and is pumped by the right ventricle into the lungs. Oxygenated blood returns from the lungs into the left atrium and is pumped by the right ventricle through the *aorta*, other arteries and very small arterioles into all the tissues of the body. After passing through the capillaries, deoxygenated blood is returned to the heart through the veins. The circulation of blood in the human body was first described by William Harvey in 1628.

The heart is constructed of specialized auto contracting muscle cells. These contractions are synchronized by an electrical signal from a specialized group of cells in the wall of the right atrium called the *sinoatrial node*. This signal flows through the two atria, causing them to contract. After about a 0.1 second delay, which allows the atria to empty, the signal flows to the ventricles, causing them to empty their contents into the *pulmonary artery* (right ventricle) and the aorta (left ventricle).

Life depends upon the ordered time response of the heart. Fibrillation is excess disorder (chaos) in this response. In atrial fibrillation, spontaneous contraction of muscle cells in the walls of the atria results in an irregular heartbeat. This condition is dangerous, but can be managed with medicines and lifestyle changes. Ventricle fibrillation prevents proper flow of blood to the brain and other body organs. It is invariably fatal after a short period and is a major cause of death in the United States.

The immune system

We are continually attacked by noxious chemicals, foreign particles and infectious organisms. Our skin provides an impervious barrier to these substances. The surfaces of our respiratory tract is less impervious, but is protected by a thick layer of mucous that is periodically expelled from the body. In addition,

> **Our immune system protects us from pathogens and removes aged and mutated cells.**

The body's immune mechanisms are specific and non-specific. Non-specific mechanisms for fighting infections include both molecules and cells. The molecules are called *complement*, a group of proteins that circulate in blood. In regions of inflammation, these proteins rapidly proliferate, producing large numbers of molecules that can destroy the integrity of bacteria cell walls. Viruses do not have cell walls and reproduce within cells of our bodies, hijacking our protein-synthesizing apparatus for this purpose. However, when cells are infected with viruses, they often produce the protein *interferon*, which blocks protein synthesis in other cells in the vicinity of the viral infection.

White blood cells are cells of the immune system that circulate in blood. They include *macrophages*, large cells with projections that kill invading bacteria by the process of *phagocytosis*. In this process the bacterium is engulfed by the macrophage and encased in a cavity called a vacuole. Powerful chemicals are released into the cavity, destroying the bacterium.[129] Blood also contains *natural killer cells* that recognize and kill, by a non-phagocytotic mechanism, cells that have undergone viral infection. In the region of an infection, blood vessels dilate and become more permeable, so that large quantities of white blood cells can be released into the infected tissues, forming the pus of a wound. After they are released into the interstitial fluid, white blood cells return to the blood through the *lymphatic system*, which includes lymph nodes and the spleen, where large numbers of white blood cells are stored.

Bacteria rapidly mutate and thus can evolve strategies for evading destruction by complement molecules or phagocytotic cells of our body. In order to counter these evasive tactics, our bodies have developed methods for highly specific identification of *antigens* (invading microorganisms and other foreign substances in the body). This identification is achieved by cells produced in the bone marrow (*B-cells*) or thymus (*T-cells*), which on their surface have *antibody* molecules that can very specifically bind to a particular antigen. When an antigen binds to a particular B-cell, the B-cell multiplies and develops into plasma cells that circulate in the blood, releasing antibodies. Each antibody contains, in addition to a segment that specifically recognizes its antigen, a non-specific

---

[129] The initial steps of phagocytosis provides a mechanism whereby organelles, such as mitochondria and chloroplasts could have initially found their way into cells.

portion that activates macrophages and killer cells for the destruction of the antigen.

Since viruses proliferate within our body cells, they are only infrequently available to be attacked by circulating antibodies. T-cells, however, are designed to destroy cells in our body that are under attack by viruses. Such cells often display on their surface small parts of the virus that act as an antigen for binding to a particular T-cell, which then activates mechanisms in the body that destroy the cell. This process also results in a proliferation of the particular T-cell.

Amazingly, the B- and T- cells in our body contain protein antibodies that specifically bind to a huge number (hundreds of thousands) of different antigens. The vast amount of information storage required to produce such an enormous variety of proteins is greatly simplified by the recognition part of these molecules being constructed from two different protein chains. This allows N such chains to generate $N \times N = N^2$ different antigen-recognition environments.

Although circulating plasma cells and T-cells have limited lifetime, the increased concentrations of particular B-cells in the bone marrow and T-cells in the thymus after removing an antigen provides a memory effect, and primes the immune system for that allergen, so that a larger and faster immune response is generated in subsequent exposure to the same antigen. Thus, the first time you are infected with a particular virus or bacteria is often the most severe; later infections are less intense because they are inhibited by the immune system. The same mechanism provides the basis for vaccination, whereby antibodies for a particular antigen are stimulated by intentionally introducing a similar antigen or part of the antigen into the body. Priming of the immune system to an antigen, however, may in some cases produce extreme reactions (hypersensitivity) upon exposure to large (or in some cases even small) concentrations of the antigen. This can lead to debilitating ***allergies***, anaphylactic shock and even death.

The task of the immune system is to perform its functions without affecting normal tissues in the body. Apparently, it is the exposure of the immune system to body tissues during fetal development that avoids an immune response to these tissues in later life. When normal tissues are attacked, any of a number of ***autoimmune diseases***, such as Lupus, Crohn's disease, multiple sclerosis or rheumatoid arthritis, can result. Individuals born with inadequate immune systems or those whose immune systems are suppressed by drugs to avoid the rejection of transplanted organs must be very careful to limit their exposure to infectious agents. One way that the immune system may be seriously compromised is by infection by HIV (human immunodeficiency virus). This virus multiplies within and destroys T-cells, rendering the individual extremely susceptible to many infections. Normal T-cell counts are 700–1000 in a standard drop of blood; counts below 200 are indicative of AIDS (acquired immunodeficiency syndrome).

The integumentary, skeletal and muscular systems

The integumentary system (our skin) stands between us and the outside world.

> **Our skin controls heat transfer and water evaporation and protects us from sunlight, microorganisms and injury.**

It consists of two layers, the *epidermis*, or outer layer, and the *dermis*. A segment of skin, including its nerves and blood vessels, is shown in Figure 19.6.

The epidermis consists of epithelial cells, rich in **keratin**, a tough protein found in nails (and claws and hoofs). Epidermal cells are being continually produced by mitosis at the inner part of this layer and dying as they migrate to the outer surface, where they slough off. This leaves the surface as an almost impervious layer of keratin. Since the epidermis does not contain blood vessels, it relies on diffusion from the dermis for nutrient and gas transfer. The skin contains granules that include the pigment *melanin*, which gives the skin its characteristic color and provides individuals of different races different degrees of protection from the ultraviolet rays of the sun.

The *dermis* consists of connective and muscular tissue, as well as epithelial tissue. It contains a rich supply of blood vessels and nerves. Control of the blood supply to the dermis is a principle mechanism for regulating body temperature. Heat loss is increased when blood vessels in the dermis dilate, increasing the flow of blood to the skin. The dermis also contains hair follicles and sweat glands and receptors of pain and pressure, such as the pancinian corpuscles. The evaporation of sweat from the skin is another mechanism for cooling our bodies.

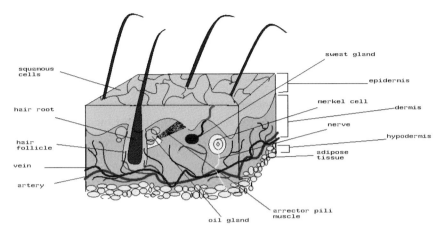

Figure 19.6. Human skin

The human body is supported by an *endoskeleton* (within the body) as compared with *exoskeletons* that support creatures such as lobsters and insects.

> **Bones support our bodies and cartilage protects our joints and organs.**

Since it is the skeletal remains of organisms that most often survive in the fossil record, there is relatively little fossil evidence for extinct creatures without skeletons. Bone is amazingly strong for its weight. Its strength is due to an extracellular matrix of very tough *collagen* fiber, hardened by crystals of calcium phosphate. The strength to weight ratio of bones is increased further by their being in the form of hollow tubes. Bone structure is maintained by cells called *osteocytes*, which are scattered through the matrix and receive nutrients from tiny blood vessels that thread through the bone. The conditions *osteopenia* and *osteoporosis* refer to bones of low and very low mineral density, respectively. The hollow parts of bones are filled with bone marrow, that produces red and white blood cells and stores fat.

Cartilage is a structural material that is more flexible than bone. It contains collagen filaments in amorphous protein and polysaccharide. Cartilage is maintained by cells called chondrocytes.

Each species has a skeletal structure uniquely designed for its survival. For example, the evolution of the hand in primates permitted the use of tools and the use of this organ in communication — allowing this branch of the evolutionary tree to prosper and multiply.

> **Muscles provide our bodies with movement and locomotion.**

Muscles work in conjunction with the skeletal system to provide animals with movement (motion of one part of the body with respect to other parts) and locomotion (movement of the entire organism from one place to another). Bones pivot around joints at which they are held in place by *ligaments*. Muscles also move soft tissues, such as eyelids and tongues, and pump blood throughout the body and food through the digestive system, as well as suck air into the lungs.

Our skeletal muscles are attached to bones through **tendons** and produce motion by contraction. Muscles come in antagonistic (working against each other) sets. Thus our biceps contract and triceps relax to bend our arms, while the triceps contract and biceps relax to straighten the arm. The skeletal muscles are under voluntary control of our nervous systems. Muscle cells are made up of long fibers composed of the proteins ***myosin*** and ***actin***. Upon receipt of their signals from nerves, muscles operate by the actin molecules moving along the myosin molecules. Muscle cells contain large numbers of mitochondria, along with fat deposits from which they can generate large amounts of ATP. Muscle movement involves doing work, which is performed with less than 50% efficiency, so that the remainder of the energy employed must be dissipated as heat.

The contractions of other muscles in our bodies are not under voluntary control, they receive their signals from the autonomic nervous system that functions below the level of consciousness. These include the muscles that move food in our digestive system and those that put pressure on our bladders. The heart muscle is unique, in that it receives it signal from electrical waves that are generated in the sinoatrial node, a small area of the right atrium. The rate of firing of this node, while typically about 70 times a second, can be accelerated or decelerated by signals from the autonomic nervous system. Such signals are based on the oxygen content of the blood and factors such as fear or stress.

The endocrine system

We respond to stimuli in two ways: by activating motor neurons, which produces a very rapid but localized response, and by releasing chemicals called *hormones* into the bloodstream, which is a slower but more general response.

> **Hormones provided by the endocrine system send signals to the organs of our body.**

It is advantageous for our body to be able to respond in both these ways. For example, seeing a dangerous animal might induce our legs to run and hormones to be released that increase our heart rate. The hormones allow us to run further and faster, but we don't want to wait around for them to work.

There are hundreds of chemicals that, at very low concentrations in the blood, play key roles in maintaining our body's homeostasis and shepherding it through the stages of life. In contrast to *exocrine glands*, such as tear, sweat and salivary glands, that release their product through a duct to a particular organ, *endocrine glands* release their products, which can be either proteins or small molecules, directly into the blood stream. The locations in our bodies of some of the most important endocrine glands are shown in Figure 19.7.

The *pineal gland*, located within the brain, produces *melatonin*, a hormone that regulates body rhythms, such as sleep. The *pituitary gland*, at the base of the brain, is a master gland that stimulates many other glands to produce their hormones. For example, upon receiving a signal from the brain, the pituitary gland increases its production of TSH, thyroid-stimulating hormone. Increased TSH levels in the blood signals the *thyroid gland* to produce more thyroxine and other thyroid hormones. Thyroxine is the most important iodine-containing compound in our bodies. Among its functions is the stimulation of the production of some proteins from DNA. It is particularly important in producing proteins in the mitochondria, and thus plays a vital role in providing energy for all the body's activities. When the level of thyroid hormone in the blood is low, a condition called hypothyroidism, metabolism is slow, and the individual feels sluggish and sleepy. Hypothyroidism at a very early age produces many

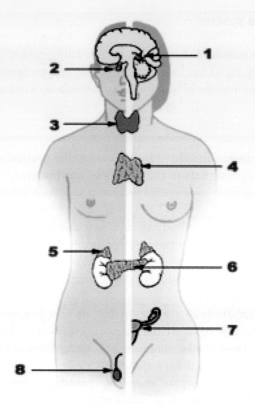

Figure 19.7. Locations of endocrine glands.
(Left, male; right, female; 1-pineal gland, 2-pituitary gland, 3-thyroid gland, 4-thymus, 5-adrenal gland, 6-pancreas, 7-ovary, 8-testes).

developmental defects, including cretinism, a form of dwarfism with severe mental retardation. Very high levels of thyroid hormone in the blood produces hyperthyroidism, a condition characterized by hyperactivity, weight loss and nervousness.

The thyroid can only respond to TSH, if there is a sufficient level of iodine stored in the gland. In the absence of sufficient iodine, a condition that often occurs in third-world countries, TSH stimulates the thyroid to produce more tissue and enlarge, a condition called *goiter*. Goiter is no longer common in industrialized nations, where iodine is routinely added to table salt.

The thymus is an important part of the immune system. In addition to making T-cells, it produces thymosin, a hormone that stimulates the production of antibodies in the immune system. The *adrenal glands*, located on top of the kidneys, produce hormones, such as *adrenaline*, that activate our body's response to stress. The *testes* and *ovaries* produce sex hormones, such as *testosterone* and *estrogen*, the levels of which determine sexual characteristics.

Most of the organs of the endocrine system have other functions in addition to producing endocrine hormones. For example, the pancreas produces digestive enzymes in addition to insulin, and testes produce reproductive cells as well as sex hormones.

The reproductive system

> The reproductive system produces gametes, provides for their fusion during fertilization, and supports embryonic development.

In sexual reproduction a zygote is formed from the fusion of two gametes (the sperm and egg). Animals use a variety of strategy to achieve this *fertilization*. We will only discuss human sexual reproduction, which employs internal fertilization of a large *ova* (egg) by a tiny *spermatozoon* (sperm cell). The resulting zygote matures in about nine months within a specialized organ, the *uterus*, of its mother. Internal fertilization is facilitated by means of the penis of the male parent, which must be rigid for insertion. Rigidity is achieved by vasodilation of the arterioles that supply blood to spongy erectile tissue in the penis.

The adult male reproductive system is shown in Figure 19.8.

Sperm are continuously produced in the *testes* at a rate of about 30 million a day. The testes are suspended outside the body in the *scrotum*, in order to keep them at an ideal temperature for sperm development, which is several degrees cooler than body temperature. In the testes, germ cells undergo meiosis, which produces immature spermatids that develop into sperm cells. As sperm moves through a long series of tubes they are mixed with the secretions of a number of glands, including the prostate gland, before being spasmodically ejected (ejaculation) through the urethra.

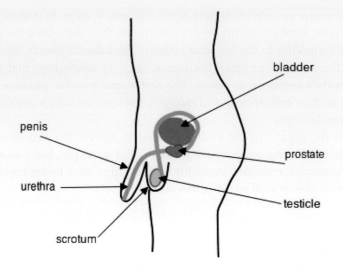

Figure 19.8. Male reproductive system.

A sperm cell is shown in Figure 19.9.

Its head contains highly compact DNA and the *acrosome*, a cap containing digestive enzymes that allow sperm to enter the ova. The tail of the sperm is a *flagellum*, which in beating moves the sperm up the reproductive canal of the female. A middle section of the sperm is richly

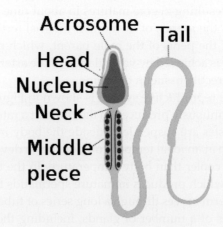

Figure 19.9. Sperm cell.

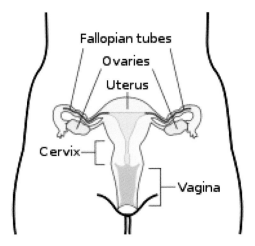

Figure 19.10. Female reproductive system.

endowed with mitochondria that supply the energy for beating of the tail. Sperm typically live only about a day after ejaculation.

The human female reproductive system is shown in Figure 19.10.

In a human female at birth, there are approximately 400,000 *oocytes* in the two ovaries, each of which can potentially form an egg. In the oocyte, the nuclear chromatin has already condensed into chromosomes, awaiting the first division of meiosis. Each oocyte is housed in a *follicle*, which surrounds it with a layer of cells that will support its development. After puberty, each month several follicles mature, during which time their oocytes grow considerably and complete the first division of meiosis, forming one large ovum and a smaller polar body. This process culminates with ovulation, when one ovum is discharged through the wall of the ovary into the surrounding *oviduct*. The ovum begins the second meiotic division during its descent to the uterus in the oviduct.

In order for fertilization of the ova to occur, a sperm cell must make its way from the vagina, through the cervix into the uterus and then into the oviduct. If a sperm enters an egg while it is in the oviduct, the second meiotic division of the ovum is completed, and one of the nuclei joins with the nucleus of the sperm. The other nucleus in the ovum forms another polar body that is ejected. The resulting zygote implants itself into the wall of the uterus and begins its development through various stages to form an embryo and eventually a newborn infant.

Summary

In multicellular organisms, cells may be organized into tissues, tissues into organs and organs into organ systems requiring a high degree of functional order. In our digestive system, foods are broken down into sugar, amino acid and lipid molecules, which are small enough to pass through the walls of the small intestine into the blood of the circulatory system. The heart pumps these molecules and oxygen attached to hemoglobin to cells throughout the body. Carbon dioxide produced by metabolism is removed from blood in the lungs, where hemoglobin is reoxidized. Other waste products are removed from blood in the kidneys.

We are protected from foreign substances by the walls of our respiratory system, by our skin, and by our immune system, which can mount a variety of attacks against invading pathogens. Our bodies are supported by our skeletal system and moved by our muscles.

The hormones that are released by our endocrine glands synchronize our organ systems. Many of these glands are controlled by signals from the brain that stimulate the master pituitary gland.

The male reproductive system produces sperm that can combine in the female oviduct with eggs to produce zygotes that, when implanted in the uterus, can become embryos and eventually newborn humans.

Questions

1. From what you know about the digestive system, what can you say about the following nutritional practices:

    (a) Taking condroitin sulfate, a large polymeric molecule, for joint health.
    (b) Eating nonsoluble fiber to move food along the colon.

2. Flatworms are multicellular organisms, but they do not have a circulatory system. Without a circulatory system, how can the cells of flatworms be provided with oxygen and relieved of carbon dioxide?

3. In men, an enlarged prostate gland can produce difficulties in urination. Using Figure 19.8, explain how this happens.

4. List two organs that are each part of more than one organ system.

5. Enteric-coated aspirin dissolves in the small intestine, not in the stomach. Speculate how this could be brought about.
6. Stomach acid is highly corrosive. How are the esophagus, stomach and small intestine protected from being digested by this acid.
7. Is the oxygenated blood, shown in blue in Figure 19.4, always carried in an artery?
8. Why is dialysis not a perfect replacement for a healthy kidney?
9. Give two examples of muscles in our body that are under voluntary control and two that are not under voluntary control.
10. What are some similarities and differences between sperm and egg cells?

20 The Nervous System

The human brain, then, is the most complicated organization of matter that we know.

Isaac Asimov

Information

In Chapter 18 we learned about information stored and transported within cells in the form of the nucleic acids, DNA and RNA. This information is present in every somatic cell and half of it in reproductive cells, so that it is passed on when cells divide and the organism reproduces. It provides an organism with its basic **Nature**. Improvements in the Nature of an organism (in the sense of giving it a greater chance to reproduce) result from evolution, a slow process requiring many generations.

In addition to information stored and transported over small distances and changed over long times, a multicellular organism also requires the ability to rapidly transport information over long distances and to store and change this information over short periods of time. The nervous system fulfills these needs. Our nervous system is made up of the *central nervous system* (CNS), consisting of the brain and spinal cord, and the *peripheral nervous system,* consisting of cells that project to other parts of our body.

The human brain is often compared with computers, which have assumed such an important role in society. Occasionally there are even formal competitions between computers and humans on tasks, such as playing chess. There are, however, fundamental differences between computers and our nervous system. Signal transmission in computers is exclusively electrical,[130] whereas

> **Signal transmission in the nervous system is both electrical and chemical.**

The electrical transmission in the nervous system is quite different from that in computers because our brains and computers are constructed from different types of materials. Computers make extensive use of excellent solid-state electrical conductors[131] and insulators. The former allows rapid transfer of information as electric current, and the latter permits it to be stored as electrical charge for long periods without requiring energy.[132] Compared to metallic conductors, the ionic solutions in our bodies do not conduct electricity as rapidly and dissipate more energy. The plasma membranes of cells do provide electrical insulation, but with much more "leakage" than solid-state insulators. *Neurons* are the cells that allow our nervous system to overcome these limitations.

The Neuron

Although there are over a hundred different types of neurons in the nervous system, most of them look similar to the one shown in Figure 20.1.

A neuron has a central cell body, consisting of a nucleus and cytoplasm (*soma*) that provide the instructions and metabolism of the cell, a multitude of **dendrites** that project from the cell body, and a single *axon* that extends from the cell body to cells in muscles, glands or the nervous system. The axon branches into a number of terminals, allowing it to simultaneously stimulate a number of other cells. Groups of cell bodies

---

[130] Some computers transmit signals optically.

[131] Some computers use **superconductors**, materials with zero resistance to the flow of electrical currents.

[132] Some computers store information as magnetic or optical domains.

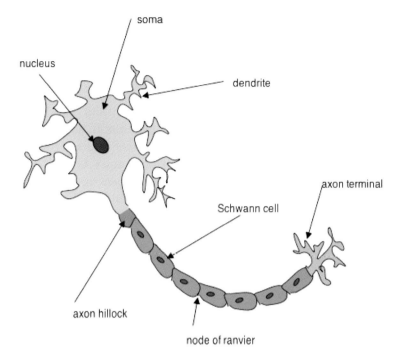

Figure 20.1. A neuron.

constitute the *grey matter* of the brain, while regions in which axons are predominant make up its *white matter*.

Signal Transmission along the Axon

Transmission of a signal along an axon of a nerve (***the axon potential***) is like the transmission of a signal along a row of upright dominoes. Each domino is in a metastable state, resting on its shortest edge, with a small barrier for knocking it over into a more stable state. As soon as the first domino makes this transition, it pushes on the adjacent domino, propagating a transverse wave pulse down the row of dominos.

The transport of electrical signals along the axon of nerve cells is also by a transverse wave, where sodium and potassium ions move across the plasma membrane of the axon. In the axon, potassium ion concentration is higher and sodium concentration lower than in the intercellular fluid. Ion pumps, which are large protein molecules imbedded in the axon cell membrane, produce these differences in ionic concentration using the reaction,

$$3Na^+(in) + 2K^+(out) + ATP \to 3Na^+(out) + 2K^+(in) + ADP + P_i,$$

where "*in*" refers to a concentration within the axon, and "*out*" refers to a concentration in the adjacent extra-cellular medium. In addition to producing a higher Na^+ concentration outside the axon and a higher K^+ concentration inside the axon, reaction I transports three units of positive charge out of the axon for every two that are pumped in, resulting in the interior of the axon accruing a negative "resting potential," which is typically −70 mV. An axon with this potential is said to be **polarized**, indicating a separation of charge across the membrane. Organic anions within the axon and chloride anions outside the axon also contribute to establishing this potential.

Differences in concentration and electric potential provide an electrochemical potential, which would normally cause ions to diffuse through the membrane to equalize their concentration on both sides. Ions, however, are not soluble in the hydrophobic environment of the lipid membrane.[133] However, the membrane also contains sodium and potassium ion channels, which leak, so that reaction I must continually pump ions. This reaction consumes the high-energy molecule, ATP, which is produced by mitochondria in the axon and requires approximately 20% of the energy used by our body (perhaps even more when we are studying for an exam). For production of this energy, the brain relies on a continual supply of glucose and oxygen in the blood stream. Interruption of blood supply, due to heart failure or obstruction of blood vessels, very rapidly leads to loss of brain function and then to death.

Ion channels have a hydrophobic exterior, which allows them to extend through the membrane and a hydrophilic interior, which allows them to dissolve ions. The dimensions of a particular ion channel are usually favorable only for the transport of one type of ion. Thus, there are channels that transport Na^+, others that transport K^+ ions and others that transport Ca^{2+} ions. Calcium ion channels are located at **synapses**, the region between neurons. Ion channels are gated, i.e., they are only

---

[133] Certain small organic molecules, called ionophores can wrap themselves around ions and facilitate this transport.

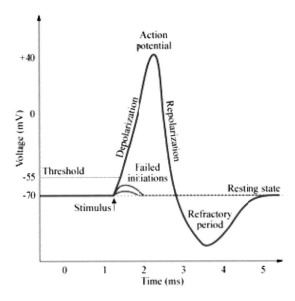

Figure 20.2. Firing of an axon potential.

open when they receive a signal, which may be either electrical or chemical.

The firing of an axon potential, shown in Figure 20.2, is initiated at the ***axon hillock***, a small region between the axon and the cell body. The axon hillock functions as an integrator, summing the electric potentials generated by charges delivered to the many dendrites of the cell body.

When the potential at the axon hillock reaches a threshold value, sodium-ion channels in the axon membrane open, and the flow of Na^+ ions into the axon briefly raises it to a positive potential, ***depolarizing*** the axon. At the positive potential, potassium-ion channels open, and the flow of K^+ ions out of the axon, restores it to a negative potential. The brief positive pulse causes ions to flow along the axis of the axon, sequentially opening channels, providing the nerve impulse. This is followed by a refractory period, during which the nerve cannot respond, during which the channels are closed and the ion concentrations are pumped back to their resting values. A nerve signal is digital; the nerve either fires or it doesn't. The magnitude of the voltage pulse is independent of the hillock potential. The size of a nerve signal measured by the rate at which the neuron fires, which is controlled by the voltage at the axon hillock. The hillock potential is determined by the inputs at all the dendrites of the cell, some of which increase and others which decrease its potential. A nerve combines and integrates these signals.

Just as the power of a computer is largely determined by the speed in which it can process data, the power of your nervous system results from the speed at which its nerve impulses travel. The flow of ions along the axis of axon is a slow process and limits the speed of the nerve impulse. However, special *glial* cells, called Schwann cells in the peripheral nervous system and oligodendrytes in the CNS, wrap around the axon and provide it with an insulating *myelin sheath*. The myelin sheath is not continuous; it is broken at the *nodes of Ranvier*. In a nerve surrounded by a myelin sheath, electrical impulses jump by a purely electrostatic effect, with no current flow, from one of these nodes to the next, at speeds up to 100 m/sec. In nerves in which the myelin has been damaged, signals move much more slowly. *Multiple sclerosis*, MS, is a disease in which the myelin coating of nerves is attacked by the body's immune system. It can cause a variety of symptoms, including muscle weakness and difficulty in moving, coordinating and balancing, and in some cases, death.

Signal Transmission at the Synapse

A nerve pulse is transmitted along the axon electrically, but from the axon to another cell it is usually transmitted chemically. Chemical stimulation provides amplification, as well as variety in the nature of the response. Nerves stimulate other cells chemically at the *synaptic cleft*, a gap of 20 to 40 nanometers that exists between the terminal of an axon and its target cell (nerve, muscle or gland).[134] Synapses have a variety of shapes; a typical one is shown in Figure 20.3.

The arrival of an action potential at the terminal of the *presynaptic neuron* (colored orange in the figure) opens calcium ions channels. Ca^{2+} ions entering the axon stimulate vesicles in the terminal to fuse with the cell membrane and release their content into the synaptic cleft, a process called *exocytosis*. Each vesicle contains a few molecules of a neurotransmitter. The most commonly employed neurotransmitter is glutamate

---

[134] The gaps at some synapses are only 3–4 nanometers, allowing signals to be transmitted electrically. Electrical transmission at a synapse is faster than chemical transmission, bur occurs without amplification.

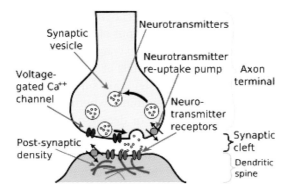

Figure 20.3. The synapse.

(glutamic acid, a basic amino acid). In Chapter 16, structures are given for serotonin, dopamine, acetylcholine, nor-epinephrine and GABA, which are among the more than 100 other neurotransmitters that have been discovered.[135] Although each axon potential pulse releases approximately the same amount of neurotransmitter, the amount of neurotransmitter released into the synaptic cleft depends on the rate of firing of the axon.

Neurotransmitters cross the synaptic cleft to the target cell, which may be a muscle cell, a gland or a knob on the dendrite of a *postsynaptic neuron*, where they bind to receptors. The small width of the gap allows rapid transmission of signals even though it occurs by the slow process of molecular diffusion.

Neurotransmitters released from the presynaptic neuron bind to receptors of the target cell. These receptors are protein molecules, which span the cell membrane and present a geometry specific to binding a particular neurotransmitter. When a neurotransmitter binds to its receptor, it induces a change in the receptor protein molecule, which may either open an ion channel or release a messenger molecule in the target cell. If the target cell is a neuron the ion flow results in either stimulation or repression of its response. These effects are only transient, however, since the neurotransmitters are rapidly destroyed by enzymes in the cleft, or taken up by a transporter in the cell membrane of the synapse of origin, a process called *reuptake*.

---

[135] Nitric oxide (NO) is a small-molecule neurotransmitter, which is synthesized as needed, rather than stored in a vesicle. Nitric oxide can diffuse through the membrane of the target cell.

If the target cell is a neuron, the effect of a neurotransmitter is to either encourage or inhibit its firing. The output of a single presynaptic neuron is not usually sufficient to cause a postsynaptic neuron to fire. However, the postsynaptic neuron receives signals at many dendritic spines from many different neurons. It is the algebraic sum of the signals received over a short period of time that determines whether the postsynaptic neuron will or will not fire, or increase or decrease its firing rate. The same neurotransmitter may have either a stimulatory or an inhibitory effect on different neurons. The firing of a neuron is a very complex function of the inputs to its synapses. Because of this, computer modeling of processes in the brain often employ "fuzzy math," a type of algebra that manipulates probabilities.

Because the nervous system controls thought and motion and regulates many other processes in our bodies, improper levels of neurotransmitters can cause a variety of illnesses. These range from those that are clearly related to thought processes (depression, anxiety and psychosis) to those involved with movement (Parkinson's disease) to malfunction of other systems in the body (high blood pressure and gall stones). Since nearly simultaneous stimulation at many synapses is required to change the firing rate of a neuron, we are able to focus our attention on tasks and ignore the huge number of stimuli with which we are usually bombarded, which usually only stimulate a few synapses.

Some of our most-often used drugs function by adjusting neurotransmitter levels. For example, serotonin is primarily an excitatory neurotransmitter and selective serotonin reuptake inhibitors (SSRIs), such as Prozac, Paxil, Zoloft and Celexa work by inhibiting serotonin reuptake. Dopamine is also excitatory, and cocaine operates by blocking dopamine reuptake.

Central and Peripheral Nervous Systems

Our nervous system is composed of two parts: the central nervous system (CNS), consisting of our brain and spinal cord,[136] and the peripheral nervous system that senses signals from the world external and internal to our

---

[136] The retinas of our eyes are outgrowths of our brain and therefore part of the CNS.

bodies and drives the responses of our bodies to these signals. Among these signals are sight, sound, smell, touch, pain and temperature.

In addition to neurons, our central nervous system contains glial cells that support, nourish and protect neurons and regulate ionic concentrations in the intercellular fluid. It has about 100 billion neurons and ten times that number of glial cells. Epithelial cells line the blood vessels of the CNS.

Neurons process and transmit information, and synapses, the junctions between neurons, store this information. There are ca. 10^{15} synapses in the CNS. The cell bodies of groups of neurons are ordered into **nuclei**, with definite organization and connections. Neurons do not generally undergo mitosis[137] in the adult brain, but glial cells are continually generated from their precursor cells. As a result, brain tumors usually occur in glial cells.

In a typical muscular response, an input signal to a sensory neuron causes the movement of a muscle. Some responses need to be exceedingly rapid — for example, the movement that enables us to regain our balance after stumbling. Such *reflexes* travel a shorter path, connections between the sensory and motor neuron being made in the spinal cord. Before our brain is aware of our stumbling, we have already acted to correct it. Other responses are routed through the brain and are part of the *somatic nervous system* if they are under voluntary control or the *autonomic nervous system*, if they occur automatically, as for example occurs with the regulation of our heart, blood pressure and digestion. In the autonomic system, the nerves are of two types, nerves of the *sympathetic system*, which are activated as response to stress, and nerves of the *parasympathetic system*, which are activated by relaxation.

Organization of the Brain

A human brain has a definite and complex structure. It is divided into two *hemispheres*, linked by a large bundle of nerves called the *corpus callosum*. The *cerebral cortex* is the thin (2–4 mm) layer of grey matter of the outer surface of the brain, where many of its higher functions are located. Its area is greatly increased by extensive wrinkling, producing ridges (gyri, singular = *gyrus*) and valleys (sulci, singular = *sulcus*). The major

---

[137] There is evidence that neurons in the hippocampus, a part of the brain involved in long-term memory and other functions, are capable of mitosis.

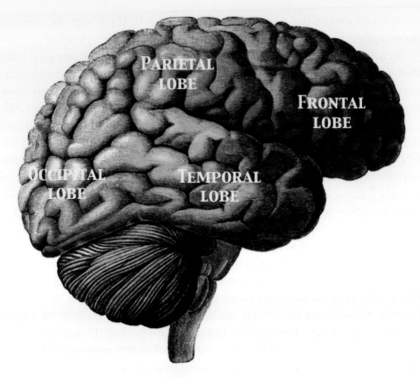

Figure 20.4. Lobes of the brain.

valleys divide each hemisphere into *lobes* (*frontal* — toward the front; *parietal* — towards the top; *temporal* — towards the sides; and *occipital* — toward the back), as shown in Figure 20.4.

The *cerebellum* and *brain stem*, located below these lobes, carry out many essential but subconscious functions, with the latter passing out of the skull and connecting to the spinal cord.

Unlike a modern computer, where most memory is flexible and can be programmed for a variety of uses, our brains are largely hard wired.

Different parts of the brain carry out particular tasks.

As an example of the specialization of the brain, the *somatosensory system* will be discussed. This system produces sensations such as touch, temperature, pressure, pain and body positions. Receptors sensitive to

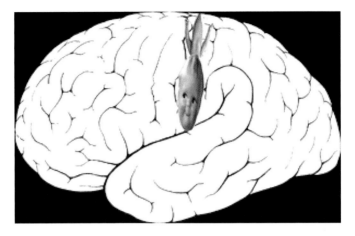

Figure 20.5. Homunculus, showing mapping of sensory regions.

these stimuli are located throughout our bodies, especially in our skin. Most of these inputs are routed to a structure deep inside the brain called the *thalamus*. There are two thalamus's, with the right thalamus receiving input from the left side of the body and the left thalamus receiving input from the right side of the body. The thalamus serves as a relay station for sensory input (and many other signals). Somatosensory signals are then sent to a particular region of the cerebral cortex called the postcentral gyrus (the first ridge) of the parietal lobe. Sensory input from each part of the body is mapped to a particular area of this region, as illustrated by homunculus (little person) shown superimposed on the brain in Figure 20.5.

As shown in the figure, in the postcentral gyrus, larger areas are devoted to regions of our bodies that are more sensitive to touch (such as our face and fingers) than to regions that are less sensitive (such as our backs).

After signals enter or before they leave the CNS, they are subjected to multiple layers of processing in different regions of the brain. For example, if a certain part of the brain is damaged, a person will report that he is blind, even though his ability to walk without bumping into things indicates that he is getting input from his brain. He can see, without being aware that he can see.

Left-right crossover also occurs for motor output signals. Language functions are largely handled by the left side of the brain, and spatial functions by the right side of the brain. Thus, depending on a person's particular strengths or weaknesses, we might label them as right or left-brain personalities.

The *hypothalamus* is a small region of the brain located just below the thalamus and above the brain stem. It is adjacent to the pituitary gland, which controls the endocrine system. The hypothalamus maintains homeostasis in the body by secreting *neurohormones* that stimulates the pituitary gland, exerting control over body temperature, hunger, thirst and fatigue.

Higher-Order Processes

Higher-order cognitive abilities, such as language, problem solving, abstract reasoning, morality and aesthetics, all involve memory.

> **The synapse is the basic element at which memories are stored in the brain.**

For example, when two closely spaced axons repeatedly fire simultaneously, synaptic connections between their stimuli are first stored in *short-term memory* and, with sufficient rehearsal or importance to us, are transferred to *long-term memory*. Short-term memory can only hold a few items, as indicated by our inability to remember much more than seven digits, whereas long-term memory capacity is very large. Certain illnesses, such as Alzheimer's disease, include an inability to convert short-term into long-term memories. Since the brain has roughly 10^{11} neurons, each of which has of the order of 10,000 synapses, the information storage capacity of the human brain is huge. Storage is achieved by increasing the strength of synaptic connections (short-term memory) and increasing the number of connections (long-term memory). Memories influence how our brains work — how we think about things. Thus, in contrast to a digital computer, in our brains, data (memories), software (how memories are processed) and hardware (synaptic connections) are all mixed together.

In order to insure that we carry out the vital functions that are necessary for our survival, the brain incorporates a *reward system*. When we eat, for example, a signal makes its way from our full stomach to the ventral tegmental area (VTA) in our brain, some of whose neurons project to the nucleus accumbens, a brain region whose stimulation provides intense feelings of pleasure. These feelings reward and reinforce our eating behavior. VTA neurons employ dopamine neurotransmitter, and pleasurable feelings increase with increased concentrations of dopamine in its synapses to the nucleus accumbens.

Cocaine is an illicit drug that functions by inactivating dopamine reuptake sites at synapses. Its use thus produces almost continual stimulation of dopaminergic neurons, including those in the reward system, producing a cocaine "high." The body compensates for such massive stimulation by reducing the sensitivity of the neurons, resulting in a "crash" when cocaine is withdrawn. In most cases, in not too long, *physiological addiction* occurs, and greater amounts of cocaine are needed to achieve the high. Vital functions no longer produce sufficient stimulation. Rats trained to press a bar to obtain a shot of cocaine will do so repeatedly and pay no attention to food, water or sex, until they are completely exhausted and die.

The VTA also projects neurons to regions in the cortex that remember the process of obtaining and using the drug. These processes are reinforced by the high. The desire to repeat these actions results in *psychological addiction* to the drug.

It's hard to believe that all our thoughts and emotions, as well as the regulation of complex processes in our bodies can be accomplished by the simple firing of axon potentials in neurons. However, as any one who has ever observed an ant farm can testify, the sum of a very large number of simple actions may be quite complex. In early computers that were slow and had a limited amount of memory, output was by a simple lighting up a selection of bulbs. Now we have computers that provide complex graphics and sometimes, even talk to the operator. Most impressive are computers that are used for artificial intelligence, i.e., attempt to "think" like humans. In such computers a degree of flexibility is incorporated into the computer's operation, so that it can learn from experience. A robot programmed this way can learn from interaction with its environment in much the same way as does the human brain.

The brain can be involved in a number of tasks at the same time. For example, we read a book while our breathing, blood glucose and hormone levels are being monitored and controlled. Vision involves simultaneously employing pathways for perceiving color, location in space, shape and movement. It is doubtful, however, whether we can *consciously* be involved in multiple tasks as is achieved in a parallel-processing computer, which is composed of a large number of central processing units.

Summary
03 80 03 80 03 80

The nervous system is composed of neurons that transmit signals and other cells that provide support services for the neurons. A neuron is comprised of a cell body that obtains input signals from a number of dendritic extensions and a single axon output. Signal transmission is electrical along the axon and chemical over the synapses through which neurons communicate with other cells. The central nervous system contains ca. 100 billion neurons, which make ca. 10^{15} synaptic connections.

Transmission of impulses along axons is achieved by the sequential opening of ion channels, dissipating the metastable ion gradients established by a pumping reaction, which is a prodigious user of energy. Axons are wrapped with a myelin sheath, permitting the impulse to jump along the axon at high speed.

Neurons stimulate or inhibit other neurons, or muscle and glandular cells at synapses, where the axon potential releases neurotransmitter molecules into a narrow synaptic cleft. Excitatory or inhibitory signals received at the dendrites are integrated at the axon hillock, which only fires the axon when a threshold potential is reached. Axon potential pulses are the same size; the nerve signal is determined by the rate of firing of the axon.

Reflexes are very rapid muscular responses to external stimuli, in which nerve connections are made in the spinal cord, rather than in the brain.

The brain has two hemispheres, which are connected by a bundle of nerves. Different structures in the various lobes of the brain carry out specific tasks. For example, different regions of the somatosensory system handle sensory input from receptors in the skin in particular parts of our body. A reward system is stimulated by the neurotransmitter, dopamine, whose uptake is blocked by drugs, such as cocaine. Short-term memories are encoded by strengthening synaptic connections, and long-term memories by increasing the number of such connections.

Questions

1. What are the strengths and weaknesses of computers and brains in playing chess?
2. Why is the signal of an axon a transverse wave?
3. What are some of the ways that the firing of an axon potential is similar to and different from the falling of a row of dominos?
4. Draw a diagram showing the motion of ions in reaction I.
5. How is the digital signal of the axon potential (either pulsing or not pulsing) converted into an analog signal that can have values in a range?
6. A person who is considered deaf is startled by a loud noise. Give a possible explanation for this.
7. It is said that an infant first gets to know the world with its lips. Why is this?
8. Cocaine acts by inhibiting the reuptake of dopamine, which is usually an excitatory neurotransmitter. Speculate on the mechanism whereby cocaine produces a "high" and its subsequent withdrawal can produce a "low."
9. Marijuana is said to be psychologically addictive, but not physiologically addictive. Explain what this means. If true, does that make marijuana any less dangerous than a physiologically addictive drug, such as cocaine?
10. What would be some strengths and weaknesses of someone who is considered to be a "right-brain person"? What about someone considered to be a "left-brain person"?

21 The Earth

The Earth was small, light blue, and so touchingly alone... The Earth was absolutely round.

Aleksei Leonov, USSR Cosmonaut

We are interested in planet Earth, not only because we live here, but also because it represents an environment in which life, especially advanced forms of life such as ourselves, has developed.

Although there may be life elsewhere in the universe, it is likely that there is none on any other body in our solar system.[138] What is it that makes the surface and near-surface regions of our planet conducive for establishing and maintaining advanced forms of life? We will see that unlike almost all the other bodies in our solar system, the surface of the Earth is continually changing — due to influences from below, from above and by processes that occur on the surface.

---

[138] The possibility of simple life forms such as bacteria living beneath the surface of Mars has not yet been ruled out, and has been suggested by hints of methane in the Martian atmosphere. Since impacting bodies result in exchange of material between Earth and Mars, the discovery of life on Mars could be consistent with life originating on either planet.

334 *Order and Disorder*

> **The continually changing nature of Earth's surface is favorable for the development of life.**

A changing environment encourages the development of advanced life forms by periodically challenging the survival of species, a challenge that is met, in part, by their evolution, which will be discussed in the next chapter.

Earth provides a variety of environments. Its surface is composed of rigid rocky parts called the *lithosphere*, bodies of liquid water, called the *hydrosphere*, and a substantial gaseous layer called the *atmosphere*. The fact that water can exist in the solid, liquid and gaseous states under conditions that exist at the Earth's surface provides one agency for changing its characteristics.

View from Above

How much do we know about planet Earth, which we call home? What is it like today and what was it like in the past? The view from space, shown in Figure 21.1, led a Russian cosmonaut to call the Earth a perfect sphere.

Figure 21.1. Earth from space.

However, more accurate measurements show that Earth's radius at the equator, 6,378 km,[139] is 0.3 percent larger than that at the poles, 6,357 km. This is as would be expected for a deformable body held together by gravity and distorted by centrifugal force due to its daily rotation around its axis.

The medieval church and most of the ancient Greeks strongly endorsed the idea of a *geocentric* universe, with the sun, moon and stars moving on spheres around the Earth. However, Copernicus, in his *heliocentric* model, introduced the idea that the Earth and the other planets revolve around the sun. This idea was heresy at the time, since it was thought that man, the most perfect creation of God, should be at the center of the universe.[140] However, after the work of Kepler, Galileo and Newton in the seventeenth century, the heliocentric model became generally accepted, because of the simplicity of its descriptions.

The Earth is at an average distance of 150 million km from the sun and moves in a slightly elliptical orbit, being ca. five million km closer to the sun in January, at its *perihelion*, than it is in July at its *aphelion*. The plane of the Earth's motion around the sun is called the *ecliptic*, and the Earth's rotation axis is tilted at an angle of 23.5° from this plane. As the Earth revolves around the sun, the sun reaches its most northerly direction at noon on June 21, which in the northern hemisphere is known as the *summer solstice* — the beginning of summer. On December 21, the *winter solstice* in the northern hemisphere, it reaches its most southerly direction. As shown in Figure 21.2, the tilt and revolution of the Earth provides more direct solar radiation on the northern hemisphere in July and more direct solar radiation on the southern hemisphere in January. This variation in the intensity (J/sec m^2) of solar radiation produces the seasons.

From space it is possible to accurately map the Earth's magnetic and gravitational fields. The Earth acts like a magnet, with its North magnetic pole displaced about 1200 km from the North geographic pole (the axis of rotation). There is considerable evidence that over time the position of the North magnetic pole has wandered, and has even "flipped" into the

---

[139] 1.0 km = 0.621 miles

[140] Giordano Bruno was burned at the stake in Rome in 1600 for a number of heresies, including his preaching the heliocentric model.

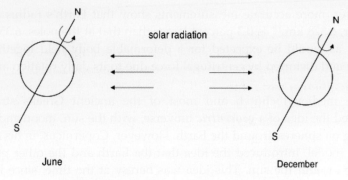

Figure 21.2. Earth in summer and winter.

southern hemisphere.[141] Recent accurate measurements of the Earth's gravitational field from satellites have provided information on the density distribution of the upper regions of the Earth and the temperature and salt-concentration of the oceans.

View from the Surface

Although processes occurring in the inner regions of the Earth largely determine the nature of its surface, we only directly experience the very top layer of our planet. Measurements in the deepest mine (4 km) and the deepest borehole (13 km)[142] indicate a marked increase in temperature and pressure with depth. However, in order to learn about even deeper regions, we must rely on remote sensing and material brought to the surface by volcanic processes.

The interior of the Earth can be probed by *seismology*, the study of the propagation of vibrations through the Earth.[143] It takes a very large disturbance, such as a nuclear explosion or an earthquake, to be detectable after traveling a long distance through the Earth from its source. Both longitudinal (*P waves*) and transverse (*S waves*) can be generated and

---

[141] From the orientation of magnetic grains at the time of their solidification from molten magma.
[142] As of 2009.
[143] A *seismograph* measures the relative motion of the surface to that of a suspended heavy mass, which does not appreciably move because of its inertia.

travel with slightly different sound velocities, which are higher in more dense materials. As these disturbances pass between two regions of different sound velocity, their path is bent, much as for light passing through a lens. From seismology we learn about the density and mechanical properties of different parts of the Earth's interior, which in turn tells us about their chemical nature, and temperature and pressure distributions. Seismology is used to verify compliance with nuclear test bans, since it can distinguish between earthquakes and nuclear explosions. The former release their energy in a very directional manner, while in the latter energy is released nearly equally in all directions.

Our present picture of the Earth's interior is shown in Figure 21.3. The thickness, density and temperature of the different regions are given in Table 21.1.

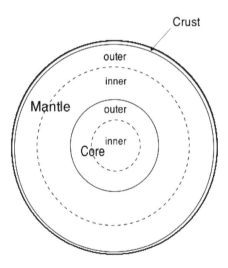

Figure 21.3. The Earth's interior.

Pressure and temperature increase with depth, with the center of the Earth being about as hot as the surface of the sun. Nevertheless, the inner core, the material within ca. 1200 km of the center, is solid, due to it being under the pressure of four million atmospheres. Iron and similar metals make up this *solid core*. Surrounding this is the *liquid core*, extending to a radius of ca. 3500 km (2170 miles), where the same materials are molten. It is the convective motion of these liquid metals in the rotating Earth that

Table 21.1. Regions of the Earth's interior

Region	State	Thickness (km)	Density (g/cm^3)	Temperature (1000 K)
Continental crust	Solid	20–40	2.7	<1.0
Oceanic crust	Solid	5–10	3.0	<1.0
Outer mantle	Solid[144]	1100	3.4–4.4	1.4–2.0
Inner mantle	Fluid	1800	4.4–5.6	2.0–3.1
Outer core	Liquid	2300	9.9–12.2	3.7–4.3
Inner core	Solid	1200	12.8–13.1	4.3–7.0

acts like a huge current passing through a coil of wire and produces the magnetic field of the Earth.

While the Earth on which we walk may seem solid and unyielding, those of us who have experienced earthquakes or tsunamis[145] or observed lava flowing from volcanoes must suspect that below its solid exterior (that we call the *crust*) the Earth is much more motile. The reason for this is that the Earth is a body in the process of cooling. When, about 4.5 billion years ago, gravity clumped together the gases, dust and rocks (the *solar nebular*) in a range of distances around the sun to produce the "proto-Earth", a tremendous amount of potential energy was converted to thermal energy, producing a planet that was initially molten. Over time, additional energy has been released in the Earth's interior by radioactive decay of heavy elements and by the latent heat released as the liquid core solidified, and at the surface by collisions with other bodies. One such collision may have been large enough to eject material into orbit around the Earth that later formed the Moon. In the liquid Earth, materials of higher density were pulled by gravity towards the center — a process called *differentiation*.

Since heat flow by conduction and convection through the rocky material of the Earth is slow, very high temperatures have been maintained in the inner regions of our planet. This is in contrast to smaller rocky bodies in our solar system, such as the planets Mercury and Mars

---

[144] Includes a rigid upper layer above a deformable layer called the asthenosphere.
[145] The tsunami of December 2004, which caused over 200,000 casualties, was caused by a sudden uplifting of up to 40 meters along a 600-mile fault below the Indian Ocean.

and our moon, which have lost most of the heat released in their formation. Venus, having a radius 95% of Earth, is an intermediate case, and probably still has a very hot interior.

The Earth's Surface

Since cooling of the Earth is by radiation at its surface, its surface solidifies first and forms a solid crust, much like what occurs when water cools in an ice tray in the freezer. The crust and the rigid upper mantle float on a deformable *mantle* layer called the *asthenosphere*. Heat transfer through the asthenosphere is by thermal convection, whereby heated material becomes less dense than its surroundings, and rises.

> **Because the Earth's solid crust sits on a moving layer in the mantle, it breaks up into sections called tectonic plates.**

The *tectonic plates* are typically several hundred km thick and several thousand km in extent. Since the velocity of the plates is on the average only a few cm a year, this motion is not usually perceptible to us. However, as the plates move against each other, sometimes they get "stuck" for a period of time, resulting in the build-up of very large shear forces, which are periodically relieved by the sudden motions that produce earthquakes.

> **Supplemental Reading: Magnitude of Earthquakes**
>
> The strength of an earthquake is usually measured by its *magnitude*. While there are a number of different earthquake magnitude scales, the two most often used, the Richter scale and the moment scale, agree quite closely. Charles Richter, a Californian seismologist, published his scale in 1930, and set the weakest movement that he could detect with his equipment equal to magnitude zero. Each increase of one unit of magnitude corresponds to about a thirty-fold increase in the energy released in the quake. The earthquake off the coast of

> Sumatra that caused the tsunami of 2004 was magnitude 9.0. Since we have more sensitive equipment today than was available to Richter, earthquakes with negative magnitudes can be detected. Worldwide, there are many earthquakes, with over a hundred with magnitude 6.0 or greater usually reported each year.

A large fraction of the heat flow to the Earth's surface is by convective flow of hot molten rock (lava) to the surface. This flow occurs near the edges of the plates where volcanoes are located. It also occurs through long cracks that exist in the thinnest parts of the crust at plate boundaries. One such crack occurs near the center of the Atlantic Ocean, at the mid-oceanic ridge. At the center of this ridge there is a deep valley called a *rift*, in which scientists have identified a *new-crust zone*, with heat release and material outflow in both directions. It is the flow of this material that drives apart the plates on either side of the rift, as it continually adds to these plates. When this molten lava solidifies, it forms a black rock called **basalt**.

Since new crust is continually being formed while the surface of the Earth retains a constant area, crust must also be destroyed. This occurs at **subduction zones**, where one plate bends down beneath another, returning material to the mantle. In this region, the downward moving plate forms a trench and lifts the upper plate, producing mountains. The subducting material becomes molten as it sinks, and lava can find its way to the surface through cracks and fissures in the upper plate. These processes are responsible for many of the features that we observe on the Earth's surface, such as the mountains and volcanoes in western parts of the US and Canada, where the Pacific Plate is colliding with the North American Plate, and the Himalayas, where the Indian-Australian Plate is colliding with the Eurasian Plate.

The hypothesis of moving continents was first proposed by Alfred Wegener in 1912. The evidence for such motion is overwhelming, and includes the "matching" contours and fossil records of the east coast of South America and the west coast of Africa. In addition, small grains of magnetic material in the molten lava flowing out of rifts tend to become oriented in the direction of the Earth's magnetic field. The periodic "flipping" of the direction of this field produces magnetic stripes in the newly formed solidified plates. Corresponding stripes have been found on the two sides of the sea floor spreading from the mid-Atlantic ridge. Tracing back the motion of the continents has led geologists to propose that they all were connected in a "supercontinent" called *Pangaea* about 200 million years ago. Since that time the continents have moved apart and pieces of the continents have separated from them to form *continental islands*, such as England, Madagascar and New Zealand, which are or have been attached to continental shelves. A similar separation is currently occurring between California and the continental United States. *Oceanic islands*, such as Hawaii or Iceland, on the other hand, are formed from volcanic activity occurring far from continental shelves.

Bombardment of the Earth

The Earth's surface is influenced by external, as well as internal factors. Objects in space continually approach the Earth. These range from high energy particles and ions emitted by the sun and higher energy cosmic rays generated elsewhere in our galaxy, to a variety of rocks (*meteoroids*) of different sizes. The latter are remnants of asteroids and comets, and material ejected from the Moon and Mars by impacts with such bodies.

The Earth's magnetic field directs charged particles toward the North and South poles, where they produce the very colorful *Aurora Borealis* (northern lights) and *Aurora Australis* (southern lights). Molecules and the smaller meteoroids are intercepted by the atmosphere, with rocky materials often producing noticeable tracks (*meteors*) in the sky as they burn up due to friction. Meteors that survive to the surface are called *meteorites*, and the larger of these can produce *impact craters* when they hit the surface. Due to their large kinetic energy, an impact crater can have a diameter ten times that of the impacting body. Several hundred such impact craters have been identified on the Earth's surface. They are characterized by their nearly circular shape and steep rim. In many cases the surrounding rocks show layers caused by shock waves and may contain very high concentrations of metals, such as platinum, iridium and rhodium that are prominent in asteroids. *Tektites*, which are beads formed from molten quartz, are usually surrounding these craters.

We have a better chance of observing an impact crater on the Earth's surface, if it is of relatively recent origin, very large size or has occurred in a dry region. Observed craters include Meteor Crater in Arizona, produced 50,000 years ago, with a diameter of one-half mile, and the Sudbury crater in Ontario, Canada, with an age of 1.8 billion years and a diameter of 125 miles. The latter is the site of extensive mining operations that extract metals deposited by the meteor. A number of large-scale features on the Earth's surface are undoubtedly due to impacts, and many rocks have also been identified as being of extra-terrestrial origin.

> **Supplemental Reading: Extinction of the Dinosaurs**
>
> The consequences of the impact of a large meteor on the Earth's surface would be dramatic. Besides the local effects, large amounts of hot material would be ejected from the surface into the atmosphere, producing worldwide forest fires when it fell to the ground. The resulting soot and ash would obscure sunlight for several years, resulting in greatly diminished plant growth and the death of animals that heavily rely on plants for food. One such impact, involving an object of 10 km in diameter, occurred 65 million years ago and is considered by most scientists to be responsible for the extinction of many species, including all the dinosaurs. Evidence of the impact crater has been found off the coast of the Yucatan peninsula in Mexico, and global deposits of iridium and tektites have been dated to that time.

Weathering and Erosion

Anyone who has seen pictures of the Moon and Mars, two of our nearest neighbors in the solar system, has noted how the surfaces of these bodies are pockmarked with a variety of craters of all different sizes. On Earth, we see evidence of only the largest or newest impacts. Also, on Mars there are mountains that are much larger than those on Earth.[146] Assuming that Earth has had a history similar to Mars and has been subjected to impacts similar to the Moon, what has happened to the order produced on Earth's surface by impacts and tectonic processes?

---

[146] Olympus Mons, a mountain of volcanic origin on Mars, is 15 miles high and its base would cover much of the states of Oregon and Washington.

Evidently, on Earth's surface, processes that break up rocks (*weathering*) and move them and their fragments from place to place (*erosion*) are much more important than they are on the Moon or Mars. These processes continually produce disorder that removes the order of the surface. The increased importance of weathering and erosion on our planet is due to the presence of liquid water on Earth throughout most of its history.

> **Weathering and erosion, enhanced by water, are continually removing surface features of the Earth.**

Recent evidence from exploration of the surface of Mars, indicates that there too, at some time in the past, liquid water has played a role in modifying the surface.

Weathering can be due to physical, chemical or biological processes. Rocks in the driest deserts, free from the influence of liquid water and life, are fractured by the large thermal stresses caused by temperatures that can vary from as high as 60°C during the day to below 0°C at night. In moist climates, rocks weather much faster, since larger thermal stresses are produced in the presence of liquid water, which flows into cracks, and expands when it freezes in either a daily or yearly cycle. Rainwater, moreover, is slightly acidic, due to carbon dioxide in the air forming carbonic acid, H_2CO_3, which dissolves carbonate rocks, such as limestone ($CaCO_3$).[147] In air that contains sulfur and nitrogen oxides, strong sulfuric and nitric acids are formed, which are very effective in dissolving a wide variety of minerals. Some plants, such as lichens (combination of algae and fungi) also produce acids that dissolve rocks. Larger plants can extend their roots into cracks in rocks, creating forces that rupture them. Tiny fragments of rock, combined with decomposed organic material, make up the *soils* that support plant life on our planet.

Once fragmented, soil and small pieces of rock can be moved by the wind. The roots of plants provide resistance to such movement, and in regions in which land has been denuded of vegetation by overgrazing,

---

[147] When solutions containing dissolved $CaCO_3$ precipitate as they drip in underground caverns, stalagmites and stalactites are formed.

massive sandstorms are common. Small fragments of rock endow the wind with the ability to further erode rocks by "sandblasting" them.

Running water, either from rain or melting snow, is the principal means of moving rock fragments and soil. The abrasive action of suspended material in running water further erodes the land. Over time, valleys, canyons and interesting rock formations, such as the Grand Canyon, are produced. Much of the sediment carried in rivers is deposited in regions where their flows are slowed in floodplains, lakes and in the deltas where they flow out to the sea. At the sea, wave action continually erodes the adjacent cliffs. The effects of water are increased many-fold in times of flood or storms.

Sedimentary deposits include, in addition to inorganic matter, the remains of microscopic, plant and animal life. Sediments are deposited in layers and become *geological strata* when they are converted to *sedimentary rock* under the influence of temperature and pressure. The ordering of layers in sedimentary rocks provides their relative age, since younger strata are deposited above older strata. However, in some regions, geological processes that tilt the strata or mudslides may confuse the ordering of layers. Radioactive dating of minerals may be used to establish the absolute age of geological layers.

Fossils are evidence of prehistoric life and include carboniferous remains of organic material, preserved bones and shells, and sometimes even entire organisms trapped in polymerized tree resin (*amber*[148]). Additionally, the structure of a living organism can be preserved by filling its cavities (e.g., of wood) or replacing its molecular constituents with inorganic ions, a process called *mineralization*. By dating the strata in which a fossil is found, the age of the organism producing the fossil can be determined. Over shorter time periods, the bodies of dead organisms can be preserved in ice. Fossils preserving the shape of soft tissues can sometimes be found in extremely dry environments.

Ice also moves across the Earth's surface with velocity much less than liquid water, but can transport both small and large rocks with its flow. During the last ice age, which peaked about 18,000 years ago, up to 30% of the Earth's surface was covered with glacial ice, and sea level was ca. 100 m lower than it is at present. The rocks and boulders carried by this

---

[148] This was the basis for the plot of the movie, Jurassic Park, in which dinosaurs were created from DNA trapped in amber.

moving ice greatly modified the landscape, sculpting mountains and creating U-shaped valleys scarred by the moving rocks. Large amounts of rocky debris were deposited in *moraines* at the limits of the extents of the *glaciers*.

Planet Earth, which seems so solid and unchanging over the period of our lives or of that of all recorded history, is very different when viewed over geological time scales ranging from ten of thousands to billions of years. New continents have been formed, old ones have disappeared and all have changed their relative positions. Mountain chains have been uplifted, impact craters have been formed and then, under the unceasing influence of wind, water and ice, these have been worn down, only to have the same process start all over again.

The Atmosphere

Probably the most significant difference between the Earth and its nearest neighbors, the Moon, Mars and Venus, is in their atmosphere. Venus, with a mass 82% that of Earth, has a surface temperature of 460°C and a surface pressure of ca. 90 atmosphere. Its atmosphere is primarily CO_2. Mars, with a mass 11% that of Earth, has an atmosphere that also is largely CO_2, but its surface pressure is less than 1% that of Earth. The surface temperature of Mars is highly variable (–135°C to 30°C). The Moon, with a mass less than 1% that of Earth, has no atmosphere to speak of and a extremely variable surface temperature (–170°C to 130°C). Only the Earth has appreciable (21%) O_2 in its atmosphere. Besides providing the chemical environment in which life on land exists, the atmosphere produces our local weather and regional climate; *climate* being the weather averaged over long time periods (decades or more).

The important factors producing the atmospheres of Venus, Earth, Mars and the Moon, all formed from similar rocky material, are their surface gravitational force, determined by their mass; their surface illumination, determined by their distance from the sun; and the presence of photosynthetic life on Earth. Surface illumination obviously affects surface temperature, making Venus hotter and Mars cooler than the Earth. The Moon, and to a lesser extent, Mars, are much less massive than the Earth. With the reduced gravitational force provided by these bodies, a greater fraction of gaseous molecules can reach *escape velocity*, the velocity required to break away from their gravitational attraction. With a lower atmospheric density (and pressure) air currents are less able to transport energy, resulting in greater temperature differences between their equatorial and polar regions. In addition, on Earth photosynthesis removes CO_2 and produces O_2.

Besides these primary creators of atmospheric properties, there is also positive feedback that can greatly magnify these direct effects. For example, on Venus, which is 72% as far from the sun as Earth, higher irradiation vaporizes more water, producing a powerful *greenhouse effect*, the interception (trapping) of emitted infrared radiation, which heats the atmosphere and the surface. At the higher surface temperature of Venus, carbonate-containing rocks decompose, releasing CO_2 to the atmosphere, further enhancing the greenhouse effect. The combined effect is called the runaway greenhouse effect and makes the surface of the planet completely uninhabitable by life as we know it. Humans should be wary about unintended consequences of any changes that they make to the atmosphere. One such change is the current transformation of the major part of the Earth's fossil fuel reserves, accumulated over hundreds of millions of years, to CO_2 in the atmosphere, in just a few centuries. Particularly frightening has been evidence collected from sea sediments indicating that there have been times in the past in which global temperatures have changed very dramatically over very short periods of time (decades to centuries).

On Earth, it was photosynthesis by certain bacteria in stagnant pools of water[149] that first injected O_2 into the atmosphere. O_2, and its photodecomposition product, ozone, shielded the surface of the planet from damaging ultraviolet rays from the sun. This allowed the development of plant life, which then injected oxygen into the atmosphere at a much larger rate.

> **Our atmosphere has evolved to produce conditions favorable for life.**

The temperature and pressure variation in the Earth's atmosphere, as a function of elevation, are shown in Figure 21.4.

---

[149] Organisms at sufficient depth in stagnant pools of water are shielded from ultraviolet rays from the sun for long periods of time, whereas in the oceans, currents periodically bring such life near the surface, where it can be destroyed by irradiation.

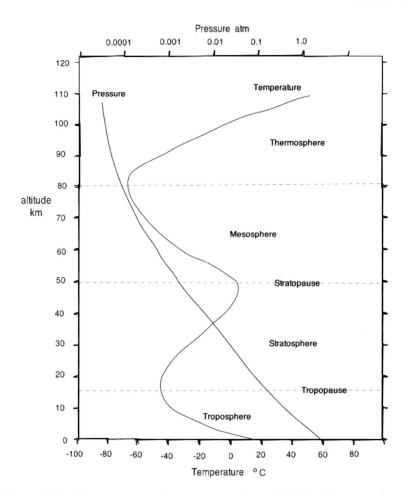

Figure 21.4. Temperature and pressure variations in the Earth's atmosphere.

Pressure continually drops with elevation, since at any level it is pressure that supports the weight of the gas above that level. In the lowest part of the atmosphere, called the *troposphere*, where we live, temperature also drops with elevation, reaching a minimum value of ca. −50°C at an elevation of ca. 15 km,[150] called the *tropopause*. Gases rapidly mix by convection in the troposphere, since they are heated at the surface,

---

[150] Temperatures and elevations of atmospheric boundaries vary with latitude and season, as well as other factors.

expand to lower density and then rise. In the *stratosphere*, which extends from the tropopause to 50 km, temperature rises. The Earth's *ozone shield* resides in the stratosphere;[151] it is the absorption of ultraviolet rays from the sun by ozone that produces the rising temperature in the stratosphere. Because of the *temperature inversion* in the stratosphere, heated gases do not tend to rise in this region, and mixing by convection is very slow. Pollutants can remain in the stratosphere for centuries.

Since the 1970s it has been realized that the concentration of ozone in the stratosphere is affected by the rate at which it reacts with trace gases, such as NO and chlorine atoms. In particular, the appearance of a massive "*ozone hole*" over the southern polar region has suggested that human activities might be damaging our ozone shield.

The Oceans

The Earth is often called the "blue planet," due to the reflectivity of liquid water and the scattering of the clear atmosphere. Brown (the continents), white (clouds and ice) and, of course, green are also prominent in views of the Earth from space.

Liquid water makes up about two-thirds of the Earth's surface and including ice, the fraction of water is ca. 80%. There is evidence that in the past, the fraction of the surface covered by ice has been both much less and much more than it is today. About 97% of the Earth's water is seawater, containing ca. 0.3% of dissolved salts (mostly sodium chloride). The remaining 3% is fresh water, of which 90% is immobilized as ice, leaving only 0.3% of the Earth's water in streams, rivers, lakes, marshes and ground water to support life on the continents.

The Earth's oceans cover two thirds of its surface and have a depth at certain places that is greater than the height of the highest mountains. Since water has a large heat capacity, the oceans, as well as other large bodies of water, function as heat reservoirs whose temperature only slowly varies. This moderates the temperature variations of cities that border on the ocean or large lakes. It also applies to the Earth as a whole,

---

[151] Lower concentrations of ozone may exist in the troposphere, especially in polluted regions, where they contribute to effects such as lung damage and plant deterioration.

whose temperature can take decades to respond to heat flux variations, such as those brought about by the increasing greenhouse effect. Ocean circulation also plays an important role in determining the climate of various locations, and is about as important as atmospheric circulation for transporting thermal energy from equatorial to polar regions.

The oceans are heated by sunlight at their upper surface, producing a temperature distribution, where upper levels are warmer than lower levels. Unlike water heated at the bottom of a pot, a temperature inversion impedes mixing by thermal convection. The wind produces waves that mix the top 100 meters of the ocean. The lower part of the ocean, however, is mainly insulated from this motion, and transfer of heat and dissolved chemicals to this region must rely on global ocean circulation. Thus, when atmospheric CO_2 dissolves in the ocean, it is limited to this upper mixed level, which rapidly becomes saturated with the gas. Additional CO_2 storage in the oceans must await currents that take it to lower levels.

Global ocean circulation is known as the ***thermohaline circulation,*** since it is driven by density changes of seawater, caused by alterations in its temperature and concentration. The density of seawater increases as its temperature decreases[152] or its salt concentration increases. In the bulk of the ocean, negligible amounts of salt are added to it, and changes in its salt concentration are due to the removal or addition of water. Removal of water is by evaporation and ice formation. Addition of water results from riverine flows and the melting of ice. As the density of seawater increases, it eventually sinks to the lower levels of the ocean, where it remains for about 1000 years. Surface water flows in to take the place of the descending water, giving rise to the global circulation shown in Figure 21.5, with water sinking in the north polar regions of the Atlantic and rising in the Pacific and Indian Oceans.

> **The oceans, as well as the atmosphere, transport heat from equatorial to polar regions.**

The thermohaline circulation is slow but enormous, having a volume equivalent to 100 Amazon rivers. It is sometimes called the great oceanic

---

[152] Unlike freshwater that reaches a maximum density at ca. 4°C.

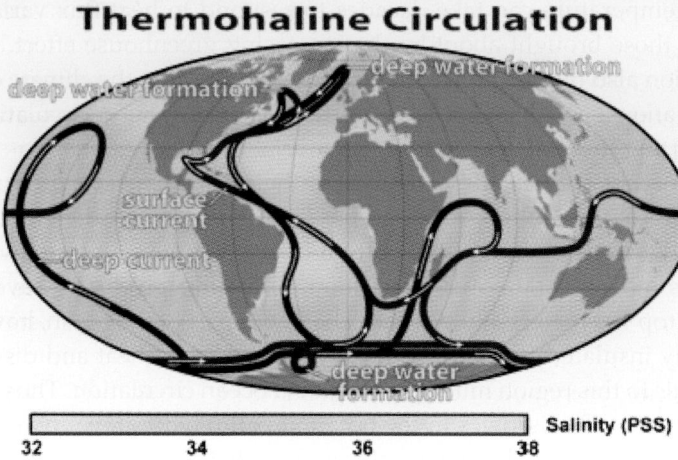

Thermohaline Circulation

Figure 21.5. Global oceanic circulation.

conveyer belt, since it transports a huge amount of thermal energy from tropical regions to the eastern North Atlantic. This transport is responsible for the warm climates of Western Europe, which for example make Sweden and Norway much more habitable than northern Canada or Alaska.

Although the oceans have a tremendous capacity for absorption of carbon dioxide, this can only happen on the time scale of the thermohaline circulation: about 1000 years. The well-mixed upper 100 meters of the ocean quickly saturates with CO_2. Many oceanic organisms use dissolved CO_2 to build their carbonate-containing shells, and when these organisms die, their shells fall to the ocean floor, where they are compressed to form carbonate-containing rocks. Ironically, atmospheric CO_2 forms carbonic acid, H_2CO_3 when it dissolves in the ocean, lowering ocean pH. In this more acidic environment, the formation of carbonate-containing shells is impeded. Increasing ocean acidity also destroys Earth's coral reefs.

Earth's History

Current theory has the Earth being formed from the agglomeration of gaseous material and rocks of a variety of sizes and chemical composition.

Although the Earth was initially molten, radiometric dating of rocks and minerals indicate that as early as 4.0 billion years ago, parts of the Earth's surface had solidified.

Gases collected by the proto-Earth were probably blown away by the burst of energy resulting from the ignition of nuclear fusion in the sun, with the atmosphere of the infant Earth then formed by outgassing of its rocky parts. This atmosphere consisted of gases such as water, carbon dioxide and nitrogen. Any oxygen present would have very rapidly reacted with reduced materials, such as elemental iron, on the surface of the early Earth, implying that the oxygen concentration at this time was very low. Many of the building blocks of life, such as amino acids and nucleotides, were formed by the addition of various forms of energy (lightning, thermal energy or UV radiation) to this early atmosphere.

With continual photo dissociation of water and escape of hydrogen atoms from the Earth's gravitational field, small amounts of oxygen were produced in the atmosphere. Atmospheric concentrations of oxygen and ozone sufficient to protect against solar ultraviolet radiation, however, would have had to wait for the development of life forms capable of photosynthesis. It is likely that the first life on Earth was formed by *methanogens*, CH_4-producing bacteria, near thermal vents deep in the ocean that were rich in high-energy compounds. However, the first photosynthetic organisms probably dwelled in stagnant pools of water, at a depth where ultraviolet light was filtered out, but visible light penetrated. As small concentrations of oxygen built up in the atmosphere and were photolyzed to produce UV-absorbing ozone, some bacteria could move closer to the surface of the water and spread throughout the Earth's oceans. In the presence of increasing numbers of photosynthetic bacteria, oxygen concentration in the atmosphere rapidly increased and surface minerals became oxidized. The concentration of oxygen in the atmosphere was probably around 10% of the present value at the beginning of the Cambrian period, 570 million years ago, when abundant fossils were first formed. Within another 50 million years, plants began to populate the solid surface of the Earth. By about 350 million years ago, atmospheric oxygen concentration had reached a value close to its present concentration, allowing animal life to emerge from the oceans onto the solid Earth.

Theories of stellar evolution predict that the radiation intensity from the sun during early Earth was 20–30% less than it is today.

However, there is no evidence that the Earth was ever completely covered by ice, a scenario called *"snowball Earth"*. This suggests that the concentration of greenhouse gases, mainly CO_2 and CH_4 produced by methanogens, was much larger during the early history of the Earth than it is today.

During the last 10,000 years global temperatures have varied very little. The conditions in this **Holocene** epoch have been ideal for the development of human civilizations. In contrast, the previous 1.5 million years, known as the **Pleistocene** epoch, was characterized by periodic dips in temperature called ice ages, during which, for example, glacial ice covered most of the northern half of the North American continent. Astronomers have correlated these temperature variations with periodic oscillations of characteristics of the Earth's motion, such as the eccentricity of its orbit and the tilt and precession of its rotation axis.

Summary
ଔ ଓ ଔ ଓ ଔ ଓ

The Earth's surface has continually been altered by new crust formation and plate subduction due to thermal convection, as heat flows from the solid and liquid core through the mantle to the surface. Meteoritic impact also produces structures on the surface. The order produced by these processes is continually being removed by disorder caused by weathering and erosion, aided by the action of liquid water and ice. Such a changing environment is ideal for the rapid evolution of advanced life forms.

Both the atmosphere and the oceans transfer heat from equatorial to polar regions and reduce temperature differences that would otherwise result from their differences in solar illumination. Components of the atmosphere that absorb and reradiate the infrared radiation emitted by the Earth, warm the Earth. This greenhouse effect was probably larger in the early Earth than it is today and is what makes the surface of Venus extremely hot. The troposphere, the lower part of the atmosphere, is heated at the bottom and is well mixed, while the stratosphere is heated by ozone absorption of sunlight and contains very little vertical motion. The oceans are heated at the top and have limited vertical circulation. They can thus only provide a reservoir for greenhouse gases, such as CO_2, over time periods of the global ocean circulation, about 1000 years.

Questions

1. Solar radiation is most direct in the northern hemisphere on June 22, but this is not usually the hottest time of year. Give an explanation for this observation.
2. What are the dates of the summer and winter solstices in the southern hemisphere?
3. Seismological methods and equipment have been greatly improved because of the commercial and military importance of this technique. Why is seismology commercially and militarily important?
4. Explain why the surface of the Moon or Mars might be considered fractal, whereas such a description would not be useful for the surface of the Earth.
5. Recently, there has been considerable controversy on whether to tap oil reserves in the North Slope of Alaska. Considering that oil is formed from the anaerobic decomposition of vegetation, and there is only sparse vegetation on the northern slopes of Alaska, how can there be large oil reserves in this region?
6. What do we mean by the Earth's magnetic field "flipping," putting the North magnetic pole in the southern hemisphere? If this happened, wouldn't it be called the South magnetic pole?
7. Since the Earth's core is 3500 km thick, while its mantle is only 2900 km thick, why does the mantle comprise most of the volume of the Earth? Calculate the volume of the Earth's mantle relative to that of its core. (The volume of a sphere is $\frac{4}{3}\pi r^3$).
8. Working your way out from the center of the Earth, compute the difference in density of materials in adjacent layers. Include the crust-atmosphere interface, where you can take the density of the atmosphere to be zero. At which boundary is the density difference the greatest?
9. Explain why, using radioactivity dating, the age (time since solidification) of rocks from the continental surface is often measured to be billions of years, while those from the oceanic surface are rarely more than a few hundred million years old.
10. Write a chemical equation showing how carbonic acid reacts with limestone to form soluble bicarbonate (HCO_3^-).
11. List as many characteristics of our planet that you can think of that make it conducive for producing and maintaining life.

12. Why do we like having ozone in the stratosphere, but not in the troposphere?
13. It is said that because of thermal and chemical lags, there is a built-in greenhouse gas temperature rise that we can do little to change. Explain some of these lags.

22 Ecology and Evolution

Nothing in biology makes sense except in the light of evolution

T. Dobzhansky

In previous chapters we have discussed single organisms. While a single autotroph, such as a photosynthetic bacterium, could probably survive and multiply,[153] a single multicellular organism is not viable. Plants need animals for pollination; animals need plants or other animals for food. Thus,

> **Multicellular organisms reside in communities.**

In this chapter, we will go beyond the individual organism, to discuss the *communities* in which they live, how individuals in the community interact and how the individuals and communities change over time.

Animal communities range from reproductive pairs to nations. In solitary species such as bears, the male and female come together for a few days each year for mating, and cubs stay with their mother for two to three years. In other species, males and females share joint responsibility for bearing children (e.g., penguins, who jointly incubate eggs) and raising young for

---

[153] In fact, the theory that all life began with a single organism would require this.

their first year. This allows one parent to protect the young, while the other forages for food. In exceptional cases, such as bald eagles, mating is for life.

Some animals, finding strength in numbers, join together in groups larger than the family. This may range from a pride of lions (a number of females and their cubs with one or a few males) to a herd of buffalo (with thousands of individuals) or a school of anchovies (with numbers in the millions). Some species can only exist when their numbers are large. This was probably the case for the North American passenger pigeon, which lived in huge migratory flocks, often numbering over a billion birds, but went extinct in the early twentieth century.

In some species, such as ants, bees and the cathedral termites shown in Figure 22.1, individual organisms band together to form hives or nests,

Figure 22.1. Cathedral termite mound.

where individuals function almost exclusively for the common good. In these communities there is irreversible morphological specialization for different functions, including reproduction, and cooperative care of young. An individual member of these communities is capable of only a limited range of activities and cannot survive by itself. Communication between individuals allows the community to accomplish more complicated tasks and persist for many generations. By analogy to multicellular individuals, such communities have been called *superorganisms*.

The millions of termites that live in the mound, shown in Figure 22.1, are not alone. Like cows and horses, they rely on microbes that live in their guts to digest cellulose, a high-energy material similar to starch, which would otherwise be useless to them. Likewise, humans harbor a wide variety of beneficial microorganisms in their saliva and intestines, and on their skin. The intestinal discomfort that we often experience after a dose of antibiotics is testament to the value provided by these natural cohabiters of our bodies. The antibiotic may kill the beneficial bacteria, allowing ill-suited bacteria to multiply. Even at the cellular level, we, along with all other animals, coexist with other species, or at least the remnants of such species. There is considerable evidence that some of the organelles of eukaryote cells were originally independent prokaryotes that were incorporated into the cells. Over time, the prokaryotes that formed our mitochondria learned to rely on us for the essentials of their metabolism, while providing us with the genetic instructions for the machinery to efficiently produce ATP.

While it is a biological necessity that humans come together for reproduction, they usually choose to live together in communities, ranging from small villages to large cities, where division of labor and specialization are the norm. Specialization among humans, such as plumber or lawyer, result from learned behaviors, and are not fixed for life by morphological changes, as in termites. In all communities, however, the reason for cooperation and specialization is *synergy*, the group doing better than if each individual acted on his own.

A group of interacting individuals of a species is a *population*. *Populations* of most species exist in a particular place at a particular time and have evolved to be uniquely suited to their biological and abiological environments. Humans are unique, in that there is sufficient mixing and cooperation between different groups, due to global travel, trade and communication, that they are rapidly becoming a single population. With the support of the greater human population, humans can exist in environments, such as Antarctica, or the space station, for which they would otherwise have little suitability.

Ecosystems

Except for autotrophs, which can satisfy their needs for energy and nutrients with only sunlight or inorganic compounds, a population of a species must exist with other organisms. A coexisting community of species with its local abiotic environment is known as an *ecosystem*. Terrestrial (land) ecosystems are classified as different *biomes* based on non-biological factors, such as temperature, moisture and light, and the predominant form of life (which are plants). For example: a temperate deciduous forest is cool and wet and populated primarily by deciduous trees, while tundra are cool, and dry and treeless, and caves are cool, dark and moldy. The biome is a general classification of a region, while an ecosystem is a more detailed description of a community and its interactions. Within a biome there may be considerable variations in ecosystems due to subtle variations in conditions, variations in migratory and evolutionary history, as well as the presence of interactions with nearby ecosystems.

In any ecosystem, *phototrophs* such as plants, algae or bacteria act as the *primary producers* of biological material.[154] For example, the roses in my garden satisfy this function. The primary producers are eaten by the *primary consumers* (aphids in the garden), which in turn are eaten by successive trophic levels of carnivores (ladybugs, then sparrows), until they are consumed by a *top carnivore* (hawks) for which there is no predator. At every level, when an organism dies, species called *decomposers* (worms, beetles and termites) ultimately return material to the abiotic environment. The successive levels in the food chain through which energy flows are called *trophic levels*, with the primary producer being the lowest level and the top carnivore the highest. In a stable ecosystem, the population of a species is held in check by limitations of its food and *prey*, the presence of *predators*, and *competition* with other species. Ecosystems are usually only stable under a narrow range of abiotic variables, such as temperature, moisture and solar illumination.

---

[154] An exception is the ecosystems deep in the ocean that derive their energy from high-energy inorganic compounds issuing from oceanic vents.

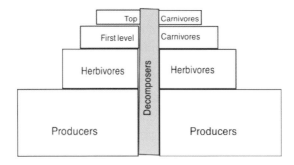

Figure 22.2. Trophic levels of an ecosystem (not to scale).

Because of waste, as well as the fundamental inefficiency required by the second law of thermodynamics, the biomass and energy content in each successive trophic level is greatly reduced. Typically, as shown in Figure 22.2, a trophic level will weigh only a tenth of the level below it and ten times as much as the level above it.

Water is often the *limiting factor* for production of biological material in an ecosystem. There is no second law for water, as there is for energy, so water can be continually recycled. The movement of water through various biomes is driven by energy from the sun, as shown in Figure 22.3.

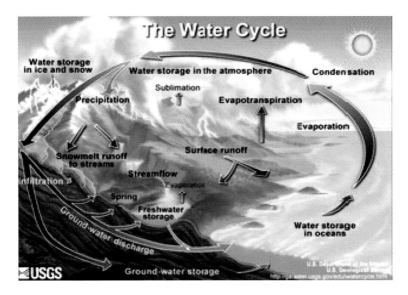

Figure 22.3. The water cycle.

Even though the quantity of water is constant, its quality can be degraded, as communities that have experienced pollution or *salination* of their water supplies have discovered. The rate at which water travels through the cycle is highly variable, with a water molecule spending only a few minutes as liquid on the surface of a leaf or thousands of years trapped in polar ice. The form in which water moves through the environment depends strongly on temperature. For example, as temperatures in the western United States have risen, precipitation has been shifting from snow to rain. As a result, surface and ground water runoff is concentrated in brief intervals after rain, rather than occurring gradually as snow melts. This has caused increased erosion and water shortages in both natural and human ecosystems.

A species can survive only over its *tolerance range* for physical variables, such as temperature, moisture and nutrient (e.g., nitrate and phosphate) accessibility, and biological variables, such as availability of food sources and limitations of predators. These define the *fundamental niche* of the species. In a given ecosystem, its competitors may limit it to a narrower *realized niche*, since no two species can occupy exactly the same niche, without one outcompeting the other and driving it to extinction. Biologists believe that diversity increases the resiliency of ecosystems to changing conditions. Diversity includes *biological diversity* and *genetic diversity*. A biologically diverse system contains a large variety of species, each occupying a slightly different niche, while a genetically diverse system includes a range of alleles, either expressed as variations or unexpressed due to *heterozygosity*. Diversity also increases the health of larger groupings, such as biomes and the entire Earth. Typically, ecosystems in tropical climates exhibit greater biodiversity than those in temperate or frigid climates.

An ecosystem contains a very high degree of interactions between species and the environment.

Small changes can produce large effects in ecosystems.

Often, if even a single species is removed or introduced or a physical factor varies beyond some limit the entire system will change. Thus, when settlers introduced a few rabbits into Australia, where they had no natural predators, the rabbit population grew to over five hundred million, practically denuding the continent of vegetation. When farmers killed the wolves in New England to protect their livestock, the deer population increased to troublesome levels. Draining of the marshes south of New Orleans contributed to the devastation caused by hurricane Katrina.

Sometimes even subtle changes in one species in an ecosystem can have a large effect on other species. This is because the symbiotic relations between species often depend on precise synchrony of life stage between these species. (e.g., larva feeding on blooming plants). Changes of external conditions of temperature, humidity etc. may affect the timing of the stages of the two species in such different manners that the symbiotic relation no longer exists.

The rate of increase of the number of members of a population is its birth rate minus its death rate. Since both birth and death rates are usually proportional to the numbers in the population (first-order kinetics), so is the rate of increase in its number. This situation is called exponential growth, and can produce massive increases in the size of a population. For example, a growth rate of only 5% will double the size of a population in 14 generations and increase it by a factor of ten in 47 generations. Exponential growth, however, is an untenable situation. Sooner or later any population will outstrip the *carrying capacity* of the ecosystem for the species. A species can surpass its carrying capacity for a short time, but then its population must fall back down to the carrying capacity, or, if the ecosystem is damaged by the overpopulation, to numbers much less than the carrying capacity. Such a phenomenon occurs with lemmings, who when their population density becomes too large, undergo mass migrations.[155]

Migrations may also occur in a population of constant size, if the parameters of its ecosystem changes. Thus, if the temperature rises, trees that thrive under cooler conditions may migrate to higher elevations in mountainous regions. If migration is blocked, such as it is when the lemmings come to the sea or the trees come to the top of the mountain, a population may die out. If this occurs for all populations of a species, the species becomes extinct. Biologists tell us that 99% of all species that have inhabited Earth have become extinct. An alternative way that populations can change, if systems parameters change slowly enough, is by evolution.

---

[155] Observations of the lemmings tumbling off cliffs into the sea during these migrations have led some to the false conclusion that they were committing mass suicide.

Evolution

In any population there are variations of characteristics among individuals. Among humans residing in the same region, some people are taller, stronger, smarter or more handsome than others. There are many other variations that are less obvious, such as resistance to different diseases or ability to digest different foods. While many of these characteristics are due to environmental factors, such as diet, education and parental nurturing, others are a direct result of differences in the alleles of individuals. Genetic variability also extends to alleles that are neutral for characteristics and those that are hidden by heterozygosity.

Even when a good racehorse is past its running prime, it can command top dollar as a breeder. This is because breeders recognize that many characteristics that make for a good racehorse are inherited and will be passed down to the next generation. This process is called *artificial selection* and involves the interceding of the breeder in the choice of individuals with desired characteristics.

The success of this process is based on, as we learned in Chapter 18, the favorable characteristics resulting from slight variations in nature or amounts of the proteins of an individual. The variations are caused by its genetic makeup, which is passed down from its parents. Thus the breeder acts as a filter that decides which characteristics will be passed on to the next generation.

As realized by Charles Darwin, Nature can also act as a reproductive filter, by providing an advantage for individuals with certain traits surviving to reproductive age and caring for their young. This process is called *natural selection*. One example that has been studied in considerable detail is the color of peppered moths in England. Originally, most peppered moths were light colored, allowing them to be camouflaged from avian predators when resting on light-colored trees and lichens. During the industrial revolution in England, pollution killed many lichen and darkened many of the trees. In this environment a darker variation of the moths was favored and soon came to dominate the population. With the introduction of more rigid environmental standards, vegetation once again became lighter and there was a resurgence of the lighter variation of moths. In these changes, the color of individual moths do not change, it is only the probability of their surviving to reproductive age and passing on their genes for color that changes.

In sexually reproducing species in which females invest a large amount of energy in producing eggs, supporting the development of embryos and perhaps nurturing young, they often have evolved to be choosy about which males they mate with. This is called ***sexual selection***.

> **Both natural and sexual selection can change the genetic makeup of populations.**

In response to female preferences males have often evolved to have elaborate plumage (peacocks), verbalizations (song birds) or behaviors (bower birds) to convince the females of their genetic superiority.

Sometimes populations of a species are geographically isolated from each other by a barrier, such as a mountain range, water in the case of oceanic islands, or even built-up areas in our modern society. If the populations are subjected to different stresses of environment, predators or food supply in these regions, they may begin to diverge in different directions. Eventually, these differences may become so large, that even if the barriers for mixing are removed, the populations can no longer interbreed and produce fertile offspring. In this case, a single species has become two different species.

> **Geographically isolated populations can evolve to become separate species.**

The evidence for the evolution of species under the pressures of natural and sexual selection, with the eventual formation of new species, is overwhelming and includes fossil, genetic and geographic-distribution data. Some of the latter was only understood after the migration of the continents was described. As stated by Dobzansky at the beginning of this chapter, it is extremely difficult to explain much of the biological world without invoking evolution. One example of this is the observation that most species are imperfect creatures, incorporating vestigial organs (such as our appendix) and less-than-optimal engineering (such as the difficulty of the head of a human fetus passing through its mother's pelvis), which is a result of evolution.

Evolution also explains why the human embryo in its development passes through stages similar to the embryos of many species in our evolutionary lineage. In addition, the system of classification of species in biology is completely evolution based, and may greatly separate species with similar appearances, based on evolutionary differences. This is because evolution may find similar solutions to survival in similar environmental niches for species with very different evolutionary backgrounds, a process called *convergent evolution*. Thus, there are substantial similarities between some marsupial species evolving in Australia and placental species found elsewhere in the world. Although there may be slight differences among biologists concerning the exact mechanism of evolution, the concept has certainly withstood the "test of time."

Evolution can produce changes in populations in just a few generations. That is why diseases are so difficult to control with antibiotics or vaccines. Production of new species, however, takes much longer, and moreover, probably doesn't occur at a uniform rate. Thus, the gene pool of a species may remain quite constant or "drift"[156] only slowly during times of relatively constant environmental conditions. In times of environmental stress, such as drought or temperature change, however, a species may change quite rapidly. Since the number of fossils found for a species are related to the time that species existed, it is not surprising that the fossil evidence of short-lived "transitional species" is sparse. Nevertheless, fossil evidence for many such species has been discovered. These include feathered dinosaurs, fish with limbs (see Figure 22.4) and ancestors of our own species, homo sapiens.

Although almost all biologists accept evolution, there is still some controversy concerning its details. One of the most important of these is

Figure 22.4. Recreation from fossils, of *Acanthostega*, a transition species between fish and amphibians.

---

[156] Genetic drift results from random variations combined with the positive feedback resulting from genes reproducing themselves.

whether natural selection can operate on the level of a group, rather than only on the level of the individual. That is, whether membership in a group increases the probability of an individual's genes being passed on to the group's progeny. Sharing of genes within a group would provide a mechanism for such an effect, as would the necessity for group survival in order for an individual's genes to be passed on. Such a mechanism would explain the development of traits such as altruism. The question is whether group mechanisms would be swamped by natural selection's effects on individuals.

Earth as an Ecosystem

We can treat the entire Earth as a continually changing ecosystem, where species are constantly evolving or becoming extinct. Usually these processes are slow and orderly, but at times when the Earth surface has been rapidly perturbed, very rapid changes of species occurred. For example, the meteor impact that occurred 65 million years ago, discussed in the last chapter, allowed many species of mammals to expand into niches vacated by the dinosaurs.

One question that has been asked is whether internally generated changes in ecosystem Earth could also produce massive and rapid changes of species. A principle called the *Gaia* hypothesis, named after the Greek supreme goddess of Earth, postulates that because it is populated by Life, Earth has self-healing capabilities for internally generated changes. If changes occur, there will exist among the species living on Earth, a sufficient number that will provide negative feedback tending to return Earth to its previous condition.

An alternative principle, called the *Medea* hypothesis, named after a Greek mythological figure who killed her children, postulates that life is basically unstable and can destroy itself. Thus, in the absence of restraints, there will be more positive feedbacks (including population growth) than negative feedbacks. Under this hypothesis, Earth can spontaneously switch between very different states. One example of this behavior occurred about a billion years ago, when photosynthetic organisms increased the oxygen concentration in the atmosphere, rapidly wiping out almost all anaerobic species. Some scientists suggest that the rapid increase of global temperatures produced by anthropogenic release of greenhouse gases, if not checked, might become another example of the Medea principle.

Summary
ଓଃ ଓଃ ଓଃ

Organisms reside in communities, ranging from families to cities, in which synergy results from specialization and cooperation. An ecosystem is a number of communities interacting with the abiotic environment in a particular place. In an ecosystem primary producers capture solar energy, which is passed on to successive trophic levels of herbivores and carnivores. An ecosystem can be extremely sensitive to small changes in conditions or species populations.

The population of individuals residing in a community has a degree of genetic variation that produces a range of suitability and adaptability to conditions. Natural or sexual selection will usually favor the reproductive success of certain members of the community, who will then pass on their particular genetic variations, causing evolution of the population. Isolated populations of the same species may evolve to a degree where they can no longer interbreed, thus becoming separate species.

Earth is an ecosystem, which can be drastically changed by externally caused events. Depending on whether positive or negative feedback predominates, it may be unstable or stable to internally generated changes.

Questions

1. What are some similarities and differences between a super organism and a multicellular individual?
2. What are some similarities and differences between a super organism and a human community?
3. What are the advantages of fish swimming in schools?
4. Choose an ecological system not discussed in this chapter and identify the members of its various trophic levels.
5. From an ecological perspective, why is it difficult to supply a diet with a large fraction of meat to the population?
6. Characterize each of the following biomes, as to temperature, moisture, light and predominant form of life: desert, freshwater marsh, savanna and coniferous forest.
7. Grasslands, characterized by cold winters, hot summers and limited rainfall, have been converted to crop production worldwide, because of their fertile soil. In the United States, dust storms during the drought of the 1930s resulted from this conversion. Explain how this happened.

8. Should latitude be an additional non-biological variable used for classifying biomes?
9. In driving along a road, you notice bunches of blue wildflowers in some places and bunches of yellow wildflowers in others. Rarely are the two flowers mixed. Give an explanation for this segregation.
10. Why might a population including many heterozygous individuals be more resilient to change than one without such individuals.
11. Discuss an ecosystem in which humans are influencing biodiversity.
12. Why does life evolve while inanimate matter does not evolve?
13. Many anthropologists believe that the almost universal belief in God (or Gods) in human societies suggests that the genetic makeup of humans may endow them with a propensity for accepting religion. How might this come about? Would it require that natural selection operate on the level of groups as well as individuals?

23 Astronomy

There are more things in heaven and earth, Horatio, than are dreamt of in your philosophy.

From Shakespeare's Hamlet

It's humbling to realize that we live on one of about ten planets orbiting one of about 300 billion stars in one of the hundreds of billions of galaxies inhabiting our universe. Nowadays, few people spend much time looking at the sky. However, questions, such as the fraction of the possible 10^{23} planets[157] that might be habitable and whether life has developed independently on some of these, probably have occurred to many of us. Is Earth special, or do we inhabit an ordinary planet, circling an ordinary star, in an ordinary galaxy of the universe?

Astronomical Observations

Astronomers, like scientists interested in the inner regions of the Earth, cannot reach out and touch most of the objects of their interest. For observation at a distance, astronomers employ electromagnetic radiation, rather than the sound waves of seismologists, because space contains no

---

[157] Obtained by multiplying the above numbers.

medium to propagate sound. Electromagnetic radiation encompasses wavelengths from extremely long radio waves to those of very short gamma rays. For simplicity we will call all of these "light" in most of this chapter.

To minimize atmospheric absorptions and emissions, astronomers usually locate their instruments in very dry regions (to minimize absorption by water vapor), on mountaintops (above clouds and away from city lights), and even on platforms orbiting the Earth.

Light can show the structure of nearby objects or extended regions that are further away. As examples, with even just a hobbyist's telescope, mountains on the moon, rings around Saturn, sunspots, and glowing gaseous nebulae can be seen. Stars, however, appear as points of light, and only the brightness, wavelength distribution and time variation of this light can be observed. In addition to light with a continuous wavelength distribution, characteristic of a black body at the temperature of the outer surface of the star, starlight contains lines of atoms and molecules, both in emission and in absorption. Lines from molecules, such as H_2, H_2^+, CH, OH, CO, H_2O and NH_3, are also observed in outer space by radio and microwave astronomy. Line spectra observed by astronomers have the same patterns as spectra observed on Earth. However, these patterns may be shifted to longer or shorter wavelengths, due to the Doppler effect resulting from the motion of the emitter or absorber with respect to Earth.

Since light spreads out from its point of origin, astronomers must know the distance to an object in order to convert its observed brightness into a measure of its inherent luminosity. For planets, this can be done by triangulation from different observation points on Earth. Distant stars are called "fixed" because they move across the sky in a manner predictable from Earth's rotation around its axis.[158] Nearby stars move with respect to fixed stars as Earth revolves around the sun. This allows their distances from Earth to be measured.

For determining distances to far-off bodies, astronomers make use of "standard candles," objects whose inherent luminosity is predicted by theory. One set of standard candles is the ***Cepheids***, pulsing stars whose period of intensity variation has been theoretically related to their luminosity. By comparing the observed brightness of a Cepheid with the

---

[158] Fixed stars may have large velocities, but their motion is too small to be discerned at their great distances from Earth.

inherent luminosity predicted by its period, the distance to it and any star in its neighborhood, can be determined.

Supernovae, exploding stars, are among the most luminous events in the universe and can be detected at great distance. The luminosities of one type of supernovae can be predicted by theory, allowing it to be used as a standard candle.

With sensitive telescopes, astronomers can record light from objects so distant that it takes a considerable time for the light to reach us. When we observe such light, we are looking at the objects at the time of their emission. This tells us about the history of the universe.

Light rays can be absorbed, reflected or bent, allowing objects that do not emit light to be studied. Light rays are bent by gravity according to Einstein's theory of general relativity, providing information about massive objects.

In addition to light, astronomers study particles. These range from protons streaming from the sun to extremely high-energy cosmic rays. Recently, using detectors deep in mines to reduce background signals, astronomers have added neutrinos to their particle repertoire.

Supplemental Reading: General Relativity

It might seem strange that a gravitational field affects light, which is composed of photons, particles with no rest mass. However, the basic hypothesis of Einstein's theory of *general relativity* is the *principle of equivalence*, which states that a gravitational field is indistinguishable from an acceleration.[159] Thus, gravity can be modeled by a warping of space by massive bodies. In turn, bodies move along paths that are everywhere straight lines in warped space, and thus curved lines in rectilinear Euclidean space.[160] Like everything else, light follows such a path. Of course, since light travels at a very high velocity, the warping of space usually has an unobservable effect on its motion, and

---

[159] If you are in a spacecraft with no windows and feel the floor push upward on your feet, you can't distinguish whether this force is due to a gravitational field or an acceleration of the spacecraft.

[160] To help understand this: consider the shortest-distance (straight-line) path between New York and Paris on the surface of the Earth, which is a curved path in space.

> negligible errors are caused by considering light rays to move in straight lines in empty space. The first experimental verification of the effect of gravity on light was made in 1919, by accurately measuring the apparent position of stars as they passed behind the eclipsed sun of that year. The agreement of theory and measurement solidified Einstein's reputation as the most brilliant scientist of his time.

Motion of Heavenly Bodies

It is not surprising that in times when the night sky wasn't "polluted" by millions of electric lights, and there was no television or internet to occupy time, observations and speculations concerning the motion of the sun, moon, planets and stars were popular. Astronomy is generally considered to be the oldest science. The description of the motion of heavenly bodies, however, depends upon the frame of reference — from where their motion is described. Until the sixteenth century, the dominant astronomical theory was that of Ptolemy of Greece (85–165 AD). The Ptolemaic theory was Earth-centered, with the stars, sun and Moon moving around the Earth on spheres successively closer to the Earth. One problem with this theory, however, was that it required very complicated motions for planets, such as Mars or Venus, which sometimes revolved around Earth in the opposite direction as the stars (retrograde motion). The endurance of the Ptolemaic theory was due to its theological correctness, placing man (and presumably God) at the center of the universe.

Although a *heliocentric* model (Earth and the other planets moving around the sun) had been proposed by Aristarchus about 400 years before Ptolemy, it wasn't until Copernicus (1473-1543) showed how such a model greatly simplified the description of the motion of heavenly bodies that it became generally accepted. While the observations of Galileo, Kepler and Tyco Brahe did much to systematize the motion of the heavenly bodies, it wasn't until Newton applied his universal laws of gravitation and motion to the problem, that the heliocentric model became generally accepted.[161] Nowadays we freely use Newton's laws to explain

---

[161] For being advocates of the heliocentric model, Galileo was censored and Giordano Bruno burned at the stake by the church.

everything from the nearly circular motion of the Earth about the sun, to the highly elliptical orbits of comets, to the somewhat chaotic motion of the rings and moons of the Jovian planets. It is only when extremely strong gravitational fields exist, such as in calculating changes in the orbit of Mercury — the planet closest to the sun — that corrections using Einstein's theory of general relativity must be made.

Order in the Universe

What principles can we use to organize our ideas concerning the types of order shown by the planets, stars and galaxies that inhabit our universe? One of these involves the shapes of these objects:

> **Spheres and disks are the dominant shapes of astronomical bodies.**

Spheres result from gravity, which being always attractive, pulls material as close together as possible, favoring spheres, which minimize the ratio of surface to volume. Disks are formed from the initial rotation of the gas clouds that condense to form objects, or by pulverization of a body orbiting the object. As the cloud loses energy (by radiating photons), it maintains its angular momentum by forming an increasingly thin and extended disk in the plane of rotation. Examples of disks are: the rings of the planet Saturn, shown in Figure 23.1a; and the spiral arms of the Andromeda Galaxy, shown in Figure 23.1b.

In addition, a ring of material, called the Kuiper belt, extends out from our solar system.

Figure 23.1a. Photo of Saturn taken by the Cassini probe.

Figure 23.1b. Photo of the Andromeda Galaxy.

In Chapter 4, we saw that only an external torque (like a collision) can change the angular momentum of a body. Thus, any astronomical object formed from material with initial rotation should show some sign of a disk, unless it is perturbed by an external influence. The inner planets of our solar system, Mercury, Venus, Earth and Mars — called the *terrestrial planets*, do not have disks, because their disk material was blown away by particles streaming out of the sun. However, the large outer planets, Jupiter, Saturn, Uranus and Neptune[162] — called the *Jovian planets*, do have disks. Among galaxies, some are elliptical, without a spiral disk, because they have undergone collisions with other galaxies. Small objects, such as asteroids and small moons, do not generate sufficient gravity to form spheres and do not attract sufficient surrounding material to form rings. These can have any form.

A Balance of Forces

The more massive a body, the more it will be compressed by gravity. Gravitational compression is accompanied by heating, as gravitational potential energy is converted to thermal energy. Compression continues until it is balanced by an opposing force. In planets and asteroids, the

---

[162] Pluto and the recently discovered Sedna, are probably better thought of as Kuiper-belt objects, rather than planets.

balancing force is the electrostatic repulsion of nuclei and overlapping electron clouds of atoms. Stars have sufficient mass that gravitational compression raises their internal temperature high enough to ignite fusion of hydrogen to helium (ca. 10^8 K). In stars, the energy released by fusion produces a pressure that balances gravity.

As a star exhausts its nuclear fuel, gravity once again takes over and compresses the star until gravity is balanced by what is called *electron degeneracy pressure*. Electron degeneracy pressure results from the Pauli exclusion principle, which requires each electron to occupy a different state of a "box."[163] Some electrons must occupy very high-energy states, producing high pressure. Stars in which gravity is balanced by electron degeneracy pressure are called *white dwarfs*.

For massive stars, even electron degeneracy pressure is not sufficient to balance gravitational force. In this case the star continues to contract until the electrons are pulled into the protons, producing neutrons. Neutrons are also fermions that obey the Pauli principle. When they occupy a small enough volume they produce neutron degeneracy pressure. Stars in which gravity is balanced by neutron degeneracy pressure are called *neutron stars*. If neutron degeneracy pressure is not sufficient to balance gravity, a star continues to contract, until the gravitational field at its surface is so strong that nothing, not even light can escape from it. Such an object is called a *black hole*.

Residents of the Universe

Hamlet's statement in the epigraph to this chapter leads us to classify the "things in the heavens."

Outer space

Outer space is a term applied to regions away from planets, stars and star-forming regions. It is not a complete vacuum, since it contains at least a few hydrogen atoms per cubic meter. Hydrogen plasmas (electrons and ions) and magnetic fields are found in regions near stars. All

---

[163] The "box" is the shrinking volume of the star.

regions of space are crossed by cosmic rays and photons of different wavelengths.

Gas

Gases consist of isolated atoms or molecules. Hydrogen and helium comprise most of the gas in the universe, due to the fact that these two (along with a small amount of lithium) were the first elements to be formed. Isolated gaseous molecules are identified by their absorption at particular wavelengths. Spectra of many small molecules have been observed in gas clouds.

In the vicinity of stars, gaseous molecules may be excited or ionized. Ions and electrons combine, forming excited states of atoms or molecules, which fluoresce, producing glowing gas. There are huge clouds of gas (and dust), called *nebulae*, in various regions of the universe, where much of the formation of new stars occurs. One such region is the Eagle Nebula, part of which is shown in Figure 23.2.

Figure 23.2. The Eagle Nebula, taken by the Hubble Space Telescope. (These towers of gas are ca. 50 trillion miles high.)

Dust

Dust is composed of microscopic solid particles that range from nanometer to micron size. These particles may consist of hydrocarbon-type molecules, or heavier elements. Dust particles absorb light over a continuous range of wavelengths, although they are more transparent in the infrared than in the visible region.

Rocks

Rocks are chunks of non-volatile minerals and/or metals, most of which orbit the sun in the *asteroid* belt between Mars and Jupiter. They range in size from a grain of sand to the asteroid Ceres, with a diameter of 950 km (590 miles).

Snowballs

Comets are bodies that orbit the Sun at distances large enough so that their volatile material, such as ice, frozen methane or other hydrocarbons, doesn't vaporize. Most of these bodies reside in the Kuiper belt that rings the solar system or in the *Oort cloud*, a sparsely populated cloud of objects that envelops the solar system at large distances. Occasionally the orbit of a snowball is perturbed by close interaction with another body, bringing it into the inner regions of the solar system. As it warms, some of the volatile material in the comet evaporates and the resulting gas is excited by photons or ions from the Sun. This produces a comet tail, which always streams away from the sun. A comet may also shed dust, which gets excited by sunlight. This second tail points back along the comet's trajectory.

A comet may collide with a planet, with the Sun, or more likely, continue in a highly elliptical orbit around the Sun, losing more of its volatile material in each orbit. Halley's comet, for example, has appeared every 75–76 years since 240 BC. In 1910 a spectacular light show was produced when the Earth passed through the tail of the comet. A very dramatic example of the impact of a comet on a planet was observed in 1994, when the Shoemaker-Levy comet,[164] having broken into 20 pieces,

---

[164] Comets are named after whoever first observes them. In this case, David Levy and Carolyn and Eugene Shoemaker, were amateur astronomers.

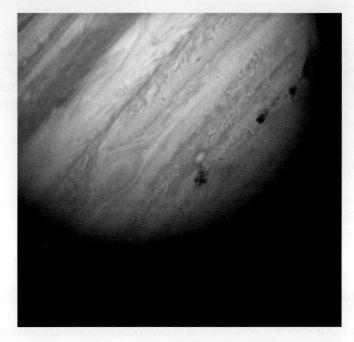

Figure 23.3. Regions of impact of Shoemaker-Levy comet on Jupiter.

impacted Jupiter. Astronomers estimate that the largest fragment, which upon impact produced a dark area as large as Earth, released energy equivalent to six million megatons of TNT. Some of the impact regions of this comet (brown spots) are shown in Figure 23.3.

We are fortunate to live in planetary system where the gravitational field of Jupiter, a giant planet, sweeps up some incoming comets before they impact Earth.[165]

Rocky planets and moons

These are rocky bodies large enough that gravity has given them roughly spherical shape. These bodies will deviate from spherical shape due to

---

[165] The effectiveness with which Jupiter protects us from such collisions has been put in doubt by recent calculations.

centrifugal distortion, external bombardment and internal processes. The four planets closest to our sun: Mercury, Venus, Earth and Mars, are of this type and called the terrestrial planets. Pluto, on the outer edge of our solar system[166] is probably part rock and part snowball. If a rocky planet is large enough, sufficient heat will be released in its formation for it to become molten and *differentiate*, i.e., for more dense materials to sink towards its center.

The planets that have been observed all revolve around a star.[167] Moons by definition are satellites of planets. In our solar system, Mercury and Venus have no moons, Earth has one and Mars has two, although the latter are so small that they are far from spherical. The Jovian planets have many more moons: 61 for Jupiter, 31 for Saturn, 27 for Uranus and 11 for Neptune — according to a recent count. Particularly noteworthy is Titan, the largest moon of Saturn, which has oceans that are comprised of hydrocarbons, such as methane and ethane and, like Earth, an atmosphere that is mostly nitrogen. Titan has been considered as a possible location for a space colony.

Gaseous planets

The four giant planets in the outer part of our solar system: Jupiter, Saturn, Uranus and Neptune, are comprised largely of hydrogen and helium, although they probably have rocky cores. Jupiter, the largest of these, has about 300 times the mass of Earth, and contains more than twice the mass of all the other planets combined.

Although it has long been suspected, it is only in the last decade that planets have been shown to orbit stars other than our sun. Several hundred of these have now been detected. Due to observational difficulties these have mainly been of the large gaseous types. However, recently, a few planets small enough to be terrestrial, have also been observed.

---

[166] Astronomers reclassified Pluto in 2006 from a planet to a dwarf planet, of which there are more than 40 known in the outer reaches of our solar system.
[167] This may be because they are detected by their effect on the motion or the light from their star.

Stars

Stars are objects that shine with their own, rather than reflected, light. Stars evolve in a predictable manner that depends on their mass. Stars can be solitary or occur in pairs (a *binary pair*). Astronomers believe that more than half the stars belong to binary pairs. It is likely that motions in close groupings of more than two stars would be unstable, so that many of these would be expelled from the group. A partial classification of types of stars is given below.

Brown dwarfs

Brown dwarfs are sufficiently massive that they are visible, at least in the infrared region, due to the thermal energy produced by their gravitational collapse. However, with mass less than about 10% that of the sun, their gravitational energy is not sufficient to ignite nuclear fusion reactions in their interior, so they glow very weakly. While the planet Jupiter is large enough to radiate a small amount thermal energy, with a mass 1/2% that of the sun, it does not get hot enough to be considered a brown dwarf. As they dissipate their gravitational energy, brown dwarfs cool and become less visible.

Protostars

These are stars in the process of forming. They are usually enveloped in a cloud of gas. When they ignite, they may blow away a large fraction of this gas. Some stars, called *T. Tauri* are unstable and periodically throw out large flares, which blow away surrounding gas.

Main-sequence stars

Main-sequence stars are those in which the contracting force of gravity is balanced by the radiation pressure created by the fusion of hydrogen to helium in the stars interior. This balance is stable, and main-sequence stars will spend up to 90% of their lifetime in this state. The conditions existing in a main-sequence star depends on the mass of the star, with heavier stars being hotter (meaning their radiation is bluer) and more luminous (producing more energy per unit time). Astronomers classify stars by the letters: O, B, A, F, G, K and M, with type O stars being the hottest, most massive and most luminous. The time a star remains in the stable conditions of the main sequence depends upon its mass. At their higher temperatures, more massive stars burn their hydrogen fuel very quickly, and may go out of balance in millions of years, while smaller stars can last for tens of billions of years. It is believed that the earliest stars formed in the universe were massive and survived only a relatively short time before they exploded and provided a suite of elements for producing subsequent generations of stars.

Our sun is an unremarkable type-G, main-sequence star that has been on the main sequence for about five billion years. The relatively stable conditions that have existed over this time are a prerequisite for the development of life on Earth. Our sun is presently 75% hydrogen and 24% helium[168] and has another five billion years remaining on the main sequence.

The inner regions of main sequence stars are so dense that they allow very little mixing, and such stars are best thought of as composed of non-mixed layers. Heat transfer in these inner regions is by radiation, rather than by convection. Only the innermost region of main sequence stars, their cores, get hot enough to sustain hydrogen fusion. In the outer regions of stars such as the sun, roiling convection occurs, producing effects such as sunspots and flares, which can easily be observed from Earth.

Red giants, planetary nebulae and white dwarfs

When the core of a main sequence star has exhausted its hydrogen fuel for fusion, it will no longer have sufficient radiation pressure to balance gravitational force, and the core will contract. The contraction will continue until the gravitational force becomes balanced by electron degeneracy pressure. The gravitational energy released on contraction increases the core temperature, and results in greater amounts of radiation energy being transferred to regions surrounding the core. At their now higher

---

[168] Other elements with appreciable concentrations in the sun are oxygen, carbon, neon and iron.

temperature, hydrogen fusion will be ignited in these regions. The energy released by this reaction will expand the outer reaches of the star, which will cool as it expands, creating a ***red giant***.[169] This will be the fate of our sun in about five billion years, when its radius expands out past the orbit of Mercury, evaporating the inner planets.

Although no one on Earth will be able to observe it, there will be further evolution of the Sun beyond the red giant stage. As mentioned above, after hydrogen fusion is extinguished in the core, it will continue to contract until the electron degeneracy pressure balances the gravitation force. As additional material falls into the core from the outer regions of the sun, its temperature will continue to rise, until new fusion reactions, forming carbon and oxygen from helium, are ignited in the core. The heat generated in these reactions and in hydrogen fusion reactions in successive layers around the core will further expand the outer regions of the star, forming a gaseous shell called a planetary nebula. With sufficient expansion of the nebula, the hot core, called a ***white dwarf***, may become visible. When the sun becomes a white dwarf, it will have about 60% of its present mass, compacted into a volume about the size of the Earth, giving an incredible density of a million times that of liquid water. A white dwarf constantly radiates energy and cools, until it is no longer visible, becoming a black dwarf. Thus the ultimate product of our sun (and its planetary system) will be such a black dwarf star surrounded by its no longer visible gaseous nebula. This is a highly disordered system.

Novae

It is probable that most stars are in binary systems. It is interesting to consider the evolution of a binary system with stars of mass not too different from that of our sun. Since more massive stars age faster, it is possible that in such a system, one star will be a red giant while the other has already become a white dwarf. In such cases, material in the outer reaches of the red giant (mainly hydrogen), being only weakly attracted to its core, may instead accrete around the white dwarf. As this material falls into the disk, it heats up, and it suddenly may become hot enough

---

[169] In some cases, the outer regions of a red giant may periodically expand and contract, producing a pulsating star.

to initiate hydrogen fusion, with the white dwarf becoming 10,000 to 100,000,000 times more luminous in just a few days. Such a sudden appearance of a bright star was first observed by Tycho Brahe in 1572, and he called it a *nova,* or a "new" star. Actually, it is not a new star, but the sudden brightening of an old star. In fact, since it is only the outer parts of the white dwarf that "explode" in this process, the binary system may be so little disturbed that after the explosion the white dwarf may continue to accrete material from the red giant, with the occurrence of recurrent novae.

Supernovae

Even more dramatic than novae are *supernovae*, in which energy is suddenly released throughout an entire star, rather than just on its surface, as in a nova. Whereas a nova might be as luminous as 100,000 suns, a supernova can be as luminous as 10 billion suns. Light from supernovae decays over a very short time, perhaps weeks or months.[170] However, during this time they may be as bright as all the other stars in their galaxy combined. Supernovae are of two types, type I and type II, the difference being that the spectrum of light from type I supernovae shows no evidence of hydrogen, while that from type II supernovae does show hydrogen lines. This suggests that type I supernovae are a product of white dwarfs, in which all hydrogen has been consumed.

The explanation of type I supernovae was developed in 1930 by Subramanyan Chandrasekhar (Nobel prize in Physics, 1983), a 19-year old Indian student, on his way to begin graduate school in England. Chandrasekhar calculated that a white dwarf more massive than 1.4 times the mass of the sun could not be stabilized against gravitational attraction by its electron degeneracy pressure. When a white dwarf exceeded this limit, it should continue to contract, getting hotter, until further fusion reactions, producing elements as heavy as iron from carbon and oxygen, ignited. The ignition of these reactions is very sudden and produces an explosion that totally destroys the star. One result of this explosion is the synthesis of elements heavier than iron. Since such elements have less

---

[170] For a supernova moving with large velocity with respect to Earth, the decay time may be extended, due to time running slower as predicted by special relativity.

binding energy per nucleon than iron, their synthesis requires the input of the massive amounts of energy generated in the supernova. Many radioactive nuclei are produced, and their decay continues to provide a source of energy to the expanding supernova.

Of particular interest to astronomers are, what is called type IA supernovae, those that release their energy in a very consistent manner. These occur in binary systems, where a white dwarf continually accretes matter from its partner, adding to its mass, until it reaches the Chandrasekhar limit. Since these events follow a predictable course, astronomers reason that they should be of uniform luminosity and act as exceedingly bright "standard candles." From their apparent brightness from Earth, the distance of these objects from Earth can be determined.

Type II supernovae are thought to result from the final stage of very massive stars, in which energy is very rapidly released in a succession of nuclear reactions, producing nuclei of mass close to that of iron (the most stable nucleus). This energy release produces a shock wave, which can synthesize even heavier elements, as well as push electrons into protons to form neutrons. Type II supernovae are comparable in brightness to type IA. However, they actually release more energy, since most of their energy escapes in the form of neutrinos that are formed in the proton + electron -> neutron reaction.

There have only been three supernovae reported in our galaxy: in 1054, 1572 and 1604. Chinese astronomers report that the first of these was bright enough to read by at night for about three weeks. It occurred about 6500 light years from Earth, and its remnant is the glowing gas of the still expanding Crab nebula, shown in Figure 23.4.

On February 24, 1987 astronomers at an observatory in Chile detected a new, very bright star in a nearby galaxy, about 170,000 light years from Earth, and quickly alerted the world's astronomy community to its presence. SN1987A (indicating the first supernova observed in 1987) was studied in great detail. Upon checking, scientists realized that neutrino detectors in deep mines had recorded neutrinos from SN1987A two hours before it had become apparent in visible light. This is because neutrinos generated at the center of the supernova leave its surface almost immediately, due to their very weak interactions, whereas light can only escape when the explosion shock wave reaches the surface. Scientists estimate that during the supernova, 10^{58}, neutrinos were released, and these carried away more energy than was provided by our sun during its entire lifetime.

Figure 23.4. The crab nebula.

Neutron stars

In a supernova the entire star may be blasted apart or, alternatively, a core consisting entirely of neutrons might remain. Such an object packs a mass greater than that of the sun into a spherical volume of about 20 km (12 miles), producing a density 10^{15} (a million billion) times that of water. Even though its surface temperature is very high, a neutron star is so small that its direct observation is very difficult. Scientists believe they have been able to indirectly detect these objects.

Since they retain much of the angular momentum of the original star, neutron stars initially rotate very rapidly. They probably have intense magnetic fields, and like the Earth their magnetic poles probably do not lie on the rotation axis and therefore rotate around the axis. Ionized matter becomes concentrated along the polar magnetic field lines and rotates very rapidly with the star. The radiation from these ions produces a "lighthouse effect," sweeping through space as the neutron star rotates. Astronomers believe that this is the explanation for *pulsars*, sources of radiation that pulse with periods of a millisecond to ten seconds. As a confirmation of this hypothesis, pulsars have been identified at the center of a number of supernovae remnants, including that of the Crab nebula.

Black holes

The minimum mass of a neutron star has been calculated to be 1.4 times that of the Sun. Is there a maximum mass for such an object? Just like there is a maximum gravitational force that can be balanced by electron degeneracy pressure, there should be a maximum (but larger) gravitational force that can be balanced by neutron degeneracy pressure. A neutron star with a mass greater than three or four times that of the sun should continue to contract, and as far as is currently known, there is nothing to stop it contracting to a point. A huge amount of mass at a single point is called a ***singularity***, and you might think that scientists would be very concerned about the possible existence of such an object. Their puzzlement is alleviated, however, by a result of Einstein's theory of general relativity: after contracting beyond a certain size, called the Schwarzschild radius, the gravitational field of such an object becomes so strong that nothing, not even light, can escape from its surface. Thus, nothing can be known about what goes on inside this *event horizon*. Its interior is completely unobservable; it is a ***black hole***. A black hole can only have properties that can be observed from its exterior, such as its mass and its angular momentum. The Schwarzschild radius is calculated to be proportional to the mass of the black hole and equal to about one kilometer for an object with the mass of the Sun. As material falls into a black hole, its mass and size increases, but all information about the material, except its mass, is lost.

Although by their very nature, black holes cannot be observed, their effects on light and other matter should be observable. A number of candidates for black holes have been identified in binary pairs, where material is drawn into the black hole from its partner, or from a surrounding region of gas called an ***accretion disk***. Due to the very high gravitational field of the black hole this material is greatly accelerated and heated, producing an X-ray source.

Conservation of angular momentum prevents material near a black hole from being drawn into its event horizon. However, collisions may disrupt this angular momentum in regions of high density and allow the hole to grow. This is certainly the case in the center of many galaxies, where super-massive black holes are generally thought to exist. These have mass of a hundred thousand to a billion times that of our sun. Early in the formation of a super-massive black hole, huge amounts of radiation are emitted, giving rise to *quasars,* quasi-stellar radio sources. The most powerful continually-emitting objects in the universe, quasars appear as points of extreme luminosity and very large red shifts, corresponding to their being at a huge distance from Earth and existing in the very young universe.

Star Clusters

Clusters range from a few hundred or thousand stars moving in the same direction, which are called open clusters, to groupings of up to several million stars, which are known as globular clusters. The open clusters are fairly young stars, while the globular clusters were formed very early in the universe. Clusters are very useful to astronomers, since the stars in the cluster all have the same age and distance from Earth. Thus if a cluster includes a Cepheid star with variable intensity, the inherent luminosity of all the stars in the cluster can be determined from their observed brightness.

Of course, there is nothing that says that the objects that we have discussed above cannot appear in combinations. These combinations can show interesting characteristics. For example, nebulae are huge diffuse clouds of interstellar gas, containing many hot stars in the process of formation. The light emitted by the stars is absorbed by the gas, which then emits the light at very distinct wavelengths. These clouds often appear reddish, corresponding to the longest wavelength emission of hydrogen atoms in the visible range. Dust clouds may form large dark regions where the emission is absorbed.

Galaxies

The diffused glow across our night sky, called the Milky Way, is seen with telescopes to be made up of billions of stars. The Milky Way also contains some fuzzy glowing patches, which were long thought to be glowing gas and sometimes called nebulae. In 1924, Edwin Hubble recognized that these patches were faraway groupings of stars, separate

388   *Order and Disorder*

Figure 23.5.   An elliptical galaxy.

galaxies, some of which were as large or larger than our Milky Way. Hubble, probably the most famous astronomer ever, would have been a shoo-in for a Nobel prize. However, until 1953, shortly after Hubble's death, the Nobel committee did not consider astronomers eligible for the prize, and it is not awarded posthumously.

Much like stars, the shapes of galaxies are usually disks (like our Milky Way and the Andromeda galaxy shown in Figure 23.1b) or elliptical, such as the galaxy shown in Figure 23.5.

While elliptical galaxies greatly outnumber spiral galaxies, many of them are dwarf companions of larger galaxies. The universe is packed with galaxies; the average distance between them is only about twenty times their size.[171] As a result many galaxies have irregular shapes resulting from collisions between galaxies.

Galaxies are not uniformly distributed through the universe. They form structures of different sizes: from groups of less than 50 galaxies, to clusters that contain many more, to the great wall of galaxies, which is the largest structure in the universe.

---

[171] As compared with stars, the average separation of which is many thousand times their size.

We reside in the Local Group, which contains about 30 galaxies, many of which are dwarf elliptical galaxies. Our Milky Way and Andromeda are the two largest members of the group and are heading for a collision with each other in about two billion years.

History of the Universe

Light emitted from other galaxies shows patterns of lines that can be identified as belonging to particular atoms. However these patterns are often shifted in wavelength from the same emissions produced on Earth. While the lines originating in nearby galaxies are either red or blue shifted (to longer or shorter wavelengths, respectively), the emissions of distant galaxies are invariably red shifted. These observations can be explained by Doppler shifting, resulting from the emitting galaxies moving with respect to Earth.[172] Nearby galaxies may be moving either towards or away from us, depending upon gravitational influences and their history of motion. However, the further from Earth a galaxy is, the more dominant is its motion away from us. In fact, the magnitude of this recession velocity is roughly proportional to the distance to the galaxy, a galaxy twice as far away is receding twice as fast. Hubble's law is the proportionality of Doppler velocities to the distance of the galaxy from Earth. This linear relationship is shown in Figure 23.6, along with the scatter due to motions of galaxies within the nearby Virgo cluster.

Does the observation that all galaxies are moving away from us mean that we are at the center of the universe? Astronomers think not; in fact a general assumption of cosmology is the *cosmological principle*, which states that,

> **All places and directions in the universe are indistinguishable.**

Rather than having the entire universe expanding away from us, we instead imagine that it is the universe itself that is expanding. A reasonable

---

[172] Since the redshifts of the most distant galaxies are greater than one (wavelengths more than doubling), special relativity must be used to calculate the Doppler velocities, to avoid predicting velocities greater than the speed of light.

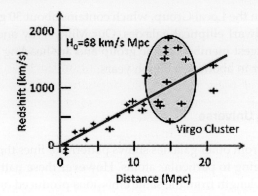

Figure 23.6. Redshift velocities vs. distance.[173]
1.0 Mpc (megaparsec) = 3.1×10^{22} m

analogy is raisin bread rising in the oven, with each raisin representing a galaxy. As time progresses, the bread rises and the raisins grow more distant from each other. Moreover, the further apart two raisins are, the faster they move away from each other.

One difference between raisin bread and the universe, however, is that the bread is expanding in the space of the oven, while there is nothing outside of the universe. According to Einstein's general theory of relativity, the redshift of distant galaxies is due to space itself expanding. Rather than deal with a Doppler shift, it explains the redshift of light from distant galaxies as due to the waves of these emissions stretching with the universe.

We can explore the early history of the universe by considering what would happen if we reversed the relative velocities of distant galaxies, corresponding to a reversal of time. Calculations show that at 13.7 billion years before the present, all the universe would be in a very small region, a singularity. The formation of the universe in this region is what is popularly called the **Big Bang**, although, since it marked the beginning of space and time, as well as mass-energy, there wasn't anyone present to hear the bang. At its formation, scientists

---

[173] H_o, the Hubble constant, is the proportionality constant between redshift velocity and distance. A more recent value determined for this constant is 74.2 ± 3.6 (km/s)/Mpc.

believe the nascent universe was inconceivably hot and tiny and consisted entirely of radiation. This radiation extended to very short wavelengths and was remarkably, although not completely, uniform, non-uniformities being produced by quantum fluctuations.[174] It is the amplification of these microscopic fluctuations by positive feedback from gravity that has resulted in all the structure that we observe today in the universe.

A confirmation of the Big Bang theory of the origin of the universe was obtained, quite accidentally, by Arno Penzias and Robert Wilson (Nobel Prize in Physics, 1978). While working at Bell Laboratories in New Jersey, this pair of scientists was plagued with an annoying background radiation picked up by a very sensitive microwave antenna that they had constructed. This signal is now explained as being due to radiation emitted from the early universe, when it first became transparent — having transformed from a plasma to a gas, mainly of hydrogen atoms. This occurred about 400,000 years after the Big Bang, when the temperature of the universe was about 3000 K. The radiation, initially in the visible and infrared regions of the spectrum, has stretched with the universe, so that it presently is in the microwave region.

Possibility of Intelligent Life Elsewhere in the Universe

It's fitting to end this book with some speculation on whether order comparable to that produced on Earth by our species, might exist somewhere else in the universe. As for any scientific question, we can approach this one both theoretically and experimentally.

Theoretically, we inquire whether any of the probable 10^{23} planets in the universe might be suitable for life. In doing this, we make the assumption that such life would be similar to ourselves, requiring liquid water and sufficient amounts of heavier elements, such as carbon, nitrogen, phosphorus and sulfur. For these stringent requirements we would need a planet at a suitable distance and following a nearly circular orbit around a star of appropriate luminosity. A gravitational field would be needed

---

[174] Quantum fluctuations are predicted by Heisenberg's uncertainty principle, which allows mass-energy to be created for very short times (as virtual electron-positron pairs).

within the range that would maintain an atmosphere, but also allow life to move around. It probably would be useful if the planet were protected from charged particles by its magnetic field and from comet impacts by the gravitational field of a much larger planet at a greater distance from its star.

A suitable star should be in a region of its galaxy that is neither too active (where the planet would frequently be exposed to massive amounts of radiation) nor too quiet (where there hadn't previously been sufficient star destruction to produce appreciable quantities of heavier elements). It should not be a binary star or one that is influenced by other stars during a lifetime of several billion years.

On such a planet, if a self-replicating system were formed, it almost certainly would evolve, although we cannot be sure that its evolution would take it in the same direction that has occurred on Earth. Life once formed would produce order, while simultaneously generating disorder in its surroundings. There might be bottlenecks that would slow the process or massive destruction, such as by impacts that would set it back. However, the number of planets is so large that it is reasonable that life with intelligence comparable to us would have formed on a few of them. Quantitative estimates of many of these factors have been made by a number of scientists, starting with Frank Drake in 1960.

Experimentally, the important question, as asked by Enrico Fermi, is, "Where are they?" That is, if there is intelligent life in the universe, why haven't we heard from it? This assumes that if intelligent life has been formed, it has survived long enough to develop technologies equal to or much greater than our own civilization's. Of course, believers in alien explanations for UFOs would say that there is a lot of evidence that we have heard from them. Most scientists remain unconvinced of this.

With interstellar and intergalactic distances being so large, it has been reasoned that an alien civilization wanting to contact us, would do so by means of electromagnetic radiation, rather than space travel. This has led to the SETI (Search for Extraterrestrial Life) program, which is exploring the heavens for signals with large information content. Nothing has been discovered to the date of writing this book, but there are so many directions and frequencies to be explored that it is certainly not yet time to abandon the search.

Summary
ଓଃ ଓଃ ଓଃ

Astronomers use different wavelengths of electromagnetic radiation to explore the universe. On its way to Earth, emitted light spreads out, is absorbed and reflected, is Doppler shifted and gravitationally bent.

Spheres and disks are dominant shapes in the universe. Disks result from angular momentum, while spheres result from gravity, which pulls matter together until it is balanced by electrostatic repulsion, fusion or electron or neutron degeneracy pressure.

The universe contains gas, dust, rocks, snowballs, planets, moons and a variety of stars. A main-sequence star with the mass of our sun, balances gravity with fusion pressure, over a period of about ten billion years. In another five billion years, our Sun will probably become a red giant, followed by a white dwarf. Larger stars can explode in novae or supernovae, and become neutron stars or perhaps black holes.

Doppler shifts indicate that distant galaxies are moving away from us. This is explained as due to the expansion of the universe, formed in a singularity, at incredibly high temperatures, 13.7 billion years ago. Radiation escaping soon after this Big Bang is seen currently as microwave background.

It is likely that of the estimated 10^{23} planets, some have conditions conducive for the development of intelligent life. SETI, a program to detect signals from such life, has not yet been successful in finding it.

Questions

1. When will Halley's comet next appear? Where does it go when we don't see it?
2. In Chapter 8, we discussed dark matter. What would be the effect of dark matter on light?
3. What is the direction of a comet's tail, as it travels toward the sun? What about away from the sun?
4. Can we expect any protection from the collision of an asteroid with Earth by the gravitational field of Jupiter?
5. How might the presence of a planet that doesn't revolve around a star be detected?
6. What might be the reason that Mercury and Venus have no moons?

7. Why do we observe planetary motions to be periodic more often than chaotic?
8. The North Star, Polaris, was long used by mariners to point to the direction of the North Pole. How could this direction be essentially the same during summer and winter, when the Earth had travelled to opposite sides of the Sun?
9. Why is the ultimate product of our sun, a white dwarf star surrounded by a gaseous nebula, a highly disordered state?
10. What is the SI unit for Hubble's constant, H_o?
11. Why would a planet on which intelligent life originated need to have a nearly circular orbit?
12. Why would a planet on which intelligent life originated need to be neither much less massive or much more massive than Earth?

Appendix

Periodic Table of the Elements

1A																	8A
1 H	2A											3A	4A	5A	6A	7A	2 He
3 Li	4 Be					8B						5 B	6 C	7 N	8 O	9 F	10 Ne
11 Na	12 Mg	3B	4B	5B	6B	7B				1B	2B	13 Al	14 Si	15 P	16 S	17 Cl	18 Ar
19 K	20 Ca	21 Sc	22 Ti	23 V	24 Cr	25 Mn	26 Fe	27 Co	28 Ni	29 CU	30 Zn	31 Ga	32 Ge	33 As	34 Se	35 Br	36 Kr
37 Rb	38 Sr	39 Y	40 Zr	41 Nb	42 Mo	43 Tc	44 Ru	45 Rh	46 Pd	47 Ag	48 Cd	49 In	50 Sn	51 Sb	52 Te	53 I	54 Xe
55 Cs	56 Ba	57 La	72 Hf	73 Ta	74 W	75 Re	76 Os	77 Ir	78 Pt	79 Au	80 Hg	81 Tl	82 Pb	83 Bi	84 Po	85 At	86 Rn
87 Fr	88 Ra	89 Ac															

Lanthanides

58 Ce	59 Pr	60 Nd	61 Pm	62 Sm	63 Eu	64 Gd	65 Tb	66 Dy	67 Ho	68 Er	69 Tm	70 Yb	71 Lu

Actinides

90 Th	91 Pa	92 U	93 Np	94 Pu	95 Am	96 Cm	97 Bk	98 Cf	99 Es	100 Fm	101 Md	102 No	103 Lr

Appendix

Periodic Table of the Elements

Glossary

abhelion	The point on the Earth's orbit where it is furthest from the sun
abiogenesis	The creation of life from inanimate matter
accretion disk	The gas surrounding a black hole that is being drawn into it
acid	A compound that releases hydrogen ions in solution
acidophiles	Bacteria that can exist in very acidic environments
acrosome	Head of sperm cell
actin	A protein that moves muscles
actinides	Radioactive elements that are filling their 5f subshell
active site	The region of an enzyme where the substrate reacts
active transport	Using energy to transport material across a membrane against a concentration gradient
adrenal glands	Located on the kidneys, produce the stress hormone, adrenaline
aerobic	With oxygen
alcohol	Containing an -OH group
aldehyde	Containing a -CHO group
alkali	An element in group 1A
alkaline earth	An element in group 2A

alkane	A hydrocarbon with only single bonds
alkene	A hydrocarbon with only single and double bonds
alkyne	Hydrocarbon with triple bonds
allele	One form of a gene
allergy	Excessive sensitivity to antigens
allosteric effect	Binding of a substance to one part of a protein changes the structure of another part
allotropes	Different molecular forms of the same element
allowed process	A process that obeys all conservation laws
alpha helix	A structural motif in proteins with hydrogen bonds along a chain
alpha particle	A high-energy helium nucleus produced in radioactive decay
alveoli	Part of the lower respiratory system where gas exchange occurs
Alzheimer's disease	A disease characterized by inability to transform short-term to long-term memory
amber	Polymerized tree resin
amide	Containing a -$CONH_2$ group
amine	Containing a -NH_2 group
amino acid	An organic molecule containing an amine and carboxylic acid group
amorphous	Without structure
amplitude	The displacement at a point in a wave
anaerobic	Without oxygen
angular momentum	Momentum times the distance from the center of rotation
anion	Negatively charged ion
annihilation	Combination of matter and antimatter to produce gamma rays
antibody	A molecule of the immune system that binds to a foreign substance
anticodon	The base sequence complementary to a codon
antigen	A foreign substance or organism in our body
antimatter	Particles analogous to those of matter, but with opposite charges
antiproton	A negatively charged proton

aorta	The artery that takes blood to all organs, except the lungs
aqueous solution	One with water as solvent
aromatic	Very stable cyclic conjugated hydrocarbon
artery	A vessel that carries blood away from the heart
artificial selection	Selection of individuals that will reproduce by a breeder
asexual reproduction	Offspring result from a single individual
asteroid	A rocky astronomical object
asthenosphere	A deformable layer of Earth's mantle
asymptotically	Continually approaching, but never reaching
atom	The smallest bit of an element
atomic theory	Atoms are the basic building blocks of matter
ATP	Adenosine triphosphate, the molecule that cells use for short-term energy requirements
atria	A chamber where blood enters the heart
Aurora Australis	The southern lights produced when solar ions get trapped by Earth's magnetic field
Aurora Borealis	The northern lights produced when solar ions get trapped by Earth's magnetic field
autocatalysis	A product of a reaction acts as a catalyst for it
autoimmune disease	Interpretation of normal body tissues as antigens by the immune system
autonomic nervous system	The part of the nervous system that is not under voluntary control
autosome	A similar pair of non-sex chromosomes
autotroph	Species that can use sunlight or inorganic compounds for energy
Avogadro's number	The number of molecules in a mole, 6.02×10^{23}
axial symmetry	Same as rotational symmetry
axon	The output projection of a neuron
axon hillock	The region between the body and the axon of a neuron
axon potential	Electric voltage signal at an axon
B-cells	Immune system cells produced in the bone marrow
balanced chemical reaction	One having the same number of each type of atom on both sides

baryon	A heavy elementary particle
basalt	A black rock formed from solidified magma
base	A compound that reacts with hydrogen ions in solution
beta cell	Insulin-producing cell in the pancreas
beta particle	A high energy electron produced in nuclear decay
beta-pleated sheet	A structural motif in proteins with hydrogen bonds between chains
bifurcation	The change in the nature of a system at some value of one of its parameters
Big Bang	The formation of the incredibly hot universe at a point, 14 billion years ago
bilateral symmetry	Same as reflection symmetry
bile salts	Substances produced in the liver that solubilize fats in the digestive system
binding energy per nucleon	A measure of the stability of a nucleus
biological diversity	The number of different species in an ecosystem
biome	An ecosystem classified by abiological factors and dominant vegetation
bit	The smallest element of information, consisting of a choice between two things
black hole	An object with gravity so strong, not even light can escape from it
blackbody	An object that absorbs all incident radiation
bladder	An organ for storing urine
boiling	Formation of gas bubbles throughout a liquid
bolus	A ball of food and saliva produced in the mouth
Bose-Einstein condensate	A group of bosons describable by a single wave function
boson	Particles whose wave function doesn't change on interchange
branched alkane	Has carbons with more than two bonds to other carbons
branched polymer	A polymer with bonds between linear chains
breeding	Artificial selection of individuals to reproduce
brittle	Property of breaking under force

bronchi	Tubes that branch off from the trachea to the lungs
bronchioli	Smaller branches of the bronchi
Brownian motion	The incessant jumping around of small particles suspended in a fluid
buckminsterfullerene	An allotropic form of carbon shaped like a soccer ball
caloric	Fluid at one time thought to be released from matter when it was heated
calorie	Energy needed to heat one gram of water one degree
capillaries	The smallest blood vessels, where exchange with tissues occurs
carbon nanotubes	An allotropic form of carbon formed of tiny tubes
carbon-14 dating	Determining the age of a plant-derived object by the decay rate of its carbon-14 nuclei
carbonyl	Containing the >C=O group
carboxylic acid	Containing a -COOH group
carnivore	A meat eater
carrying capacity	The number of individuals of a species that can be supported by an ecosystem
cartilage	A somewhat flexible structural material of our bodies
catalyst	Substance that changed the speed of a chemical reaction without itself being consumed
cation	Positively charged ion
cecum	The region where the small and large intestines are joined
cell	The simplest form of life that is capable of independent existence
cellulose	A non-digestible glucose polymer found in the walls of plant cells
center of symmetry	The point about which an object has inversion symmetry
centi	One hundredth
central nervous system (CNS)	The brain and spinal cord

centromere	The place on sister chromatids where they are held together by proteins
Cepheid	A pulsing star
cerebral cortex	The thin layer of grey matter on the outer surface of the brain
chaos	Motion, which while bounded, is too complicated to predict
charged amino acid	One with a normally ionized acid or base side group
chemical energy	Energy due to the electrostatic interactions of electrons and nuclei
chemical formula	A schematic representation of information about the atoms in a molecule and perhaps their arrangement
chemoautotroph	Species that can use inorganic species for energy
chiral	A molecule that has a non-superimposable mirror image
chromatin	Stretched-out DNA
chromosome	A strand of DNA with associated proteins
cilia	Hair-like projections that wave in the upper respiratory system to expel mucous
circular waves	Waves that emanate from a point on a surface
cis-, trans-	Isomers differing only by the placement of groups around a double bond
climate	Weather averaged over an appreciable time
clone	Having the same genetic material as another individual
close-packed	A crystal structure in which atoms have 12 nearest neighbors
coagulation	Sticking together
cocaine	A drug that provides extended stimulation of the reward system of the brain
codon	The ordered set of three bases on mRNA that code for a particular amino acid
coenzyme	A small organic molecule that is a cofactor for an enzyme
cofactor	A molecule that participates in the reaction of the substrate at the active site of an enzyme

coherent	Monochromatic radiation that is in phase and polarized
collagen	A tough protein fiber
colloid	A long-lived suspension
colon	The part of the large intestine where water reabsorption occurs
comet	An icy astronomical object
common name	How a compound is generally referred to
community	A group of interacting organisms sharing an environment
competition	The struggle between species for resources
complement	The molecular part of the non-specific immune system
complementary	The specific base to which a base of DNA or RNA binds
completion	When one of the reactants of a chemical reaction is completely consumed
component	A chemical species
compound	Matter consisting of combinations of two or more elements
Compton effect	Conservation of energy and momentum in the scattering of radiation by matter
condensation	Formation of a liquid from a gas
condensation reaction	A joining of molecules by producing water
conduction	Heat transfer by interaction of stationary molecules
conformers	Species differing only by rotation around a single bond
conjugated	Hydrocarbon with a region of alternating double and single bonds
connective tissue	The material of bones, cartilage, tendons and ligaments
conservation law	A law stating that a quantity doesn't change with time
conservation of mass	Theory that mass doesn't change in chemical transformations
conservative replacement	Replacement of an amino acid in a protein with a similar one

continental islands	Small pieces of land that have broken off from major continental plates
continuous spectrum	Emission or absorption of all wavelengths over a range
control	A comparison group not subjected to intervention in an experiment
convection	Heat and mass transfer by small currents in a fluid
convergent evolution	The tendency of very different species to find similar solutions to environmental challenges
coordinate covalent bond	A bond in which both of the bonding electrons are provided by the same atom
coordination number	The number of nearest neighbors
core electrons	Electrons in inner shells
corpus callosum	The bundle of nerves that joins the two hemispheres of the brain
cosmic rays	Very high-energy particles produced in outer space
Coulomb	The SI unit of charge
Coulomb's law	Law for the force between two charged particles
covalent bond	A bond between two atoms resulting from their sharing electrons
crest	A position of maximum displacement in a wave
crossing over	Exchange of genetic material between homologous chromosomes during meiosis
crosslink	Bonds joining linear chains in polymers
crust	The most outer layer of the solid Earth
crystallization	Forming crystals, usually from a cooling solution
cyclic order	A system that repeats after a period of time
cycloalkane	An alkane in which the end carbons have been joined
cytoplasm	Material not in the nucleus of a cell
d state	A state with orbital angular momentum of two
damped motion	Motion in which velocity slows down
dark energy	Proposed repulsion of matter at large distances that accounts for the expansion of the universe.

decomposer	An organism that breaks down other dead organisms
degeneracy	The number of different states with the same value of energy
delocalized	Electrons are spread over many atoms in a molecule
denaturation	Changing from its natural form
dendrite	A projection from the body of a neuron that receives input signals
deoxy sugar	One in which an OH has been replaced by an H
depolarization	The brief rise of the potential of an axon when the sodium ion channels open
dermis	Inner layer of the skin
deuterium	A hydrogen isotope with one proton and one neutron in its nucleus
diabetes	A disease in which glucose is not properly regulated
dialysis	Passage of ions and small molecules out of solution through pores in a membrane
diaphragm	A muscle whose movement alternately expands and compresses the lungs
diastolic	When the heart is relaxed
differentiate	The settling of heavier materials towards the center of a fluid massive object due to gravity
diffraction	The bending or spreading of waves
diffusion	Movement of matter by molecular motion due to a difference in concentration
digestive system	The system that provides nutrients from food to the blood
diploid	Having two sets of chromosomes, one from each parent
dipole moment	A molecule has positively and negatively charged ends
direction of polarization	Direction of the electric field of electromagnetic waves
directly proportional	One variable rises the same percentage as another
disorder	Lack of restriction

displacement	The distance that material in a wave is moved
dissolving	When a compound enters solution as either individual molecules or ions
distribution	The fraction of individuals having different values of a property
disturbance	Movement of material from its quiescent position
disulfide	Containing the -SS- group
DNA	The molecule that stores the genetic information of life
dominant	An allele that determines a phenotype
Doppler effect	A change in the frequency of a wave due to motion of the wave source
double bond	A covalent bond involving two pairs of electrons
double helix	The two complementary chains of DNA held together by hydrogen bonds
double layer	A layer formed by back-to-back stacking of surfactant molecules
duality	Being thought of as both particles and waves
ecliptic	The plane in which the Earth travels around the sun
ecosystem	A coexisting community of species with its local abiotic environment
egg cell	Large gamete
elastomer	Can be stretched reversibly up to some maximum amount
electron donor	An atom that easily loses electrons
electric field	Electrostatic force per unit charge at a point in space
electric motor	A device for converting electrical energy into mechanical energy
electrical generator	A device for converting mechanical energy into electrical energy
electromagnetic force	Force between charged or magnetic particles
electromagnetic radiation	Time varying electric and magnetic fields propagating in space
electromagnetic spectrum	The complete wavelength range of electromagnetic radiation

electron	The negatively charged constituents of atoms
electron configuration	A designation of how electrons occupy orbitals
electron degeneracy	Since two electrons can't be in the same state, some must have very high energy
electron volt	Energy of an electron accelerated through a potential of one volt
electronegativity	The electron attracting ability of an atom in bonds
electrostatic force	Force between charged particles
element	Fundamental type of matter in chemistry
emphysema	A lung disease characterized by destruction of alveoli
empirical formula	One that just gives the ratios of constituent atoms of a molecule
emulsify	Solubilization of a nonpolar substance in water by incorporating it in a micelle
enantiomers	A chiral pair
endocrine gland	A gland that releases its product into the blood
endoskeleton	A support for the body from the inside
endothermic	Requiring energy
energy	That which is required to do anything useful
energy barrier	An amount of energy required before a change can occur
energy distribution	The fraction of molecules that have different energies
energy surface	Energy as a function of position in a multi-dimensional space.
entangled	Distant particles described by a single wave function
entropy	A measure of disorder
enzymes	Biological catalysts, usually proteins
epidermis	Outer layer of the skin
epigenetic	Modifications beyond the genetic sequence
epithelial tissue	The material of the surfaces of vessels, cavities and structures
equilibrium	A state with no further tendency for change
erosion	Moving of soil and rocks
escape velocity	The velocity needed to escape from the gravitational attraction of a body

esophageal sphincter	The muscle that keeps the acidic contents of the stomach from refluxing into the esophagus
essential amino acid	One that cannot be synthesized in our bodies
ester	Containing a -COO- group
ether	Containing a -O- group
eukaryote	An organism comprised of cells with nuclei
eV	An electron volt
evolution	The genetic change of a population over time
excited state	State of a microscopic system with more than the lowest possible energy
exocrine gland	A gland that releases its product through a duct
exocytosis	Release of contents of a vesicle at a cell membrane
exoskeleton	A support for the body from the outside
exothermic	Releasing energy
exponent	The number of times a quantity is multiplied by itself
expression	When mRNA is being produced from a gene
extensive property	A property of an entire system
extracellular matrix	A polysaccharide-filled region that supports animal cells
extremophiles	Organisms that can exist at extreme conditions
falsifiable	The ability to be shown to be wrong
fat	An ester of glycerol with three fatty acids.
fatty acid	A carboxylic acid derivative of a long-chain hydrocarbon
fermion	Particles whose wave function change sign on interchange
fertilization	Union of a sperm and an egg
fibrous	Stretched out in fibers
first law of thermodynamics	Energy is a conserved quantity
first order	Rate of consumption is proportional to amount remaining
fission	A process in which a large nucleus breaks into smaller pieces
flagellum	A tail-like projection
fluctuation	Transient deviation from most probable arrangement

Glossary   409

fluorescence	Emission of light of changed wavelength after absorption
follicle	Organ in which an oocyte develops
forbidden process	A process that violates a conservation law
force	Something that can bring about change
fossil	Preserved trace or remains of an ancient organism
fossil fuels	Primarily, coal, petroleum and natural gas; buried plant material that has been converted to fuels over millions of years
fractal	An object showing symmetry of scale
fractal dimension	How a fractal object fills space
free energy	Systems change in the direction that decreases this quantity
free radical	A molecule with an odd number of electrons
free-electron model	A sea of electrons moving through cations
frequency	The number of disturbance per unit time at a point
frontal	Toward the front
functional group	A group of atoms that can take the place of hydrogen in hydrocarbons
functional order	How well different parts of a system work together
fundamental niche	The range of all environmental factors needed to support an organism
fusion	A process in which small nuclei combine
Gaia	The theory that life buffers Earth against change by negative feedback
gallbladder	An organ that stores bile salts until they are needed
gametes	Cells from one parent for sexual reproduction
gamma rays	Very short wavelength electromagnetic radiation produced in nuclear processes
gene	A portion of DNA that codes for a protein
general relativity	Einstein's theory that includes gravitation
genetic code	The three bases in mRNA that code for the amino acids
genetic diversity	The number of different alleles in a population
genotype	The genetic material of an individual

geocentric	Characterized by Earth being at the center of the universe
geological stratum	A layer of soil that has been converted into rock by heat and pressure
geometry of a molecule	Specification of the arrangement of its atomic nuclei
geothermal energy	Energy tapped from the heat of Earth's interior
giga	A billion
glacier	An accumulated, moving mass of ice
glial cell	A cell in the nervous system that supports the neurons
global minimum	Lowest value anywhere
global warming	Rising temperature of the Earth's surface
globular	Balled up
glucagon	A hormone that stimulates the release of glucose from glycogen
glucose	A simple sugar used by the body for energy
gluon	Boson that transmits the strong force
glycogen	The polymerized form in which glucose is stored in the liver and muscles
glycoprotein	A protein molecule with one or more carbohydrates attached
goiter	An enlarged thyroid produced by iodine deficiency
Golgi complex	An organelle where molecules are packaged in vesicles
gradient	The rate of change of a quantity in a given direction
gravitational field	Gravitational force per unit mass at a point in space
gravitational force	Attraction between masses
graviton	Boson that transmits gravity
greenhouse effect	Adsorption and reradiation of infrared radiation by atmospheric constituents
grey matter	Regions containing primarily cell bodies of neurons in the brain
ground state	State of a microscopic system with the lowest possible energy

group	Elements in the periodic table with similar properties
group theory	The mathematical theory dealing with symmetry.
gyrus	A ridge on the surface of the brain
hadrons	Elementary particles that interact by the strong force
half life	The time for removal of half of the initial particles
halide	Containing a halogen atom
halogen	An element in group 7A
haploid	Having a single set of chromosomes
heartburn	Leakage of stomach acid through the esophageal sphincter.
heat	Thermal energy transferred from system to surroundings
heat death	A state of maximum entropy in which nothing useful can be achieved
heliocentric	Characterized by the sun being at the center of the universe
hemisphere	Either the left or right side of the brain
hemoglobin	The oxygen transport protein in blood
herbivore	An animal that eats only plants
heterocycle	Having rings containing more than one atom
heterogeneous catalysis	Catalysis by surfaces
heterotroph	Species that directly or indirectly get their energy from autotrophs
heterozygous	Having two different alleles for a gene
hexose	A sugar containing six carbon atoms
histone	A protein molecule around which DNA is coiled
HIV	Human immunodeficiency virus
Holocene	The last 10,000 years, during which Earth has had a fairly constant temperature
homeostasis	Ability of an organism to control its internal conditions
homo sapiens	Humans as a species

homogeneous catalysis	Catalysis by molecules in solution or the gas phase
homologous pair	A set of chromosomes, one inherited from each parent
homozygous	Having identical copies of the gene for a protein
hormone	A chemical messenger in the body
hydrocarbon	Compound containing only carbon and hydrogen
hydration	Combining or interacting with water
hydroelectric power	Energy derived from flowing water
hydrogen bond	A weak interaction between a positively charged hydrogen atom and a lone pair
hydrophilic	Strongly attracted to water
hydrophobic	Not strongly attracted to water
hypertension	Higher than normal blood pressure
hypotension	Lower than normal blood pressure
hypothalamus	The part of the brain involved in maintaining homeostasis
idealization	Simplification that maintains the essence of a situation
ileum	The small intestine
impact crater	The depression formed when a meteor hits Earth's surface
incomplete dominance	A phenotype transmitted in a blended manner
inertia	Resistance to a change of the motion of a body
information	How much we know about something
information-coding base	The basic unit of the genetic code
infrared radiation	Electromagnetic radiation with wavelength somewhat longer than 700 nm
inhibition	A compound slows down a reaction
inorganic chemistry	The chemistry of all atoms except carbon
insulin	A glucose-regulating hormone produced in the pancreas
integumentary	Having to do with skin
intensive property	A property of each bit of matter of the system
interference	The adding of amplitudes of two waves
interferon	A substance excreted by some cells infected with viruses

intermolecular	Between molecules
internal energy	Energy of microscopic parts of a body
intramolecular	Within a molecule
intrinsic property	Properties common to each piece of a system
intron	A section of DNA that does not code for a protein
inversely proportional	One variable decreases by the same percentage that another rises
inversion symmetry	An object is indistinguishable after all points are moved to an equal distance on the opposite side of a point.
ion pump	A protein channel in a cell wall that can move ions towards higher concentrations
ionic bond	Attraction between positive and negative ions
ionizing radiation	High energy particles that produce large numbers of ions when interacting with matter
ions	Molecules or atoms with electrical charge
isoelectronic	Having the same number of electrons.
isolated	Not interacting with surroundings
isomers	Molecules of different structure with the same molecular formula
isotopes	Atoms of the same element with different mass
Joule	The SI unit of energy
Jovian planets	The gaseous planets: Jupiter, Saturn, Uranus and Neptune
Kelvin	SI unit of temperature
ketone	Containing the -C-CO-C- group
kidney	An organ that removes urea and excess salts from blood
kilo	One thousand
kilogram	SI unit of mass
kinetic energy	Energy of motion
kinetic theory	An explanation of properties of matter based on the motion of its constituent particles
kinetics	Having to do with the speed and detailed mechanism of a process
l quantum number	A measure of the shape (or angular momentum) of an orbital
lanthanides	Elements filling their 4f subshell

laser	Device for producing coherent radiation in a very narrow wavelength range
lattice	An object that when repeated can fill up space
lattice energy	The electrostatic energy of interaction of ions in a crystal
law	A summary of observations
law of definite proportions	All samples of a compound contain unvarying ratios of elements
law of multiple proportions	In different compounds of the same elements, the ratios of these elements are in proportion of small whole numbers
lepton	A light fermion elementary particle
Lewis structure	Structures with octets of electrons around atoms except for hydrogen
ligament	Fibrous tissue that holds bones in place at joints
light ray	A description of light focusing on its path
light waves	A description of light focusing on the oscillations of its component electric and magnetic fields
limit cycle	The path of a periodic motion
limiting factor	That whose availability controls the magnitude of a process
limits	Allowable values of variable
line spectrum	Emission or absorption of only discrete wavelengths
linear alkane	Each carbon has no more than two bonds to other carbons
linear polymer	A polymer without branching
lipid	A water-insoluble molecule
liquid core	A liquid region surrounding Earth's solid core
local maximum	A displacement tends to increase the displacement
local minimum	Lowest value in a range, but there are lower values outside the range
logarithm	The exponent to which ten must be raised to give a number
lone pair	A pair of electrons not involved in a covalent bond

long-term memory	Long-term increase of number of synaptic connections
longitudinal wave	A wave in which disturbances are in the direction of propagation of the wave
loop	A structural motif which changes the direction of protein chain
luminiferous ether	Medium through which light was thought to be propagated by 19th century scientists
lymphatic system	Tissues and vessels through which lymph travels
macromolecule	A large molecule
macrophage	An immune system cell that engulfs foreign matter in out bodies
magnetic force	Force produced by magnets or electric currents
magnetic moment	The magnet produced by spin
magnetic resonance imaging	Images based on changing orientations of nuclear spins
magnitude	Measure of the strength of an earthquake
main group elements	Elements in groups 1A through 8A in the periodic table
malleable	Property of being distorted by forces
mantle	The largest part of the Earth, lying directly below the crust
mass	A property of matter that gives it inertia
mathematical model	A representation of relationships between physical quantities by mathematical equations
Maxwell's equations	Mathematical equations that summarize the laws of electromagnetism
mechanism	The detailed steps of a process
Medea	The theory that Earth is unstable due to positive feedback
medium	Material through which a disturbance propagates
mega	One million
meiosis	The process of forming gametes by two successive divisions of a cell
melatonin	A sleep-regulating hormone
meson	An elementary particle of intermediate mass
messenger-RNA	The molecule that transports genetic information from the nucleus to the cytoplasm

metabolism	Chemical reactions in an organism that support life
metastable equilibrium	Small displacements tend to shrink, some large displacements tend to grow
meteor	A "shooting star" produced when a meteoroid burns up in the atmosphere
meteorite	The solid remnant of a meteor on the ground
meteoroid	Rocky or icy material traveling through space
meter	The SI unit of mass
methanogen	An anaerobic, methane-producing bacteria
methylene	A -CH_2- group
micelle	A tiny droplet, hydrophobic on the inside, hydrophilic on the outside, suspended in water
micro	One millionth
micro RNA	Small pieces of RNA
microwave radiation	Electromagnetic radiation with wavelength longer than infrared
milli	One thousandth
mineralization	Replacement of organic with inorganic molecules with retention of shape
mitochondria	Organelle where ATP is synthesized
mitosis	Splitting a cell into two identical cells
m_l quantum number	A measure of the orientation of an orbital
mole	The standard amount of matter in the SI system of units
molecular formula	One that gives the actual number of each type of atom in a molecule
molecular machine	A number of different protein molecules bound together to efficiently produce sequential changes in substrates
momentum	The product of mass times velocity
monatomic	Molecules containing only a single atom
Monera	The kingdom of bacteria
monochromatic	Electromagnetic radiation of a single wavelength
monolayer	A single layer on a surface
monomer	A small molecule that is linked together to form a polymer

Glossary

MRI	Magnetic resonance imaging
mRNA	Messenger RNA
m_s quantum number	Indicates the orientation of a spin
multiple bond	Either a double or a triple bond
multiple sclerosis	A disease in which the immune system attacks the myelin sheath on neurons
mutagen	Something capable of producing a mutation
mutation	An unrepaired change in DNA
myelin sheath	An insulating coating on axons, which increases the speed of nerve pulses
myosin	A protein that moves muscles
n quantum number	A measure of the size of an orbital
nano	A billionth
natural killer cells	Cells that recognize and kill infected cells
natural selection	Favoring of reproductive success due to the environment; selection of individuals that will reproduce by natural factors
nearest neighbors	The atoms in a crystal that are at the closest distance to a given atom
nebula	A cloud of gas and dust
negative feedback	Any displacement tends to reduce the displacement
negentropy	A measure of order
neurohormone	A hormone released by a neuron
neuron	A cell that transmits signals in the nervous system
neurotransmitter	Molecule used to transfer information between neurons
neutral	Neither acidic nor basic
neutralization	A reaction between an acid and a base
neutral equilibrium	Equilibrium at a position with no force in a displacement from the position
neutrino	A particle with zero charge and very small mass produced in some nuclear transformations
neutron	A particle with zero charge and mass approximately that of the proton
neutron star	A star made of neutrons
new-crust zone	A region from which magma outflow is producing new crust
Newton	The SI unit of force

Newton's first law	In the absence of forces, objects move with constant velocity
Newton's second law	Force is the rate of change of momentum
Newton's third law	Objects exert equal and opposite forces on each other
nitrogen fixation	Conversion of atmospheric N_2 to forms that can be used by plants
noble gases	Helium, neon, argon, krypton and xenon
node	Position of a wave at which displacement is zero
nodes of Ranvier	Breaks in the myelin sheath on a neuron between which the signal jumps
non-metals	Elements towards the right-side of the periodic table
non-polar amino acid	One whose side group has little attraction to water
nova	An explosion of the outer part of a star
nucleation center	A bit of material on which a crystal starts to grow
nucleon	Either a proton or a neutron
nucleotide	A base-sugar-phosphate combination
nucleus	The positively charged part of an atom, containing almost all its mass
nucleus	An organelle storing the genetic material of a cell
nucleus	The cell bodies of a group of neurons that work together
nuclide	The nucleus of a particular isotope of an element
occipital	Toward the back
oceanic islands	Islands formed by volcanic activity far from continental plates
octet	Structures in which elements other than hydrogen are surrounded by eight electrons
omnivore	An animal that eats both plants and animals
oocyte	An immature egg cell
Oort cloud	Material that orbits the solar system at large distances from the sun
optical isomers	A chiral pair

orbital	A one-electron wave function
order	Restriction
order of magnitude	A measure of something to within a factor of ten
organ	A number of tissues joined in a structural unit
organ system	A group of organs that work together to perform a particular function
organelle	A region where specific reactions are carried out in a cell
organic chemistry	The chemistry of carbon compounds
organism	Something that is living
oscillating	Changing back and forth
osmolality	The concentration of species in a solution
osmosis	Passage of solvent through a semipermeable membrane
osteocytes	Cells that nourish and renew bone
osteopenia	Having bones of low density
osteoporosis	Having bones of very low density
ovary	Gland that produces eggs and female sex hormones
oxidation	The loss of electrons
oxidation state	The charge state an atom is considered to have in a compound
oxidizing agent	A substance that removes electrons from other species
oxyacid	An acid containing oxygen
ozone hole	Recent depletion of the ozone shield in the spring over Antarctica
ozone shield	A region of the stratosphere containing ozone that shields Earth from short UV rays
p state	A state with orbital angular momentum of one
P waves	Longitudinal seismological waves
pair production	Conversion of gamma rays into matter plus antimatter
Pangaea	Earth's single land mass that supposedly existed 200 million years ago
parasympathetic system	The part of the autonomic nervous system that is activated by relaxation
parietal	Toward the top

particle	A discrete entity
particle-waves	Phenomena that can be described as both particles or waves
pentose	A sugar containing five carbon atoms
peptide bond	An amide bond between two carbons
perfect observer	One limited only by the laws of physics
perhalo	Hydrocarbons in which all hydrogen atoms have been replaced by halogen atoms
perihelion	The point on the Earth's orbit where it is closest to the sun
period	The time for repeating a repetitious motion
period	A number of successive elements in the periodic table
periodic motion	Motion that repeats in time
periodic table	An arrangement of elements showing their similarities
peripheral nervous system	Nerves that project from CNS to other parts of our body
permanent magnet	A magnet that does not require an electric current
peroxy	Containing the -OO- group
peristalsis	A wave of muscle contraction
pH	A measure of the hydrogen ion concentration in solution
phagocytosis	The process of engulfing foreign particles
pharynx	The part of the respiratory tract containing the vocal cords
phase	Position of the crest of a wave at a given time
phase difference	Difference in the positions of two waves at a given time
phenotype	The outwardly observable characteristics of an organism
phospholipid	A fat with one fatty acid replaced by a phosphate group
phosphoryl	Containing a $-OPO_2-$ group
photo-electric effect	Ejection of electrons from a surface by radiation of sufficiently high frequency
photo-thermal power	Conversion of solar energy to heat and then to electric current

photoautotroph	Species that can carry on photosynthesis
photo electricity	Direct conversion of solar radiation into electric current
photon	Energy package of electromagnetic radiation
photosynthesis	Production of oxygen and carbohydrates from CO_2 and water using the energy of sunlight
phototroph	An organism that uses sunlight as its primary source of energy
physiological addiction	A condition in which a larger amount of drug is required for the same stimulation
pi bond	Bond with electrons above and below the line between the atoms
pineal gland	Located within the brain, produces melatonin
pituitary gland	A master gland located at the base of the brain
planar wave	Waves that move in a line on a surface
plasma	A state of matter containing an appreciable fraction of ionized species
plasma membrane	The double layer that encloses a cell
Pleistocene	The time of the ice ages
pluripotent	A cell that can form any type of cell
point group	The group of symmetry operations that characterize an object
polar amino acid	One whose side group is attracted to water
polar bonds	Covalent bonds in which the electrons are not centrally located
polarized	The charged state of an axon ready to transmit a signal
polycyclic	Containing multiple connected rings
polygenic	A trait determined by more than one gene
polymer	A large molecule with many subunits
polysaccharide	A sugar polymer
population	The individuals of a species residing in a particular location
positive feedback	A displacement produces forces that increase the displacement
positron	A positively charged electron
post-translational modification	Modification of an amino acid after it is incorporated into a protein
potential energy	Energy of position

power	Energy per unit time
predator	An organism that consumes another
predictable motion	Motion that can be calculated from Newton's laws
presynaptic neuron	The neuron of the axon at a synapse
prey	Animals that a carnivore consumes
primary consumer	An organism that consumes the primary producer
primary producer	An organism that uses sunlight and CO_2 to produce carbohydrates
primary structure	The sequence of amino acids in a protein
principle of equivalence	A gravitational field is indistinguishable from an accelerating reference frame
prion	A misfolded protein
probability density	A function giving the probability of an electron being at different locations
probability theory	Considerations of random events
product	The compounds on the right side of a chemical reaction
prokaryote	Cell without a nucleus
promoter	A piece of DNA that when attached to a transcription factor, activates transcription of a gene
protein sequence	The ordered listing of the amino acids in a protein
Protist	The kingdom of microscopic eukaryotes
proton	A positively charge constituent of an atomic nucleus
psychological addiction	A state where external cues produce desire for a drug
pulmonary artery	Artery that takes blood to the lungs
pulsed laser	A laser that emits its energy in very short bursts
purine	An information-coding base with two rings
pyloric sphincter	Muscle that keeps food in the stomach until it is digested
pyrimidine	An information-coding base with one ring
qbit	Information given in terms of probabilities
quantum dots	Particles in which electrons are confined to very small regions

Glossary 423

quantum fluctuations	Transient appearance of particle-antiparticle pairs in vacuum.
quantum mechanics	The mathematics of very light particles
quantum number	A number that can only take on discrete values that describes a state
quark	Particle with charge of 2/3 or –1/3 of which baryons are composed
quaternary structure	How multiple subunits are connected in a protein
racemic mixture	Contains equal amounts of each enantiomer
radio waves	Very long wavelength electromagnetic waves
random motion	Motion described by the laws of probability
reactant	The compounds on the left side of a chemical reaction
realized niche	The range of conditions a species can survive in when competing with other species
recessive	An allele that is masked by another
red giant	A large but fairly cool star
reduction	The gain of electrons by a species
reductionist	Explaining complex systems based on their component parts
reflection plane	The plane through which an object has reflection symmetry
reflection symmetry	An object is indistinguishable after all points are moved to an equal distance on the opposite side of a plane
reflex	A rapid response to a sensory input
renewable energy	Energy used soon after it arrives on Earth from the Sun
replication	The act of a system reproducing itself
resonance	Different arrangements of the electrons in a molecule
rest	The state of no motion
rest energy	The energy equivalence of the mass of an object
reuptake	Taking up of neurotransmitter molecules by the presynaptic neuron
reversible	A reaction that has a tendency to go in either direction
reward system	Parts of the brain whose stimulation provides intense pleasure

ribosome	A grouping of proteins and RNA that synthesizes proteins in a cell
ribozyme	An RNA-based enzyme
rift	A deep long valley in the Earth's crust
RNA	A molecule that works with the genetic information of life
rotational symmetry	An object is indistinguishable after a rotation around an axis
s state	A state with zero orbital angular momentum
S waves	Transverse seismological waves
SI units	Set of units generally used by scientists
salination	Increase in concentration of inorganic ions in water
saliva	Lubricants and digestive juices produced in the mouth
saturated hydrocarbon	An alkane
saturated solution	A solution containing the maximum amount that will dissolve
scientific notation	Expressing numbers using powers of ten
scrotum	Sac in which a testicle is suspended
second	The SI unit of time
second law	Entropy always increases in isolated systems
secondary structure	Short-range structural motifs in a protein
sedimentary rock	A rock that was formed from sediments converted by heat and pressure
seismology	The study of the propagation of vibrations through the Earth
semiconductor	A material whose electrical conductivity is between that of a conductor and an insulator
semipermeable	Allows the passage of water, but not of solutes
sex chromosomes	The very different X and Y homologous pair of chromosomes
sex-linked	A trait carried on the X or Y sex chromosome
sexual reproduction	Offspring result from two individuals
sexual selection	Selection of individuals that will reproduce by sexual preferences of the opposite sex
shell	Electrons with the same n quantum number
short-term memory	Short-term enhancement of synapse strength

side group	The third group attached to the alpha carbon in an alpha amino acid
sigma bond	Bond with electrons near the line between atoms
sinoatrial node	Group of cells that generate the electrical signal for the heart muscles
single bond	A covalent bond involving one pair of electrons
sister chromatids	Identical copies of a chromosome
skeletal formula	Structural formula of an organic compound showing only the carbon backbone and functional groups
small intestine	The organ in which nutrients pass into blood
snowball Earth	The idea that Earth was once completely covered with ice
soil	Rock fragments combined with decaying organic material
solar energy	Energy from the sun
solar nebular	The gas and rocks that condensed to form the solar system
solid core	The part of the Earth closest to its center
solubility	The amount of a substance that will dissolve in a solvent
solute	Components added to solvent to make a solution
solvent	The component of largest amount in a solution
soma	The central body of a neuron
somatic nervous system	The part of the nervous system that is under voluntary control
somatosensory system	The part of the brain that produces sensations, such as touch and pain
sound wave	A wave composed of alternating compression and rarefaction of a medium
special relativity	Einstein's theory that deals with objects moving at constant velocity
species	Individuals that can interbreed and have fertile offspring
spectroscopy	Probing a molecular system with emitted or absorbed light
speed of sound	Velocity with which a longitudinal wave moves in a medium

spermatozoon	The smaller gamete
spherical waves	Waves that emanate from a point in a three-dimensional medium
sphincter	A muscle that normally keeps a passage closed
spin	An internal angular momentum
spin orbital	A designation of both spin and orbital of an electron
spindle apparatus	Filaments that pull sister chromatids apart during mitosis
stable equilibrium	Equilibrium with a restoring force at any displacement from the equilibrium position
stable motion	Motion that approaches rest or a limit cycle
standard model	The quarks, leptons and bosons that comprise all matter
standing wave	Stationary wave
starch	A digestible glucose polymer
state function	Something that depends only on the present state of a system
static electricity	Transfer of charge due to friction
stationary wave	A wave in which the positions of nodes do not move
steady-state	An unchanging condition having flows of matter and/or energy
stem cell	A cell that can divide and change type
steric hindrance	An increase in energy due to the close approach of two nonbonded atoms in its structure
steroid	A lipid based on a particular 4-ring organic system
stoichiometry	The information about a chemical reaction provided by its balanced chemical reaction
stratosphere	The part of the atmosphere directly above the troposphere
strings	According to some physicists: the ultimate particles of which everything is composed
strong acid	A compound that releases all its ionizable hydrogens in solution
strong base	A compound that releases all its ionizable OH- ions in solution

strong force	Force between nucleons
structural formula	One that gives information about the arrangement of atoms in a molecule, as well as their numbers
structural isomers	Isolatable different species with the same molecular formula
subduction zone	A region in which one tectonic plate is moving under another
subshell	Electrons with the same n and l quantum numbers
substrate	Molecules acted on by an enzyme
sulcus	A depression on the surface of the brain
summer solstice	The day when the Earth's axis is tilted closest to the sun
supernova	A violent implosion of a large star
superorganism	A number of organisms that live together, differentiated for specialized functions
super saturated	Containing more than the maximum that will dissolve
surface energy	Energy required to increase the surface area of a liquid
surface tension	Surface energy
surfactant	A molecule having a hydrophilic and a hydrophobic end
surroundings	Everything in the universe except the system
suspension	Small regions of one component dispersed throughout another
symmetry axis	A direction around which a system can be rotated by some angle without discernable change
symmetry operation	A procedure that does not produce a distinguishable change in a system
sympathetic system	The part of the autonomic nervous system that is activated under stress
synapse	A region where a signal is transmitted between two neurons
synaptic cleft	The gap between an axon and its target cell

synergy	Advantageous cooperation between individuals
system	The part of the universe we focus our attention on
systematic name	The naming of a compound according to its constituent atoms following definite rules
systolic	When the heart is pumping
T-cells	Immune system cells produced in the thymus
taxonomy	The science of classifying life forms
tectonic plate	A moveable section of Earth's crust
tektite	A bead of solidified molted quartz
temperature	A measure of thermal energy per unit mass
temperature inversion	Temperature increasing with elevation
temporal	Toward the side
tendon	Fibrous tissue that connects muscles to bones
terrestrial planets	The rocky planets: Mercury, Venus, Earth and Mars
tertiary structure	How a protein chain is folded in three dimensions
testes	Glands that produce sperm and male sex hormones
tetrahedron	A 3-D figure with four faces and four vertices
thalamus	A part of the brain that relays signals
theory	A hypothesis that has stood the "test of time"
thermal energy	Kinetic energy on the microscopic scale
thermodynamics	Considerations of energy and entropy
thermohaline circulation	The global circulation of the oceans, driven by temperature and salinity variations
thermophile	An organism that can exist at very high temperatures
thermostasis	Maintaining constant temperature
thiol	Containing a -SH group
thymus	The gland located in the chest that produces T-cells
thyroid gland	Located in the neck, regulates energy level
tissue	An aggregate of cells and intercellular material
tolerance range	The range of an environmental factor over which an organism can survive
top carnivore	A meat eater for which there is no predator

torque	Force times the distance from a center
trachea	The windpipe
transcription	Converting DNA to messenger-RNA
transcription factor	A protein that activates transcription of a gene by attaching to its promoter sequence
transfer-RNA	Small RNA molecules that attach amino acids and their anticodons
transition elements	Elements in groups 1B through 8B in the periodic table
translation	Moving the entire body in the same direction
translation	The process of generating proteins from mRNA
translation group	The movements of objects that leave an infinite collection of those objects indistinguishable
translational symmetry	A repeating object is indistinguishable if all its parts are moved in the same direction by the same distance
transverse wave	A wave in which the disturbance is perpendicular to the direction of propagation
triple bond	A covalent bond involving three pairs of electrons
tritium	A hydrogen isotope with one proton and two neutrons in its nucleus
trophic levels	The successive levels in the environment through which energy flows
troposphere	The lowest level of the atmosphere
tunneling	Go through an energy barrier
Turing pattern	A spontaneously developed spatial pattern in a system
ultra-violet light	Electromagnetic radiation with wavelengths somewhat shorter than 400 nm
uncertainty principle	Limits on simultaneously measuring position and momentum of a particle
unit cell	A small area or volume with which two or three-dimensional space can be completely filled
unsaturated hydrocarbon	A hydrocarbon with double or triple bonds
unstable equilibrium	A displacement from this equilibrium tends to grow

urea	The compound that nitrogen wastes are converted to in the liver
urethra	The tube through which the bladder is emptied
uterus	Organ in which embryo matures
vaccination	Priming the immune system by adding antigens
valence electrons	Electrons in the outermost shell of an atom
vapor pressure	The pressure in a gas, in equilibrium with its liquid
variable	A quantity that can take on different values
vein	A vessel that carries blood toward the heart
velocity	A vector indicating the magnitude and direction of speed
velocity of light	An invariant 3×10^8 m/s for all electromagnetic radiation
ventricle	A chamber from which blood leaves the heart
vesicle	A volume completely enclosed by a double layer
vestigial	A no-longer used organ left over from earlier stages of evolution
virtual particles	Bosons that transmit forces
virtual photon	Photons that are responsible for transmitting the electrostatic force
visible light	Electromagnetic radiation with wavelength of 400–700 nm
vitamin	A necessary coenzyme that cannot be synthesized in our bodies
volt	Joule per Coulomb
W and Z boson	Bosons that transmit the weak force
watt	Joule per second
wave	Undulating motion in space and/or time
wave function	The ultimate quantum-mechanical description of matter
wave length	Repeating distance of a periodic disturbance
wave velocity	The velocity with which a disturbance travels through a medium
wavelength	The distance between identical parts of a wave, such as its crests
weak acid	A compound that releases some of its ionizable hydrogens in solution

weak force	Force causing change of fundamental particles
weathering	The breaking up of rocks
white dwarf	A small star in which gravity is balanced by electron degeneracy pressure
white matter	Regions containing primarily axons of neurons in the brain
WIMP	Weakly interacting massive particle
winter solstice	The day when the Earth's axis is tilted furthest from the sun
work	Force times the distance moved in the direction of the force
X-rays	Electromagnetic radiation with wavelengths shorter than ultra-violet
zygote	The cell formed from combination of two gametes

Credits

Cover: with the permission of the King Library, San Jose, Ca.

Chapter 2

Figure 2.6.	From the "Minerals in Your World" project, a cooperative effort between the United States Geological Survey and the Mineral Information Institute
Figure 2.7.	Drawn with a program of Michael Barnsley
Figure 2.8.	Uploaded by AVM to Wikimedia Commons

Chapter 3

Figure 3.1.   Image by Erwin Rossen

Chapter 4

Figure 4.1.   From Practical Physics, by Newton Henry Black and Harvey N. Davis, London, Macmillan, 1914

Chapter 7

Figure 7.1.   Uploaded by P. Oystein to Wikipedia
Figure 7.3.   With the permission of Ocean Nanotech Co.

Chapter 8

Figure 8.1.   C. D. Anderson, Physical Review **43, 491 (1933)**

Chapter 9

Figure 9.2.   Jan Homann in WikiProject Physics/Archive October 2009
Figure 9.4.   Uploaded by en:User:SeeSchloss to Wikimedia Commons

Chapter 13

Figure 13.1. With the permission of the King Library, San Jose, Ca.

Chapter 14

Figure 14.4. On Wikipedia, created by SuperManu
Figure 14.6. On Wikipedia, created by Dr. H. U. Bodecker

Chapter 15

Figure 15.1. Paul Verkade, Joke Owendijk, Michele Solimena Med. Fac., University of Technology, Dresden, Germany

Chapter 16

Figure 16.6. On Wikipedia, created by Tom Murphy VII

Chapter 17

Figure 17.3.	Uploaded by J. W. Schmidt to Wikipedia
Figure 17.4.	Uploaded by G3pro to Wikipedia
Figure 17.5.	Jason Koval & Kevin Cartwright on Wikipedia
Figures 17.7. and 17.13.	Uploaded by Richard Wheeler to Wikipedia
Figures 17.6. and 17.8.	With the permission of Professor Richard Bowen, Colorado State University

Chapter 18

Figure 18.1.	Created by Madeline Price Ball

Chapter 19

Figure 19.1.	Created by Mariana Ruiz (LadyofHats)
Figure 19.3.	Created by Patrick J. Lynch, medical illustrator and C. Carl Jaffe, MD, cardiologist
Figure 19.5.	From Wikimedia Commons
Figure 19.6.	Created by Crystal Mason
Figure 19.7.	Figure prepared for the U. S. Federal Government
Figure 19.9.	Created by K. H. Maen
Figure 19.10.	Figure prepared for the U. S. CDC

Chapter 20

Figure 20.2.	Prepared by Chris73 for Wikimedia
Figure 20.3.	Created by Nrets for Wikimedia
Figure 20.4.	From *Manuel de L'anatomiste*, by Charles Morel and Mathias Duval, published in 1883
Figure 20.5.	Created by User:jimhutchins for Wikipedia

Chapter 21

Figure 21.1. From NASA's Visible Earth website
Figure 21.4. Created by Robert Simmon, NASA; modified by Robert A. Rohde

Chapter 22

Figure 22.1. Photo by Brian Voon Yee Yap
Figure 22.3. From U. S. Geological Survey, U. S. Department of the Interior
Figure 22.4. Created by Nobu Tomura

Chapter 23

Figure 23.1a. Carolyn Porco, Cassini Imaging Team Leader, NASA
Figure 23.1b. John Lanoue, from BedfordNights.com
Figure 23.2. Image taken by the Hubble Space Telescope
Figure 23.3. Hubble Space Telescope Comet Team and NASA
Figure 23.4. Image taken by NASA's Hubble Space Telescope
Figure 23.5. Image taken by NASA's Hubble Space Telescope
Figure 23.6. Patterned after William C. Keel (2007), *The Road to Galaxy Formation*, Berlin, Springer

Index

abiogenesis 8
acid 168–170, 183, 185, 186, 188, 191, 210, 211, 212, 229, 242, 243, 245–248, 250, 254, 256, 258, 259, 261, 271, 273, 274, 278–283, 288, 291, 295, 314, 315, 323, 343, 350, 353
    carboxylic 183, 184, 186, 190, 191, 242, 245
    oxyacid 168
    strong 46, 48, 51, 52, 56, 58, 67, 75, 87, 96, 115–117, 120, 123, 124, 127, 133, 142, 144, 147–149, 151, 152, 164, 166, 168, 169, 185, 212, 235, 247, 261, 272, 274, 277, 307, 335, 343, 360, 362, 373, 375, 386
    weak 41, 46, 47, 51, 52, 56, 58, 97, 115, 116, 123, 124, 142, 145, 146, 148, 152, 166, 169, 184, 186, 210, 226, 243, 262, 273, 277, 322, 328, 331, 339, 380, 382, 384
alkali 146, 147
alkaline earth 146, 147
alkanes 176, 177, 179, 190
allosteric effect 264
allotropes 148, 149, 151, 152
    of carbon 33, 40–42, 90, 91, 136, 150–152, 156, 158, 175–177, 181, 190, 214, 235, 236, 314, 350
    of oxygen 33, 89, 148, 155, 191, 297, 300, 302, 351
allowed process 113, 114
amino acid 188, 190, 223, 226, 235, 242–248, 255, 256, 258–261, 263, 268, 270, 273–275, 278, 279, 281–283, 288, 291, 295, 314, 323, 351
    charged 17, 35, 36, 41, 46, 47, 48, 51, 76, 115, 117, 123, 127, 128, 143, 146, 160, 162–164, 166, 209, 210, 219, 243–246, 252, 253, 255, 263, 266, 273, 313, 341, 392
    essential 13, 24, 27, 95, 116, 146, 149, 179, 216, 223, 226,

228, 235, 245, 255, 261, 270, 277, 326, 357, 394
  nonpolar   179, 208, 209, 255, 259, 263, 270
  polar   159, 167, 173, 208–210, 242–245, 255, 259, 263, 266, 288, 313, 346, 348, 349, 352, 360, 385
anion   38, 146, 147, 156, 160, 169, 173, 183, 191, 271, 320
  polyatomic   160
antibody   304
antigen   270, 274, 304, 305
antimatter   117, 120, 125
asteroids   51, 58, 341, 348, 374
astronomical   10, 57, 58, 369, 372, 373, 374
  dust   55, 122, 205, 338, 366, 376, 377, 387, 393
  gas   4, 5, 13, 16, 18, 26, 27, 36, 90, 122, 123, 125, 137, 139, 143–147, 155, 172, 197, 198, 208, 212, 213, 228, 299, 306, 347, 349, 354, 373, 376, 377, 380, 384, 386, 387, 391, 393
  observations   2–8, 12, 33, 70, 80, 95, 97, 111, 148, 232, 234, 361, 369, 372, 389
  order   1, 15, 31, 45
    disks   373, 374, 388, 393
    spectra   370, 376
      redshifts   389
    spheres   161, 335, 372, 373, 374, 393
atmosphere   13, 16, 26, 40, 41, 89–91, 117, 143, 144, 148, 149, 153, 155, 171, 183, 198, 202, 204, 236, 255, 256, 333, 334, 337, 341, 342, 345–349, 351–353, 365, 379, 392
  ozone layer   171, 172, 348
  stratosphere   171, 172, 183, 348, 352, 354
  troposphere   347, 348, 352, 354
atomic theory   33, 34, 169
  law of definite proportions   33, 157
  law of multiple proportions   33
atoms   3, 16, 31, 69, 80, 98, 127, 139, 157, 199, 245, 348, 370
ATP (adenosine triphosphate)   252
autotrophy   235, 295, 355, 358

baryons   116, 117, 120, 121
base   1, 11, 12, 82, 89, 105, 127, 133, 154, 168–170, 173, 184, 190, 197, 212, 230, 248, 250–252, 255, 256, 271–274, 276–281, 283–285, 288, 290, 291, 298, 308, 309, 342, 358, 362, 364
  information coding   251, 252, 271, 276
beta cell   222–226, 264, 294
Big Bang   15, 29, 56, 390, 391, 393
black hole   56, 108, 375, 386, 387
blackbody   94
Bohr theory   129, 133
boiling   213
bond   143, 148, 149, 151, 159, 164–167, 177–180, 186, 187, 199, 201, 250, 258, 260, 261

Index 439

coordinate covalent 166
covalent 143, 152, 155, 158, 159, 164–166, 168, 172, 173, 277
  disulfide 243, 259, 261, 263, 273
  double 4, 14, 46, 148–151, 157, 166, 170, 173, 176, 179–181, 186, 188, 190, 191, 210, 219, 251, 272, 273, 277, 279, 280, 290, 291, 361
  hydrogen 16, 32, 86, 106, 107, 108, 116, 127, 139, 157, 175, 201, 212, 229, 242, 261, 277, 351, 375
  ionic 146, 147, 155, 157, 160–162, 164, 168–170, 172, 175, 318, 320, 325
  metallic 47, 95, 146, 147, 153–56, 164, 318
  polar 159, 167, 173, 208, 209, 242, 243, 245, 255, 259, 263, 266, 313, 346, 348, 349, 352, 360, 385
  triple 149, 150, 166, 173, 176, 179, 180, 190
bone 40, 69, 106, 146, 261, 262, 274, 293, 294, 304, 305, 307, 308, 344
  osteocytes 307
Boson 107, 108, 111
brain 3, 10, 204, 224, 227, 230, 234, 267, 268, 294, 297, 302, 303, 309, 314, 317–320, 324–331
  addiction 329
  cocaine 324, 329–331
  corpus callosum 325
  grey matter 319, 325
  lobes 326, 330
  memory 217, 305, 325, 326, 328, 329
  reward system 328–330
  somatosensory system 326, 330
  white matter 319
Brownian motion 95, 198

cartilage 106, 261, 294, 307
catalyst 170–173, 217, 223, 253, 265, 290
cation 38, 39, 136, 137, 146, 153–156, 160, 166, 169
cell 22, 29, 70, 143, 161, 187, 204, 221–225, 227–230, 235, 237, 238, 241, 242, 251, 255, 260, 266, 269, 270, 274, 276, 277, 279, 283–286, 287, 290, 291, 293, 294, 295, 297, 300, 304, 305, 311–313, 317–319, 321–323, 325, 330,
  cytoplasm 223, 225, 229, 237, 238, 241, 276, 279, 281, 284, 318
  gamete 230, 231, 276, 279, 286, 287, 291, 311
  mitochondria 225, 238, 282, 304, 308, 309, 313, 320, 357
  nucleus 17, 36–40, 42, 46, 51, 52, 58, 59, 69, 86, 87, 105, 114, 119, 124, 125, 127–130, 132, 133, 136, 141, 142, 144, 146, 147, 167, 170, 223, 224, 225, 227, 237, 238, 241, 276, 279, 280, 284, 285, 290, 291, 313, 318, 329, 384
  plasma membrane 210, 223, 225, 237, 241, 318, 319

ribosome   224, 273, 281–283, 291
stem   177, 182, 230, 284, 326, 328
cepheids   370
chemical reactions   33, 81, 84, 169, 170, 173, 201, 210, 212, 215, 216, 218, 225, 228, 252, 275
   completion   170, 172, 212
   equilibrium   78, 82, 85, 139, 170, 207–212, 214, 215, 218, 230, 238, 248, 260, 274
   neutralization   212
   patterns in   214–216, 275
   reversible   170, 173, 357
   stoichiometry   160, 161, 170
chromatin   279, 285, 286, 313
chromosome   279, 285, 287, 288, 289, 291, 313
circulatory system   241, 250, 294, 299, 300, 314
   aorta   303
   arteries   300, 303
   capillaries   260, 299, 300, 302, 303
   heart   13, 58, 186, 219, 226, 241, 295, 300, 302, 303, 308, 309, 314, 320, 325
      atria   302, 303
      sinoatrial node   303, 308
      ventricles   302, 303
   veins   300, 303
climate   343, 345, 349, 350, 360
   greenhouse effect   346, 349, 352
collagen   261, 262, 307
comet   55, 377, 378, 392, 393
   Halley's   55, 377, 393

comets   45, 341, 373, 377, 378
community   5, 355, 357, 358, 366, 384
complexity   1, 117, 127, 136, 216, 221, 222, 225, 227, 230, 235, 237, 257, 275, 293
compounds   16, 23, 32, 33, 97, 136, 144–147, 152, 157–158, 160, 161, 164–166, 169, 170, 172, 173, 175, 176, 181–188, 190, 191, 202, 209, 227, 235, 236, 254, 255, 257, 294, 295, 297, 351, 358
   aromatic   181, 182, 184, 190
   carbonyl   184, 185, 187, 190
   common name   158, 164, 173, 178, 180, 182
   conjugated   180
   empirical formula   158, 160, 161, 172, 173
   hydroxyl   187, 243, 248
   inorganic   157, 158, 175, 235, 255, 344, 358
   molecular formula   158, 164, 173, 177, 178, 256
   organic   157, 158, 169, 175, 176, 184, 185, 187, 190, 235, 236, 254, 255, 266, 295, 320, 343, 344, 358
   systematic name   158, 160, 164, 173
configurations   134–136, 140, 142, 146, 147, 197, 198, 205, 208, 270
convection   213, 214, 218, 338, 339, 347–349, 352, 381
coordination number   161, 162
cosmic ray   117
Coulomb's law   47, 114

Index 441

crystals 22–24, 45, 70, 161–164, 247, 248, 262, 307
cycloalkanes 178, 179, 190

dark energy 56, 57
dark matter 122–125, 393
degeneracy 131, 132, 137, 375, 381–383, 386, 393
diabetes 226
digestive system 294–297, 300, 308, 314
    esophagus 295, 315
    large intestine 297
    small intestine 295, 297, 314, 315
    stomach 269, 295, 315, 329
dinosaurs 342, 344, 364, 365
    extinction of 342
dipole moment 159, 167, 173, 179
disorder 1, 15, 31, 45, 163, 193, 208, 274, 303, 343, 382
Doppler effect 63, 64, 74, 370
duality 100

Earth's 3, 16, 53, 54, 57, 79, 86, 90, 117, 143, 144, 146, 147, 149, 150, 153, 171, 236, 255, 334–344, 346–348, 350–353, 370
    core 3, 16, 86, 144, 157, 164, 337, 338, 352, 353, 379, 381, 382, 385
    crust 143, 144, 146, 147, 150, 338, 339, 340, 352
    erosion 16, 342, 343, 343, 352, 360
    impact craters 341, 345
    magnetic field 48–51, 67–69, 73, 100, 105, 106, 109, 117, 118, 130, 131, 338, 340, 341, 353, 375, 385, 392
    mantle 338, 339, 340, 352, 353
    orbit 16, 53, 54, 55, 57, 128, 137, 335, 338, 352, 373, 377, 379, 382, 391, 394
    seasons 55, 335
    tectonic plates 339
    tilt 54, 335, 344, 352
    weathering 205, 342, 343, 352
ecosystem 28, 237–361, 365–367
    carrying capacity 361
electric motor 50
electromagnetic 46–48, 51, 56–67, 68–70, 73, 74, 81, 83, 90, 93–95, 105, 115, 123, 142, 369, 370, 392, 393
    field 11, 41, 47–51, 56, 68, 69, 76, 79, 85, 100, 105, 106, 109, 117, 118, 123, 130, 131, 151, 152, 162, 213, 247, 248, 336, 338, 340, 341, 351, 353, 371, 375, 378, 385, 386, 391–393
    force 5, 24, 34, 45, 46, 47, 48, 51, 52, 53, 54, 55, 56, 57, 58, 59, 75, 77, 78, 79, 82, 87, 102, 108, 115, 116, 117, 123, 124, 127, 142, 175, 335, 345, 371, 374, 375, 380, 381, 382, 386
electromagnetic radiation 67, 69, 70, 73, 90, 93, 94, 105, 123, 369, 370, 392, 393,
    polarization 69, 71, 73, 247, 248
    spectrum 67–69, 74, 104, 105, 129, 133, 136, 383, 391

electron configuration   133, 134, 136, 137, 141, 153, 155
electronegativity   159, 173
electrostatic   47, 48, 51, 52, 57, 79, 87, 114, 123, 142, 146, 151, 154, 158, 161, 162, 164, 172, 252, 322, 375, 393
   energy   5, 16, 32, 48, 71, 94, 113, 128, 143, 161, 177, 194, 207, 225, 248, 266, 276, 297, 318, 337, 357, 371
   force   5, 24, 34, 45–48, 51–59, 75, 77–79, 82, 87, 102, 108, 115–117, 123, 124, 127, 142, 175, 335, 345, 371, 374, 375, 380–382, 386
endocrine system   309, 311, 328
   adrenal glands   311
   pituitary gland   309, 310, 314, 328
   thyroid gland   309, 310
energy   5, 16, 32, 48, 71, 94, 113, 128, 143, 161, 177, 194, 207, 225, 248, 266, 276, 297, 318, 337, 357, 371
   barrier   85–88, 152, 171, 179, 223, 253, 303, 319, 363
   conservation   34, 53, 55, 59, 82, 84, 96, 101, 109, 113–115, 118, 124, 125, 129, 194, 195, 209, 386
   generation   8, 88, 121, 214, 232–234, 239, 288, 290, 362
      aerobic   236
      anaerobic   236, 353, 365
   internal   34, 76, 80, 84, 90, 105, 116, 120, 131, 225, 227, 229, 311, 324, 341, 375, 379
   kinetic   5, 13, 76, 80–84, 88, 90, 91, 102, 114, 170, 199, 200, 202, 204, 341
   potential   76–85, 87, 88, 90, 91, 111, 199, 200, 202, 204, 221, 230, 237, 319–323, 330, 331, 338, 374
   resources   88–90
   rotational   19, 148, 177
   thermal   75, 80–83, 86, 88, 89–91, 145, 146, 178, 199, 200, 202, 203, 205–207, 213, 266, 338, 339, 343, 349–352, 354, 374, 380
   translational   22, 27, 81, 261
   vibrational   81, 212
energy surface   78, 79
   stability of   78, 175, 183
entropy   75, 163, 193–208, 212, 217, 218, 268, 274
enzymes   170, 187, 209, 223, 224, 248, 259, 265, 266, 269, 275, 277, 284, 285, 290, 295, 296, 311, 312, 323
equilibrium   78, 82, 85, 139, 170, 207–212, 214, 215, 218, 230, 238, 248, 260, 274
evolution   6, 28, 53, 54, 203, 234, 235, 255, 258, 289, 290, 308, 317, 334, 335, 351, 352, 355, 361–364, 366, 382, 392
   convergent   364
   selection   234, 289, 329, 362, 363, 365–367
      artificial   229, 329, 362
      natural   24, 25, 38, 42, 90, 144, 150, 179, 194, 200, 227, 234,

267, 289, 290, 304,
357, 360, 362, 363,
365–367
    sexual   230–233, 235, 238,
276, 287, 291, 292, 311,
363, 366
  species formation   363, 364
Exclusion   108, 134, 137, 375
  force   5, 24, 34, 45–48, 51–59,
75, 77–79, 82, 87, 102, 108,
115–117, 123, 124, 127, 142,
175, 335, 345, 371, 374, 375,
380–382, 386
  principle   50, 51, 75, 85, 98,
101, 102, 108, 134, 136, 137,
195, 306, 365, 371, 375, 389,
391
extremophiles   229

Fermion   111, 116
fission   42, 86–90, 227
fluorescence   104
forbidden process   113
free energy   201, 202, 252, 266,
267, 274
functional group   176, 182, 183,
187, 188, 190, 273
  amide   183, 187, 190
  amine   183, 187, 188, 245
  carbonyl   184, 185, 187,
190
  halide   183
fusion   38, 40, 86

gaia hypothesis   365
galaxies   1, 7, 10, 16, 55, 122, 369,
373, 374, 386–390, 393
genetic code   281, 282, 288, 291
genetics   234

  heterozygous   233, 260,
267
  homozygous   233, 260
  mendelian   234, 287
genotype   231, 233
glial cell   322, 325
glucagon   223–225, 270
gravitational   34, 46–48, 51, 53,
55, 56, 58, 59, 76, 77, 79, 82, 84,
85, 88, 114, 115, 122–124, 142,
202, 213, 235, 335, 336, 345, 351,
371, 373–375, 378, 380, 381, 383,
386, 389, 391–393
  energy   5, 16, 32, 48, 71, 94,
113, 128, 143, 161, 177, 194,
207, 225, 248, 266, 276, 297,
318, 337, 357, 371
  field   11, 41, 47, 48, 49, 50,
51, 56, 68, 69, 76, 79, 85,
100, 105, 106, 109, 117,
118, 123, 130, 131, 151,
152, 162, 213, 247, 248,
336, 338, 340, 341, 351,
353, 371, 375, 378, 385,
386, 391, 392, 393
  force   5, 24, 34, 45–48, 51–59,
75, 77–79, 82, 87, 102, 108,
115–117, 123, 124, 127, 142,
175, 335, 345, 371, 374, 375,
380–382, 386

hadron   117, 121
halogen   147, 183
heat   67, 79, 83, 88, 89, 91, 95,
139, 162, 185, 199, 200, 201, 204,
206, 207, 212, 213, 218, 227, 306,
308, 338–340, 348, 349, 352, 379,
381, 382
heliocentric model   335, 372

hemoglobin   182, 259, 264–266, 300, 314
homeostasis   227–229, 235, 257, 309, 328
hydrates   164

idealization   19, 57
immune system   226, 270, 300, 303, 304, 305, 306, 311, 314, 322
   allergies   305
   autoimmune diseases   305
   B-cells   304, 305
   non-specific   304
   specific   8, 257, 274, 290, 295, 304, 323
   T-cells   304–306, 311
information   3, 16, 17, 28, 69, 79, 99, 109, 110, 120, 140, 150, 151, 158, 159, 172, 193, 197, 199, 204, 205, 217, 222, 227, 231, 242, 248, 251, 252, 254, 256, 257, 259, 271, 274–280, 282, 285, 290, 305, 317, 318, 325, 328, 336, 371, 386, 392
inheritance   287
insulin   184, 223–226, 258, 259, 263–265, 270, 289, 290, 292, 294, 311
isomers   159, 173, 178, 180, 190, 191, 247, 248, 256
   optical   247, 248, 256, 318
   structural   146, 147, 159, 165–180, 188, 189, 242, 248, 256, 259, 261, 267, 269, 294, 307
isotopes   38, 39, 42, 86, 120, 125, 132, 154, 248

kidneys   226, 300, 301, 311, 314
kinetic theory of gases   5, 13

kinetics   170, 361
lattice energy   162, 163
lepton   116, 117, 121, 122
Lewis structures   147–149, 165, 173
life   8, 16, 32, 75, 114, 146, 157, 175, 201, 207, 221, 241, 257, 275, 293, 333, 355, 369
   definition of   18, 208
ligaments   259, 294, 308
lightning   48, 58, 64, 73, 148, 236, 255, 351
limits   18, 28, 29, 45, 98, 195, 322, 345
lipid   250, 270, 314, 320
   phospholipid   210, 211, 250, 251, 256
liver   209, 222, 223, 224, 227, 270, 286, 294, 297, 300
luminiferous ether   70, 123
lymphatic system   300, 304

magnetic moment   105, 131
magnetic resonance   105, 106, 144
magnets   14, 47–50, 59
main group element   140, 144, 146
Mars   57, 58, 107, 238, 242, 333, 338, 341–343, 345, 353, 372, 374, 377, 379
mass   3, 7, 9, 33, 34, 35, 36, 38, 39, 42, 47, 53, 55, 80, 84, 85, 86, 90, 91, 96, 97, 99, 116, 117, 119, 122, 123, 127, 140, 141, 144, 173, 242,336, 345, 361, 371, 375, 379, 380, 381, 382, 383, 384, 385, 386, 387, 393

conservation of   34, 55, 59,
    82, 84, 96, 101, 114, 115, 124,
    125, 129, 195, 386
  rest   52, 57, 82–85, 90, 91, 96,
    111, 118, 276, 371
mathematical model   12
Maxwell's equations   51, 68, 73, 83
medea hypothesis   365
meiosis   286, 311, 313
mesons   116, 117, 120, 121, 124, 125
methanogens   351, 352
micelle   209–211, 218, 219
mitosis   285, 286, 291, 306, 325
molecular machines   273
momentum   91, 96–99, 105, 107, 109, 110, 113, 114, 116, 124, 125, 130, 131, 194, 373, 374, 385, 386, 393
  conservation of   34, 55, 59,
    82, 84, 96, 101, 114, 115, 124,
    125, 129, 195, 386
  of particle   101, 116
  of wave   73
moon   46, 57, 71, 80, 137, 335, 338, 339, 341, 342, 343, 345, 353, 370, 372–374, 378, 379, 393
motion   24, 28, 34, 45, 46, 48, 50–59, 63–65, 69, 70, 74, 76, 80–83, 90, 95, 98, 101, 107, 128, 130, 194, 198, 199, 214, 252, 259, 298, 308, 324, 331, 335–337, 339, 340, 341, 349, 352, 370–372
  chaotic   59, 373, 394
  conservation of   34, 55, 59,
    82, 84, 96, 101, 114, 115, 124,
    125, 129, 195, 386

constant   13, 46, 47, 52–54, 55, 57, 59, 62, 68, 69, 81, 82, 94, 103, 114, 129, 202, 203, 204, 212, 214, 227, 228, 239, 240, 360, 361, 364, 390
  damped   57
  Newton's laws of   82, 194
  orbital   55, 59, 60, 116, 130, 131, 133, 134
  periodic   22, 40, 57, 58, 120, 131, 133, 136, 140–142, 147, 155, 159, 161, 165, 216, 219, 221, 340, 352
  random   28, 58, 59
  angular momentum   54, 55, 59, 75, 105, 107, 109, 113, 114, 116, 124, 125, 130, 131, 194, 373, 374, 385, 386, 393
  spin   54, 55, 59, 60, 105–107, 109, 110, 116–118, 121, 125, 131, 132, 134, 137, 206
muscles   223, 259, 270, 295, 308, 314, 315, 318
mutation   278–289, 291

nebula   255, 376, 382, 384–386, 394
negative feedback   78, 91, 238, 365, 366
nervous system   254, 256, 294, 308, 317, 318, 322, 324, 325, 330
  autonomic   308, 325
  central   69, 148, 198, 225, 247, 259, 261, 285, 317, 318, 324, 325, 330
  peripheral   317, 322, 324
  somatic   317, 325
neuron   10, 204, 254, 294, 309, 318–325, 328, 329, 330

axon 318, 319, 320, 321, 322, 323, 328, 329, 330, 331
axon potential 319, 321, 323, 329, 330, 331
dendrite 318, 321, 323, 330
postsynaptic 323, 324
presynaptic 322–324
Schwann cells 322
neurotransmitters 254, 256, 323, 324
neutrino 40, 114–116, 121, 124, 125, 384
neutron 57, 39–41, 46, 52, 56, 85–87, 91, 107, 108, 114–116, 119–121, 123–125, 127, 375, 384–386, 393
niche 360, 364, 365
noble gas 145–147
novae 382, 383, 393
nucleation 164
nucleotide 255, 258, 271, 276, 285, 291, 351
nucleus 17, 36–40, 42, 46, 51, 52, 58, 59, 69, 86, 87, 105, 114, 119, 124, 125, 127–130, 132, 133, 136, 141, 142, 144, 146, 147, 167, 170, 223–225, 227, 237, 238, 241, 276, 279, 280, 284, 285, 290, 291, 313, 318, 329, 384
in brain 234
of atom 33, 58, 139
of cell 242, 256, 270, 284, 286, 294, 318

oceans 88, 89, 90, 229, 236, 255, 336, 346, 348, 349, 350, 351, 352, 379
circulation 213, 303, 349, 350, 352

order 1, 15, 31, 45, 69, 75, 94, 113, 135, 163, 177, 193, 208, 221, 247, 257, 275, 293, 328, 336, 365, 370
organ 222, 232, 237, 241, 294, 297, 308, 309, 311, 314
system 9, 10, 16, 18–22, 26, 28, 29, 45, 53, 58, 80, 81, 85, 86, 97, 100, 101, 102, 104, 107, 109, 117, 137, 193, 195–208, 212, 214, 216–218, 221, 226, 227, 238, 241, 242, 254, 256, 270, 294–301, 303–306, 308, 309, 311–314, 317–319, 321–331, 333, 338, 342, 360, 364, 366, 373, 374, 377, 378, 379, 382, 383, 392
origin 65, 120, 255, 323, 341, 342, 370, 391
of building blocks 255
of life 8, 16, 28, 146, 149, 150, 175, 190, 203, 207, 221, 222, 228, 230, 235, 236, 238, 241, 251, 252, 255–258, 275, 276, 290, 293, 309, 333, 334, 351, 358, 361, 366, 381
osmosis 228

pair production 118
particle in a box 102, 103
periodic table 40, 120, 131, 133, 136, 140–142, 147, 155, 159, 165
phenotype 231–234, 260, 287, 289
photoelectron effect 94–96, 110, 111

photosynthesis   41, 89, 155, 235, 295, 346, 351
plasma   87, 88, 210, 223, 225, 237, 241, 300, 304, 305, 318, 319, 391
polymers   187–189, 250, 257, 268, 271
   branched   178, 189, 268, 270
   linear   36, 120, 166, 167, 178, 180, 189, 190, 248, 258, 259, 268, 270, 271, 273, 277, 279, 389
   vinyl   182, 189
polysaccharides   250, 258, 268, 274
population   3, 231, 234, 238, 260, 268, 289, 357, 358, 360, 361–367
population   3, 231, 234, 238, 260, 268, 289, 357, 358, 360–367
positive feedback   78, 91, 214, 346, 364, 391
positron   116–118, 120, 124, 125
   annihilation   118–120
   emission tomography (PET)   120
power   10, 14, 38, 76, 79, 90, 92, 107, 110, 225, 227, 322,
prion   267, 268
   disease   11, 149, 167, 186, 226, 232, 259, 260, 261, 262, 267, 268, 284, 288, 289, 299, 300, 305, 322, 324, 328, 362, 364
probability   3, 11, 96, 98, 100, 103, 107, 109, 110, 129, 131, 134, 197, 198, 206, 208, 287, 362, 365
proteins   149, 187, 188, 190, 210, 223, 224, 226, 229, 241–243, 245, 248, 255, 258–260, 262, 263, 265, 267, 268, 270, 273–275, 278–285, 290, 291, 295, 296, 304, 305, 308, 309, 362
   fibrous   259, 361, 274
   globular   243, 259, 262, 263, 265, 266, 274, 387
   primary structure   258, 273
   quaternary structure   261
   secondary structure   260, 262, 273
      alpha helix   260, 261
      beta sheet   262, 263, 267
   tertiary structure   260, 263, 265, 266, 273
proton   35, 38, 39, 41, 43, 46, 52, 56, 59, 79, 85, 86, 91, 107, 114, 116, 118, 120, 121, 123–125, 127, 131, 133, 167, 384

qbit   110
quantum dot   103, 104
quantum mechanics   7, 101, 102, 105–109, 111, 117, 122, 127, 133, 136, 141, 165
quantum numbers   102, 108, 110, 129, 131, 132, 137
quarks   17, 113, 121, 122, 124, 125

radioactivity   41, 87, 235, 353
   alpha particles   35, 36
   beta particles   40, 42
   dating by   41
   gamma rays   67, 69, 70, 118, 120, 124, 370
   half life   40, 41, 56, 89, 114, 120
relativity   71, 83, 85, 95, 117, 133, 371, 373, 383, 386, 389, 390

general 6, 8, 12, 17, 58,
    78, 82, 85, 102, 130, 159,
    177, 186, 187, 193, 202,
    242, 246, 250, 271, 309,
    358, 371, 373, 386, 389, 390
  special 71, 83, 95, 114,
    117, 133, 230, 322, 369,
    383, 389
replication 28, 285, 286, 288,
  290
reproduction 204, 222, 228,
  230-233, 235, 238, 276, 285, 287,
  291, 292, 311, 357
    asexual 230, 231, 235, 238,
      285, 291, 292
    cell 22, 29, 70, 143, 161, 187,
      204, 216, 221-225, 227-230,
      235, 237, 238, 241, 242, 251,
      255, 256, 260, 269, 270, 274,
      276, 277, 279, 283-287, 290,
      291, 293, 294, 300, 304, 305,
      311-313, 317-319, 321-325,
      330
    sexual 230-233, 235, 238,
      276, 287, 291, 292, 311, 363,
      366
reproductive system 311-314
  egg cells 311, 313
  fertilization 311, 313
  sperm cells 231, 311
resonance 105, 106, 144, 148, 149,
  151, 181, 191
respiratory system 297, 298, 300,
  314
  alveoli 298, 299, 300
  emphysema 299
RNA 225, 249, 251, 271, 273, 274,
  277, 279, 280-282, 284, 290, 291,
  317

messenger 225, 279-281,
  291, 323
transfer 47, 96, 97, 157,
  159, 172, 199, 200,
  212-214, 218, 227, 251,
  253, 256, 266, 276, 279,
  282, 300, 306, 318, 328,
  339, 349, 352, 381
world 1-4, 10, 18, 24, 31,
  32, 34, 48, 87, 93, 94, 98,
  101, 102, 107, 127, 139,
  141, 144, 150, 189, 195,
  197, 216, 234, 235, 257,
  290, 297, 306, 310, 324,
  331, 363, 364, 384

scientific method
  falsifiable 6
  hypothesis 1-6, 8, 13, 14, 39,
    70, 85, 94, 95, 98, 340, 365,
    371, 386
  law 4, 5, 14, 33, 47, 52, 53,
    82, 83, 113, 114, 118, 157,
    194-197, 200, 203-206, 217,
    359, 389
  measurement 2, 3, 5, 12, 26,
    33, 34, 42, 52, 58, 74, 84, 97,
    98, 101, 105, 109, 148, 166,
    335, 336, 372
  observation 2-8, 12, 13,
    25, 26, 33, 36, 56, 66, 70,
    80, 95, 97, 102, 111, 148,
    191, 232, 234, 353, 361,
    363, 369, 370, 372, 379,
    385, 389
  theory 4-7, 11-14, 17, 21,
    33, 34, 39, 51, 56, 65, 71,
    83, 85, 93, 95, 96, 101,
    115, 119, 120, 122, 128,

129, 133, 141, 169, 197,
232, 350, 355, 370–373,
386, 390, 391
scientific notation   10, 11, 13
scientific units   10
second law of thermodynamics
  194, 195, 205, 206, 217, 359
seismology   336, 337, 353
  earthquake   63, 193, 194,
    336–340
shell   64, 70, 114, 134, 136, 137,
  142, 143, 144, 145, 147, 149, 150,
  153, 155, 382
skin   37, 96, 97, 106, 169, 171,
  230, 234, 237, 259, 261, 288,
  294, 303, 306, 307, 314, 327,
  330, 357
  epidermis   306
  dermis   306
solubility   163, 250, 255, 300
solutions   129, 133, 163, 164, 169,
  210, 218, 242, 244, 245, 248, 318,
  343, 364
spectroscopy   104
spectrum   67–69, 74, 104, 105,
  129, 133, 136, 383, 391
  continuous   31, 73, 88, 94,
    104–106, 111, 114, 218, 322,
    370, 377
  hydrogen atom   35, 38, 107,
    127–129, 131–133, 137, 153,
    167, 182, 201, 249
  line   3, 21, 24, 26, 49, 64, 81,
    104, 105, 129, 133, 143, 147,
    148, 159, 165, 180, 285, 287,
    325, 370, 371
standard model   121, 122
stars   10, 16, 32, 45, 86, 87, 88,
  108, 122, 206, 335, 369, 370, 371,

372, 373, 375, 376, 379, 380,
381–385, 387, 388, 392, 393
  brown dwarfs   16, 380
  clusters of   122
  main-sequence   380, 381,
    393
  neutron   16, 39–41, 46, 52, 56,
    85, 86, 87, 91, 107, 108,
    114–116, 119, 120, 121,
    123–125, 127, 375, 384–386,
    393,
  red giant   382, 383, 393
  white dwarfs   108, 375, 381,
    383
steady state   207, 212, 214–218,
  230
steroid   250, 251
strong force   46, 52, 58, 87, 115,
  117, 124, 127
subshell   134–136, 142, 144, 146,
  153, 154
sugar   89, 164, 235, 242, 248–250,
  252, 255, 268, 271, 273, 274, 276,
  277, 280, 291, 295, 314
  deoxyribose   249, 271, 273,
    277, 280
  glucose   223–227, 235, 238,
    250, 256, 268, 269, 270, 296,
    320, 329
  hexose   248, 255, 256
  pentose   348, 255
  ribose   249, 252, 273, 277,
    280
Sun   10, 16, 34, 53–55, 57, 59, 66,
  70, 86–90, 104, 115, 116, 125,
  144, 194, 204, 206, 217, 218, 236,
  238, 306, 335, 337, 338, 341, 345,
  346, 348, 351, 359, 370, 371–374,
  377, 379, 380–387, 393, 394

super organism   237, 366
supernova   383-385
supernovae   87, 371, 383, 384, 386, 393
surfactant   209, 210, 218, 219, 250, 297
symmetry
  averaged   28, 96, 345
  fractal   24-26, 29, 353
  functional   18, 28, 29, 176, 182, 183, 187, 188, 190, 237, 273, 284, 287, 293, 314
  group theory   21
  inversion   21, 24, 27, 29, 348, 349
  of the universe   114
  reflection   20, 22-24, 27, 29, 146
  rotational   19, 148, 177
  translational   22, 27, 81, 261
    unit cell   22, 23, 29, 70, 161, 173
synapse   320, 322-325, 328-330
  neurotransmitter   254, 256, 322-324, 329-331
    reuptake   323, 324, 329, 331
  receptor   263, 265, 270, 306, 323, 326, 330
  release   48, 71, 83, 84, 86, 87, 90, 95, 119, 168, 169, 173, 185, 201, 213, 224, 228, 245, 304, 309, 314, 322, 323, 330, 337-340, 365, 375, 378, 379, 381-384

taxonomy   236
teeth   146, 261, 262, 274

tendons   259, 294, 308
tetrahedron   150, 176, 177
torque   54, 55, 59, 374
transcription   279, 280, 281, 284, 290
transformations   25, 56, 58, 84, 115, 119
transition   140, 153, 154, 160, 319, 364
translation   9, 22, 24, 27, 29, 81, 279, 282, 283, 284
transition element   140, 153, 154
trophic levels   358, 359, 366

uncertainty principle   98, 102, 391
universe
  creation of   119, 335
  expansion of   89, 90, 225, 382, 393

vapor pressure   208, 219
virtual particles   123, 124

water   16, 18, 29, 32, 33, 37, 43, 48, 51, 61, 62, 63, 65-67, 70, 76, 81, 83, 89, 91, 117, 143, 146, 150, 157-159, 163, 164, 167-170, 173, 175, 179, 184-187, 199-202, 206-213, 215, 218, 225, 228, 229, 231, 236, 242, 243, 244, 250, 255, 258, 266, 268, 274, 294, 297, 301, 306, 329, 334, 339, 343, 344-346, 348, 349, 351, 352, 359, 360, 363, 370, 382, 385, 391
  cycle   57, 58, 60, 183, 231, 343, 359, 360

wave function 99–103, 107, 108, 110, 111, 129, 130, 133, 134, 159
    mixed 109, 110, 216, 291, 295, 311, 328, 349, 352, 367
    of hydrogen atom 132
    of multi-electron atoms 133, 141
    of particle in a box 102–104

wave
    amplitude 62, 63, 65, 68, 69, 73, 95, 100, 111
    diffraction 65, 66, 70, 97, 99, 100
    frequency 24, 62, 63, 65, 67, 68, 69, 73, 74, 94, 96, 101, 104, 105, 106, 110, 129, 214, 234
    interference 65, 66, 67, 70, 72–74, 95, 96, 99
    nodes 63, 66, 304, 322
    of pulse 71–73
    phase 62, 65, 71, 73, 143, 161, 172, 202, 208, 255
    velocity 7, 18, 29, 40, 52–55, 57, 59, 62–64, 68–70, 73, 74, 80–84, 97–99, 102, 110, 114, 124, 128, 137, 199, 337, 339, 344, 345, 371, 383, 389, 390
    wavelength 62, 63, 65–67, 69–74, 93–99, 101, 103, 104, 110, 111, 137, 171, 370, 376, 377, 387, 389, 391, 393
        of particle 17, 95, 97, 98, 101, 116, 119, 121, 122, 124, 197–199

weak force 46, 47, 52, 56, 58, 115, 116, 123, 124, 142

work 3, 28, 50, 51, 77, 79, 98, 141, 194, 216, 225, 228, 234, 237, 293, 308, 309, 324, 328, 335